Asymmetric Reactions and Processes in Chemistry

Asymmetric Reactions and Processes in Chemistry

Ernest L. Eliel, EDITOR
University of North Carolina at Chapel Hill

Sei Otsuka, EDITOR
Osaka University

Based on a U.S.–Japan seminar cosponsored by the Japan Society for the Promotion of Science and the National Science Foundation and held at Stanford University, Stanford, California, July 7–11, 1981.

ACS SYMPOSIUM SERIES **185**

AMERICAN CHEMICAL SOCIETY
WASHINGTON, D. C. 1982

Library of Congress CIP Data

Asymmetric reactions and processes in chemistry.

(ACS symposium series, ISSN 0097-6156; 185)

Includes index.

1. Chemistry, Organic—Synthesis—Congresses. 2. Stereochemistry—Congresses.

I. Eliel, Ernest Ludwig, 1921- . II. Otsuka, Sei. III. Nippon Gakujutsu Shinkokai. IV. National Science Foundation (U.S.). V. Series.

QD262.A78 547'.2 82-3908 ASCMC8 185 1-300
ISBN 0-8412-0717-8 AACR2 1982

PRINTED IN THE UNITED STATES OF AMERICA

ACS Symposium Series

M. Joan Comstock, *Series Editor*

FOREWORD

The ACS SYMPOSIUM SERIES was founded in 1974 to provide a medium for publishing symposia quickly in book form. The format of the Series parallels that of the continuing ADVANCES IN CHEMISTRY SERIES except that in order to save time the papers are not typeset but are reproduced as they are submitted by the authors in camera-ready form. Papers are reviewed under the supervision of the Editors with the assistance of the Series Advisory Board and are selected to maintain the integrity of the symposia; however, verbatim reproductions of previously published papers are not accepted. Both reviews and reports of research are acceptable since symposia may embrace both types of presentation.

CONTENTS

PREFACE

The possibility of an asymmetric synthesis was foreseen in Le Bel's pioneering 1874 paper on the asymmetric carbon atom and the idea was reduced to practice by Emil Fischer, W. Marckwald, and A. McKenzie around the turn of the century. Although there seems to have been a long lasting air of mystery about asymmetric induction, the process was put on a firm mechanistic basis by the classical studies of W. Doering, L. M. Jackman, H. S. Mosher, V. Prelog, and D. J. Cram around 1950. However, it was only in 1961 that a practical asymmetric synthesis proceeding in high optical yield—the hydroboration–oxidation of *cis*-2-butene by tetrapinanyldiborane—was achieved by H. C. Brown and G. Zweifel. By 1971, when the definitive book in the area, *Asymmetric Organic Reactions* by J. D. Morrison and H. S. Mosher appeared, innumerable asymmetric reactions were on record and the mechanism of a number of them was quite well understood, but optical yields much in excess of 40–50 percent were surprisingly rare.

Perhaps as a result of the publication of the Morrison–Mosher book or simply because the time was ripe, the situation has changed drastically in the last ten years. Recent reviews by H. B. Kagan and J. C. Fiaud, D. Valentine and J. W. Scott, and J. W. ApSimon and R. P. Seguin, as well as the current literature indicate large numbers of methods, probably now well over a hundred, by which chiral products may be obtained with enantiomeric excess in the 80- to 90-percent region. Although such methods have been reported from all over the chemically active world, a substantial number have originated either in Japan or in the United States.

It therefore seemed timely to arrange a joint U.S.–Japan Seminar on the topic of asymmetric reactions and processes—the title including not only asymmetric syntheses, both conventional chemical and enzymatic, but also certain separation methods involving the same kind of diastereomeric interactions that are involved in asymmetric synthesis. The seminar was held July 7–11, 1981, at Stanford University with the editors and Professor Harry S. Mosher as co-organizers. It was supported jointly by the Japan Society for the Promotion of Science and the National Science Foundation and featured 19 plenary speakers: nine from Japan and ten from the United States. The names and brief biographies of all but one of these speakers are given on pp. xi–xiii, and their subjects are listed in the Table of Contents.

Approximately half the chapters (1–10) deal with what might be termed "classical asymmetric synthesis": a chiral adjuvant is combined with a prochiral reagent, thus producing diastereotropic ligands or faces. Stereoselective replacement of the ligands or addition to the faces followed by removal of the chiral adjuvant leads to chiral products. The second largest block of chapters (11–13 plus parts of 7 and 8) deals with asymmetric catalysis, involving, in several instances, chiral organometallic reagents. It is remarkable that, not only in the stoichiometric but also in the catalytic reactions presented, optical yields frequently exceed 90%. Chapters 14 and 15 are concerned with large-scale commercial applications of asymmetric enzymatic synthesis and Chapters 16 and 17 deal with biochemical applications of enzyme chemistry.

Despite the fascination of asymmetric synthesis, separation methods for enantiomers even today compete, often successfully, with direct synthetic routes. Thus it is appropriate that the topic of separation was included in the seminar. Chapter 18 is concerned with this topic, as was a paper presented by D. J. Cram on host–guest complexation [J. Am. Chem. Soc., *103,* 3929 (1981); J. Org. Chem., *46,* 393 (1981); and J. Chem. Soc., Chem. Comm., 625 (1981)]. Substantial parts of Chapters 1 and 10 also deal with the role of resolution and separation methods in the synthesis of chiral compounds, including enantioconvergent syntheses.

In addition to the 19 main papers, two short papers and nine posters (preceded by short oral presentations) were featured at the Seminar. Nine of these eleven short communications are summarized in the form of abstracts on pages 261–286. The titles of these communications are listed in the Table of Contents. Abstracts of two communications are not included: that of D. Valentine, Jr. (Catalytica Associates, Inc.) on phosphines having both chiral phosphorus and chiral ligands has been published in J. Org. Chem., *45,* 3691 (1980) and that by B. Sharpless (MIT), on kinetic resolution, has appeared in J. Am. Chem. Soc., *103,* 6237 (1981).

By the judgement of most of the participants, the Seminar was a success in that it brought together information on a wide variety of diverse, although often fundamentally related, methods for efficient synthesis of chiral organic compounds. We hope that this written record will prove of equal interest to the reader.

ERNEST L. ELIEL
University of North Carolina
Department of Chemistry
Chapel Hill, North Carolina 17514

SEI OTSUKA
Osaka University
Department of Chemistry
Toyonaka, Osaka, Japan 560

CONTRIBUTING AUTHORS

ICHIRO CHIBATA is Director of the Research Laboratory of Applied Biochemistry, Tanabe Seiyaku, Co., Ltd. Born Osaka, Japan, 1926; B.S., 1948, Ph.D, 1959, Kyoto University. With Research Laboratories, Tanabe Seiyaku, Co., Ltd. since 1948, Director since 1971.

ERNEST L. ELIEL is W. R. Kenan, Jr. Professor of Chemistry, University of North Carolina, Chapel Hill. Born Cologne, Germany, 1921, D.Phys.-Chem.Sci., University of Havana, Cuba, 1946; Ph.D. University of Illinois, 1948. University of Notre Dame, 1948–72; University of North Carolina since 1972.

HEINZ G. FLOSS is Lilly Professor of Medicinal Chemistry at Purdue University, West Lafayette, Indiana. Born Berlin, Germany, 1934. Diplom., Technical University, Berlin, 1959; Ph.D., Technical University, Munich, 1961; Postdoctoral, University of California at Davis (Conn) 1964–65. At Purdue University since 1966.

KAORU HARADA is Professor of Chemistry, University of Tsukuba, Ibarati. Born Toyonaka, Osaka, Japan, 1927, B.S., 1952, Ph.D., 1961, Osaka University. Research Associate, Florida State University, 1956–64. University of Miami, 1964–74; The University of Tsukuba since 1974.

TAMIO HAYASHI is Instructor, Department of Synthetic Chemistry, Faculty of Engineering, Kyoto University. Born Gifu, Japan, 1948. B.Eng., 1970, Ph.D., 1975, Kyoto University; Postdoctoral, Colorado State University (Hegedus), 1976–77. At Kyoto University since 1975.

CLAYTON HEATHCOCK is Professor of Chemistry, University of California at Berkeley. Born San Antonio, Texas, 1936. B.S. Abilene Christian College, 1958; Ph.D., University of Colorado, 1963; Postdoctoral, Columbia University (Stork), 1963–64. At University of California at Berkeley since 1964.

KENJI KOGA is Professor, Faculty of Pharmaceutical Sciences, University of Tokyo. Born Aichi, Japan, 1938; B.S., 1960, Ph.D., 1966, University of Tokyo; Postdoctoral, University of California at Los Angeles (Cram), 1971–73. At University of Tokyo since 1976.

ALBERT I. MEYERS is Professor of Chemistry, Colorado State University, Fort Collins. Born New York City, New York, 1937. A.B., 1954, Ph.D., 1957, New York University. With Cities Service Research & Development, 1957–58; Louisiana State University, 1958–70; Wayne State University, 1970–72; Colorado State University since 1972.

TERUAKI MUKAIYAMA is Professor, Department of Chemistry, University of Tokyo. Born Nagano prefecture, Japan, 1927. B.S., Tokyo Institute of Technology, 1948; D.Sc., University of Tokyo, 1956. Assistant Professor, Gakushuin University, 1952–58; Professor, Tokyo Institute of Technology, 1962–73; University of Tokyo since 1974.

HITOSI NOZAKI is Professor, Department of Industrial Chemistry, Kyoto University. Born Okayama prefecture, Japan, 1922. B.Eng., 1943, Dr.Eng., 1949, Kyoto University; Postdoctoral, Cornell University (Meinwald), 1956–57. At Kyoto University since 1963.

ATSUYOSHI OHNO is Associate Professor, Institute for Chemical Research, Kyoto University. Born Kure, Hiroshima, Japan, 1936. B.S., Kyoto University, 1958; Ph.D., Osaka City University, 1963; Postdoctoral, Massachusetts Institute of Technology (Swain), 1963–65; Purdue University (Davis) 1965–66. At Kyoto University since 1974.

IWAO OJIMA is Senior Research Fellow and Group Leader, Sagami Chemical Research Center, Sagamihara. Born Yokohama, Japan, 1945. B.S., 1968, M.S., 1970, Ph.D., 1973, University of Tokyo. At Sagami Research Laboratories since 1970.

SEI OTSUKA is Professor, Department of Chemistry, Osaka University. Born Tchingtao, China, 1918. B.S., 1941, D.Sc., 1955, Osaka University; Postdoctoral, Ohio State University (Newman), 1955–57; Max–Planck Institut für Kohleforschung, Mülheim (Wilke), 1958. With Japan Synthetic Rubber, Co. Ltd, 1957–1964. At Osaka University since 1964.

WILLIAM H. PIRKLE is Professor, School of Chemical Sciences, University of Illinois, Urbana. Born Shreeveport, Louisiana, 1934. B.S., University of California, Berkeley, 1959; Ph.D., University of Rochester, 1963; Postdoctoral, Harvard University (Corey), 1964. At University of Illinois since 1964.

GARY H. POSNER is Professor, Department of Chemistry, Johns Hopkins University. Born New York City, New York, 1943; B.S., Brandeis University, 1965; Ph.D., Harvard University, 1968; Postdoctoral, University of California at Berkeley (Dauben), 1969. At Johns Hopkins since 1969.

GABRIEL SAUCY is Associate Director, Chemical Research Department, Hoffmann–La Roche, Incorporated, Nutley, New Jersey. Born Schaffhausen, Switzerland, 1927; Diplom, 1951; Ph.D., 1954, Federal Institute of Technology, Zurich, Switzerland. With Hoffmann–La Roche, Basle, 1954–64, Nutley since 1964.

BARRY M. TROST is Evan P. and Marion Helfaer Professor of Chemistry, University of Wisconsin, Madison. Born Philadelphia, Pennsylvania, 1941. B.A., University of Pennsylvania, 1962; Ph.D., Massachusetts Institute of Technology, 1965. At University of Wisconsin since 1965.

GEORGE M. WHITESIDES is Professor of Chemistry, Massachusetts Institute of Technology, Cambridge, Massachusetts. Born Louisville, Kentucky, 1939. A.B., Harvard University, 1960; Ph.D., California Institute of Technology, 1964. At Massachusetts Institute of Technology since 1963.

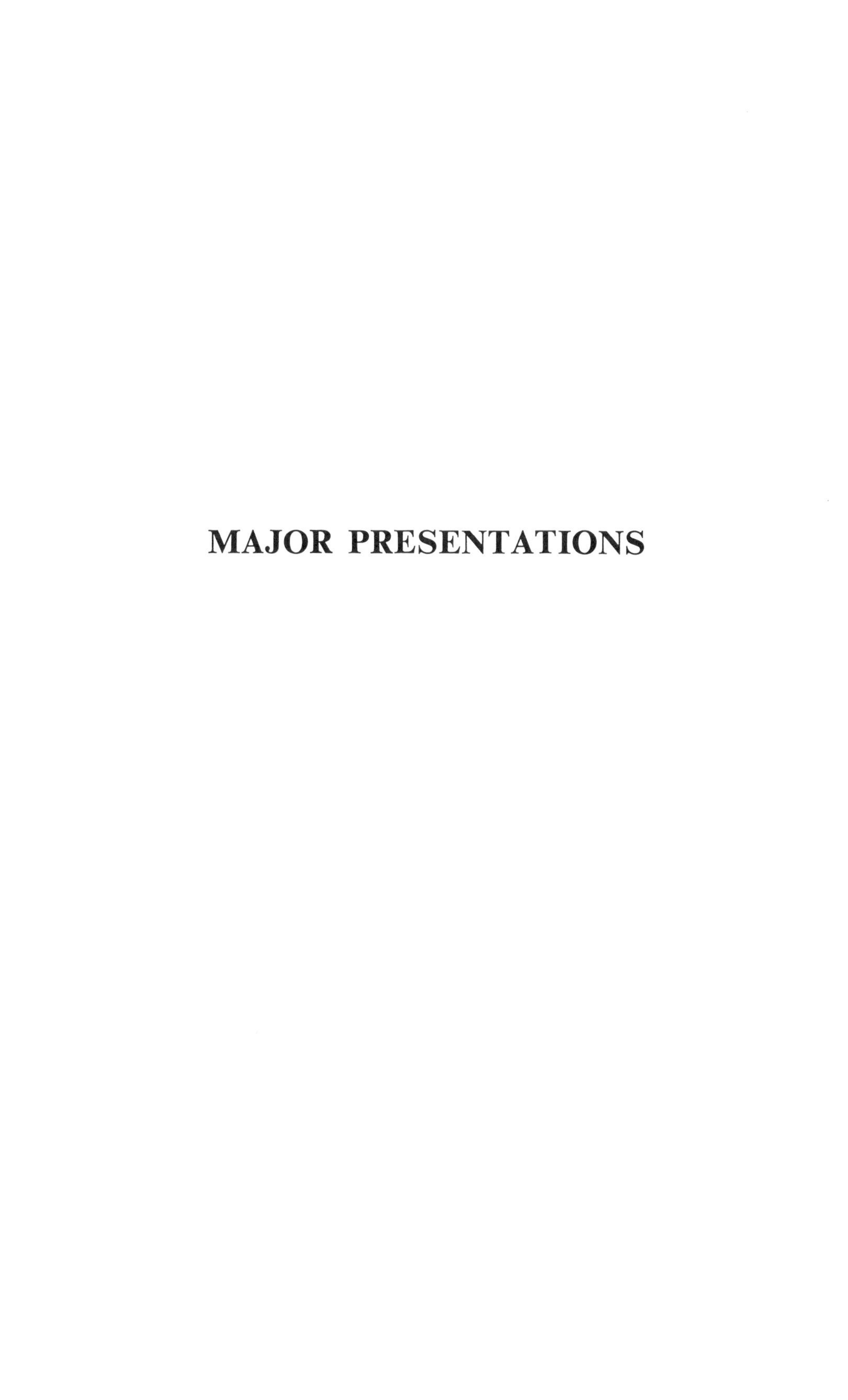

MAJOR PRESENTATIONS

1

Approaches for Asymmetric Synthesis as Directed Toward Natural Products

BARRY M. TROST

University of Wisconsin, Department of Chemistry, McElvain Laboratories of Organic Chemistry, Madison, WI 53706

Asymmetric synthesis of natural products embraces one of four different strategies - 1) resolution of a convenient intermediate or final product, 2) utilization of enantiomerically pure starting materials, 3) asymmetric induction at the stage of an achiral intermediate, and 4) enantioconvergency. Each of these is illustrated. A new type of enantioconvergency embodying a [3.3] sigmatropic rearrangement is employed for the synthesis of prostanoids. Asymmetric induction strategy is examined in the context of a model for asymmetric induction in the Diels-Alder reaction. The enantiomerically pure and partially pure adducts are employed in the synthesis of iboga alkaloids and pillaromycinone. Erythrynolides are the framework for strategies embodying resolution and enantiomerically pure building blocks. For the former, both enantiomers of a key building block are utilized to synthesize different halves of the molecule. In the latter, a single enantiomerically pure intermediate is utilized for the synthesis of the two halves of the molecule in a convergent approach. Emphasis is placed on the general utility of O-methylmandelic acid as 1) an enantiomeric inducing agent, 2) a resolution agent *via* HPLC, 3) an analytical tool to determine % ee, and 4) a tool for deducing absolute configuration.

The total synthesis of complex natural products offers challenges in the construction of the carbon framework, adjustment of the oxidation pattern, control of relative stereochemistry and control of absolute stereochemistry. While all of these areas offer exciting opportunities, the last remains the least considered and most perplexing in developing particular synthetic strategy. To a very large extent, total synthesis of natural products still implies the synthesis of a racemate which, by definition, contains only 50% of the natural product and may be resolved at the end or along the way.

0097-6156/82/0185-0003$05.00/0

In attempting to specifically consider the problem of absolute stereochemistry in developing strategy, four options emerge. First and most common is the resolution of some convenient intermediate or final product. Second and increasingly popular is the utilization of optically pure starting materials. Third is the dissection of the target molecule into an achiral intermediate in which asymmetry can be induced in a subsequent step. Fourth is the design of an intermediate which allows easy interconversion of two enantiomers (enantioconvergency). In this presentation, I wish to consider some aspects of each of these strategies in the context of several problems in the total synthesis of natural products.

Enantioconvergence Strategy

The concept of enantioconvergent synthesis has heretofore been virtually restricted to cases in which the chiral center is directly epimerizable such as in α-amino acids. In an alternative view, the separate transformation of two enantiomers via stereochemically complementary pathways into a single enantiomeric series represents a case of enantioconvergence.(1) For example, conversion of the enantiomeric alcohols **1a** and **1b** to **1c** via the *Z* and *E* olefins respectively converge to the same enantiomer of the product derived via a Claisen ortho ester rearrangement (equation 1). (2,3)

(1)

1a **1b** **1c**

An alternative conceptual approach to enantioconvergent synthesis involves intermediates whose enantiomers may be readily interconverted by simple chemical reactions. Compound **2** potentially represents such a species since it can be reasoned that a [3.3] sigmatropic rearrangement commutes the *S*,*S* isomer **2a** into the R,R isomer 2b (equation 2). (1) Attempts to equilibrate 2a

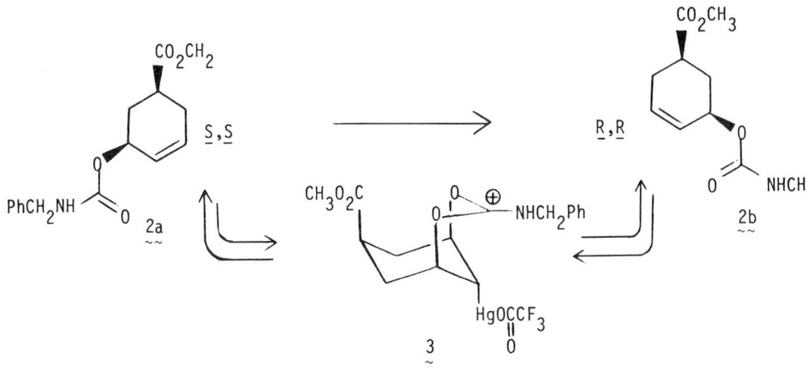

(2)

(56% ee, $[\alpha]_D^{25}$ + 16.3° (c 5.25 ($CHCl_3$)) thermally led to recovered 2a with no change in rotation. However, addition of 0.2 equiv. of mercuric trifluoroacetate to a refluxing dioxane or THF solution of 2a for 6-10 hours leads to complete racemization and no competing *cis*-*trans* isomerization. A mechanism invoking the intermediacy of 3 rationalizes this observation.

Although in its simplest form this process would seem merely to lead to racemization, it can, in fact, be adapted to optical enrichment. Thus, use of an optically active urethane such as 4 combined with fractional crystallization of one diastereomer creates a "resolving machine." Not only optical rotation, but NMR analysis as well allows determination of optical purity with

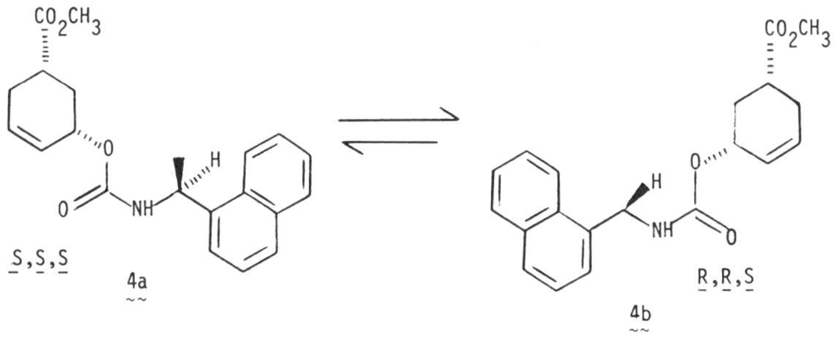

the S,S,S isomer 4a showing the methyl ester absorption upfield (δ3.58 to that in the R,R,S isomer 4b (δ3.68)). Indeed, reacting the racemate of the hydroxy ester with the isocyanate derived from S-α-naphthethylamine gave the urethane 4 as a 1:1 mixture of 4a and 4b with $[\alpha]_{436}^{25}$ -13.2° (c 2.46, PhH) and two singlets of equal intensity in the NMR spectrum at δ3.68 and 3.58. Mercury catalyzed equilibration converts this mixture to an approximately 1:2 ratio of 4a to 4b as determined by the NMR spectrum, with the signal at δ3.68 larger than that at δ3.58, and a rotation

$[\alpha]^{25}_{436}$ -11.8° (c 1.95, PhH). To verify that the urethanes could be succesfully employed in synthesis, their conversion to the corresponding hydroxy esters without racemization must be demonstrated. In fact, treatment with trichlorosilane in hot benzene containing triethylamine converts optically active 4a back to its hydroxy ester with no loss of optical activity.

With the phenomenon established, a strategy for prostanoid synthesis emerges as shown in equation 3. In the major simplifi-

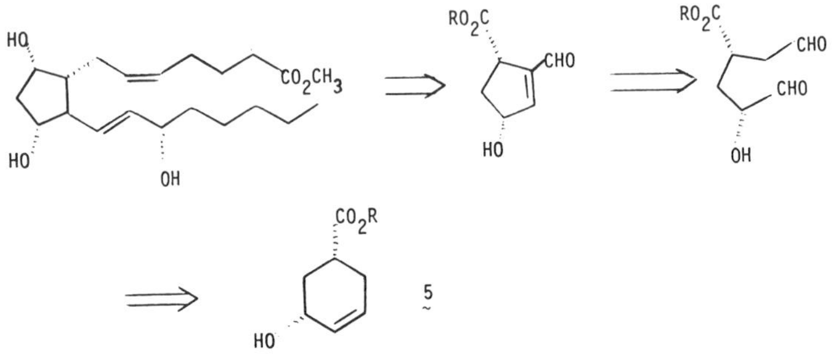

(3)

cation, two key points must be recognized 1) the ability of an α,β-unsaturated aldehyde to lead to introduction of both side chains and 2) the ability of a carboxy group to serve as a precursor to an alcohol by a carboxy inversion procedure. The ready accessibility of a cyclopentene-1-carboxaldehyde by a directed aldol condensation and the generation of a dialdehyde by the oxidative cleavage of an olefin leads to the alcohol 5 which relates to 4. This approach also intrinsically differentiates the C(9) and C(11) oxygens to provide an entry into a number of PG compounds. Scheme 1 illustrates the utilization of this strategy for an analogue.

Asymmetric Induction Strategy

While a great deal of emphasis has been placed upon searching for asymmetric induction, strikingly successful results remained elusive until very recently. One of the most powerful approaches for the synthesis of natural products has been the Diels-Alder reaction but asymmetric induction in this reaction remained disappointing. (4) This reaction is particularly important since the cycloaddition normally creates chiral products from achiral reactants. A working model for inducing chirality could greatly enhance the power of the approach.

Although the mechanism of the Diels-Alder reaction is still controversial, it is practically useful to envision a relationship between its transition state (t.s.) and a charge transfer complex. (5) In its simplest version, the two enantiomeric t.s.'s 6a and 6b become diastereomeric t.s.'s if a chiral "solvator" is present. Preferential solvation of one of the two enantiotopic

SCHEME 1. An Enantioconvergent Approach to Prostanoids

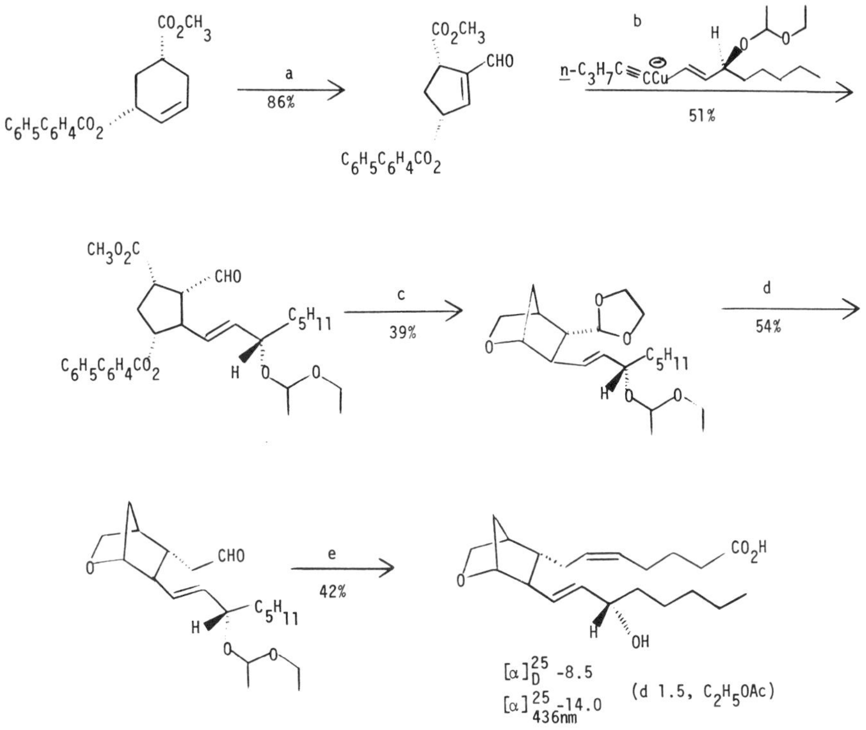

$[\alpha]_D^{25}$ -8.5

$[\alpha]_{436nm}^{25}$ -14.0 (d 1.5, C_2H_5OAc)

(a) i) 1 mol% OsO_4, $NaClO_3$, H_2O; ii) $NaIO_4$, THF, H_2O; iii) $C_5H_9\overset{+}{N}H_2\overset{-}{OAc}$, PhH, 50°.
(b) HMPA, THF, ether, -78° (1 h), -20° (3h).
(c) i) 2-(dimethylamino)-1,3-dioxolane [drawn as a ring structure followed by] $N(CH_3)_2$, HOAc, CH_3CN; ii) LAH, THF, 0°; iii) TsCl, C_5H_5N, 0°; iv) NaH, DMF, rt.
(d) i) CH_3COCO_2H, THF, BF_3·ether; ii) $C_2H_5OCH=CH_2$, $POCl_3$, $(C_3H_5)_3N$; iii) $Ph_3PCH_2OCH_3Cl$, t-C_4H_9Li, THF, 0°; iv) $Hg(OAc)_2$, THF, H_2O then KI, H_2O.
(e) i) $Ph_3\overset{+}{P}CH_2(CH_2)_3CO_2^-$, CH_3SOCH_2Na, DMSO; ii) HOAc, THF, H_2O.

faces of the diene will then lead to preferential reaction via **6a** or **6b**, i.e. to asymmetric induction. Reasoning that a π-stacking type of interaction might be an effective "solvator," that such an interaction would be more favorable with the diene than

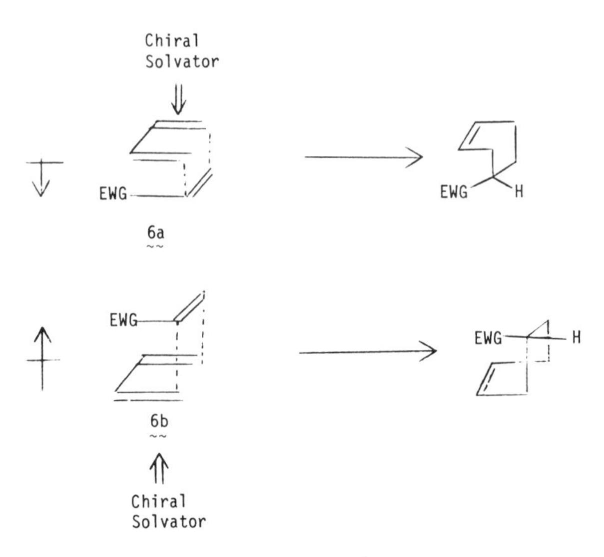

the dienophile (for electronic reasons), and that internal solvation would be more favorable than external solvation led us to explore incorporation of a mandelic acid group into the diene. In such a mandelate system, a large substituent L either projects toward (i.e. 7a) or away (i.e. 7b) from the diene and would thus bias the sense of stacking of the system to want the latter inter-

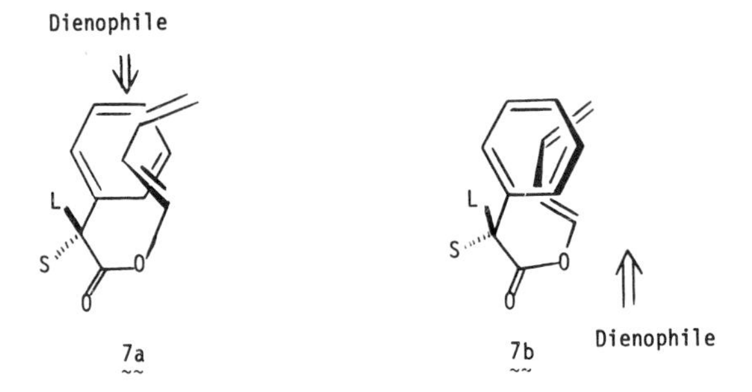

action. Choice of L=OCH_3 and S=H has the advantages that the chiral inducing agent is readily available in optically pure form and offers a direct method for analyzing the degree of asymmetric induction by NMR spectroscopy. In addition, it offers a method for the direct determination of absolute configuration. Mosher suggested a model, represented in 8a and 8b, in which the extended dihedral angle of 0° between R' and Ph in 8a and R and Ph in 8b leads to shielding of R' in 8a and R in 8b (in these extended Newman projections, the circle represents the CO_2 group) (6). An alternative model depicted in 9a and 9b suggests a dihedral angle of about 60° between R and Ph in 9a and R' and Ph

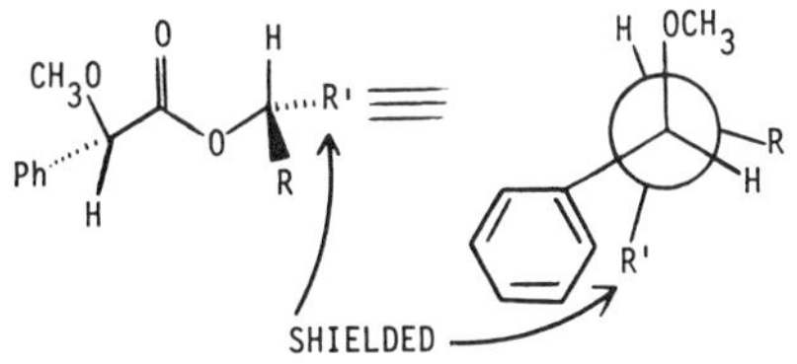

8a

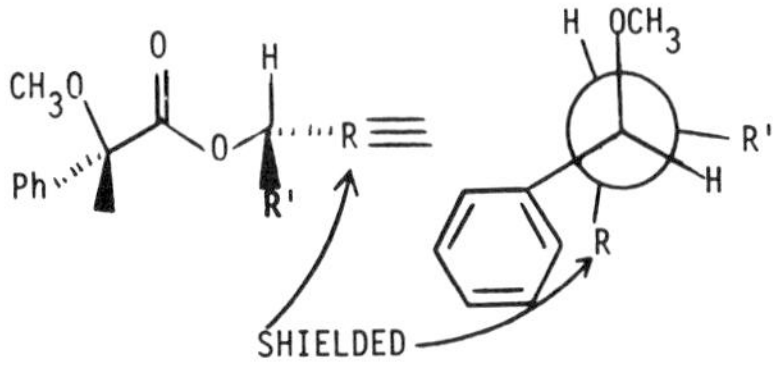

8b

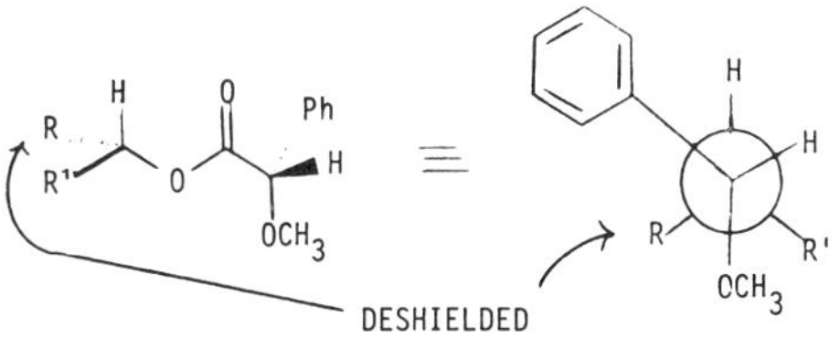

9a

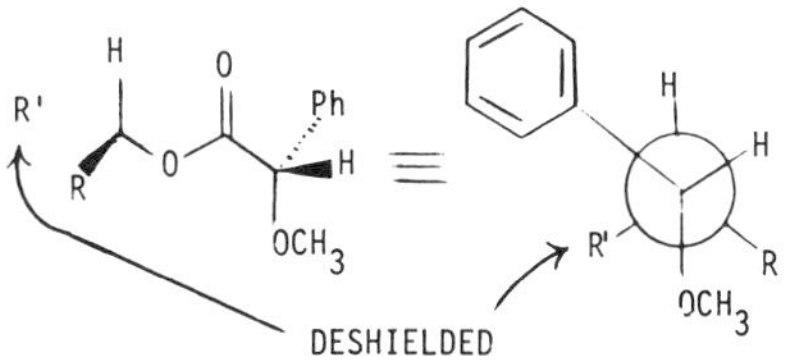

9b

in 9b which should lead to deshielding of R in 9a and of R' in 9b due to the anisotropy of the phenyl ring. Dipolar effects and extended eclipsing interactions tend to favor conformations 9a and 9b while eclipsing interactions around the carboxylate function tend to favor 8a and 8b. Fortunately, both conformational models predict the same trend - the S-mandelate ester of enantiomer 10a should exhibit the NMR signal for R downfield of the corresponding signal for the S-mandelate ester of enantiomer 10b

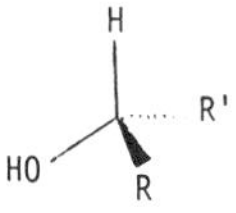

10a

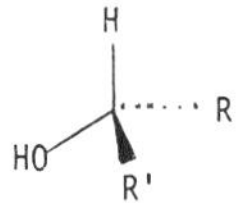

10b

and the converse situation holds for the signals for R'. We have found that this little used method has correctly predicted every absolute configuration we were able to verify independently. Chart 1 exemplifies some of the molecules with key NMR absorptions.

Diels-Alder condensation of diene 11 with acrolein and juglone, catalyzed by BF_3·ether and $B(OAc)_3$ respectively, led to adducts 12 and 13 (5). For 12, the ratio of the absorptions for the aldehydic protons at δ9.65 and 9.20 of 82:18 represents the degree of asymmetric induction and assigns the 1R,2R configuration to the major product - an interpretation confirmed by correlation with the known 1R,2R-2-hydroxymethylcyclohexanol.

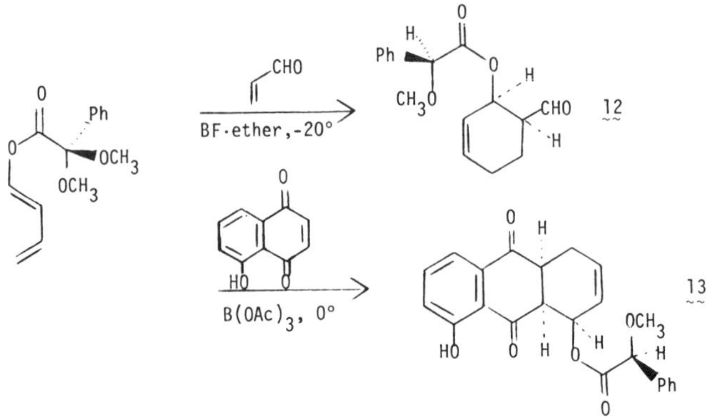

The 64% ee rises to ∿100% ee for the juglone adduct 13 whose enantiomeric purity and absolute stereochemistry were assigned by the NMR method (see Chart 1). Both results agree with a t.s. invoking folding as represented in 7b - in good accord with the model. Most interesting is the enhancement of the ee as a function of dienophile. Again the model presented accommodates this observation. For juglone, charge transfer interactions in the t.s. for cycloaddition should be more important than for acrolein - a fact that should lead to tighter complex formation at the t.s. and thus enhanced chiral recognition.

A strategy for synthesis of iboga alkaloids evolves from the acrolein cycloaddition (9,11). Focusing on the simplest conceptual approach to iboga alkaloids via Diels-Alder chemistry, a double bond must be introduced into the existing cyclohexyl ring. Such a retrosynthetic analysis, represented in equation 4, rapidly dissects the problem to the cycloaddition of acrolein to a 1-acyloxy-1,3-hexadiene. Scheme 2 summarizes the synthesis to

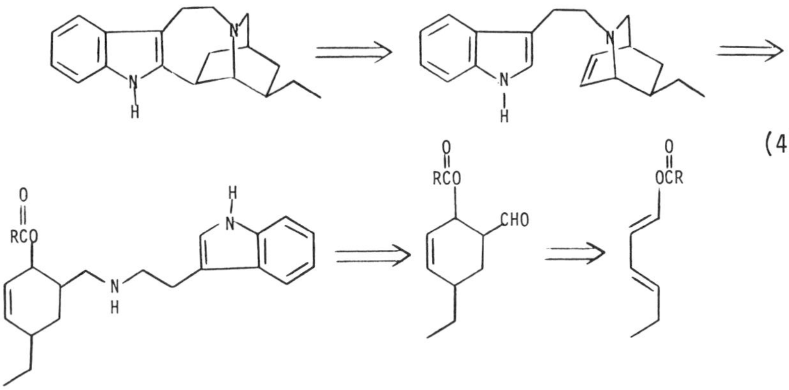

(4)

give 80% of 3R,4S,6R-ibogamine and 20% of the 3S,4R,6S isomer(9).

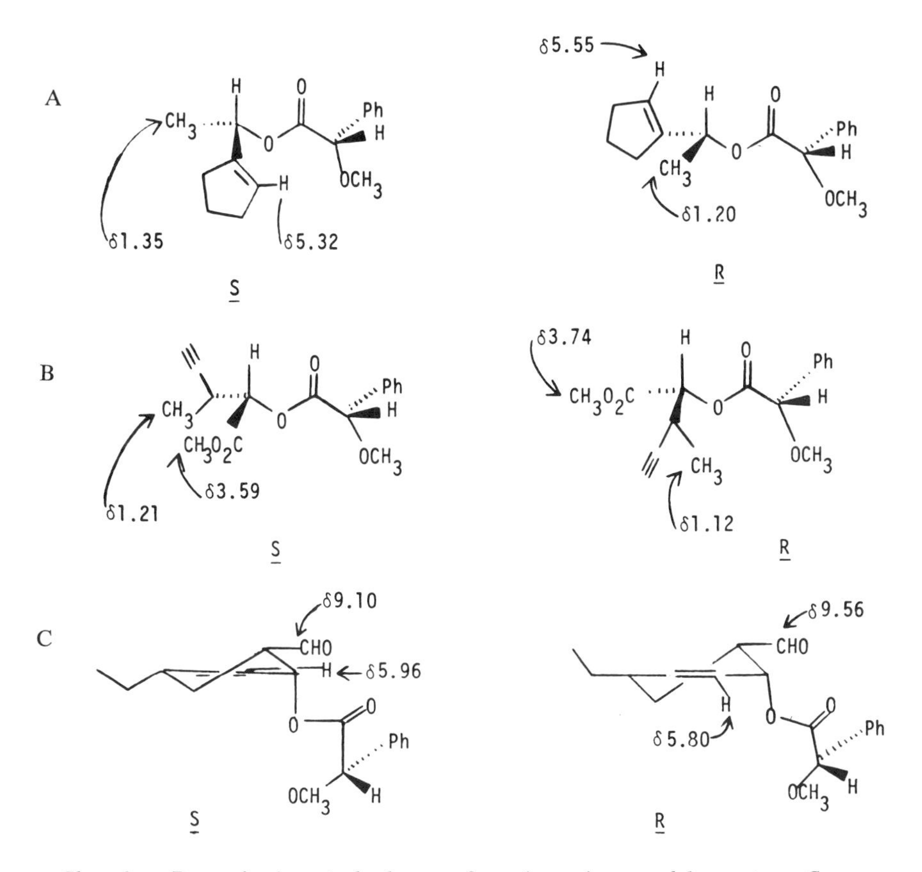

Chart 1. Determination of absolute configuration using mandelate esters. Configurations denoted refer only to carbinol carbon atom. Key: A, Ref. 7; *B, Ref.* 8; *C, Ref.* 9; *D, Ref.* 10; *and E, Ref.* 5. Continued on next page.

D

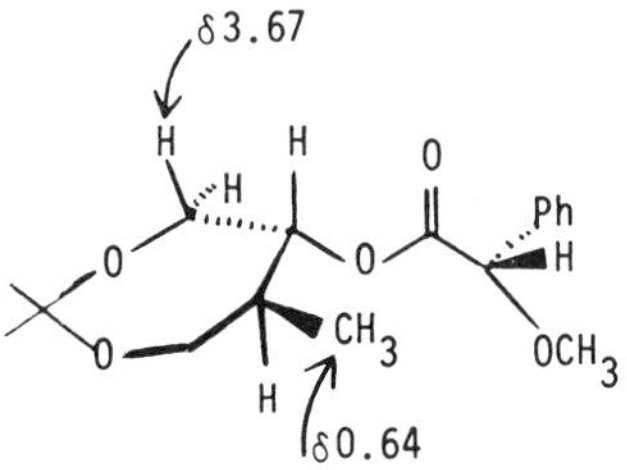

δ3.67 δ0.64 S

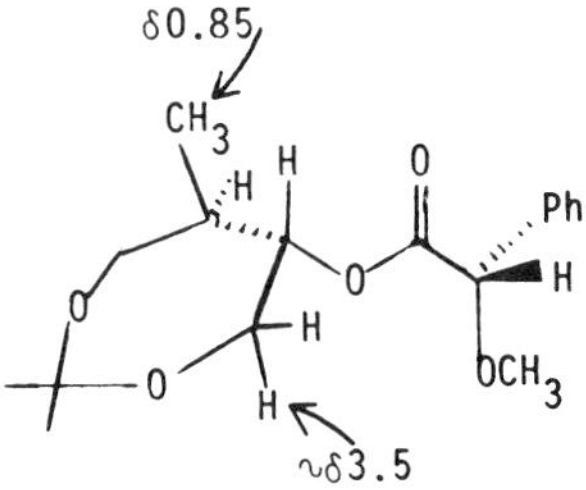

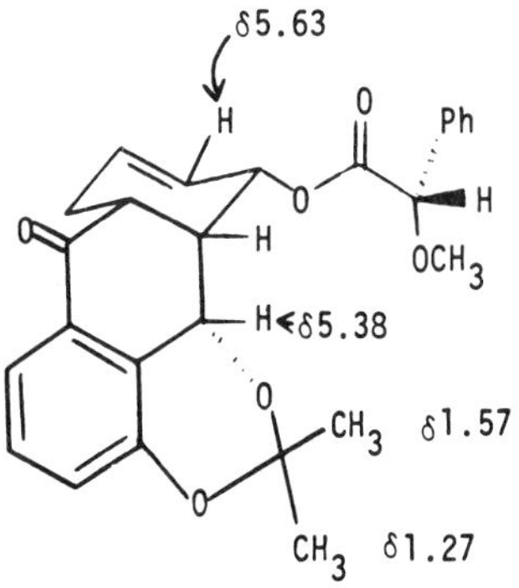

δ0.85 ~δ3.5 R

E

δ5.63 δ5.38 δ1.57 δ1.27 S

δ5.46 δ5.46 δ1.67 δ1.61 R

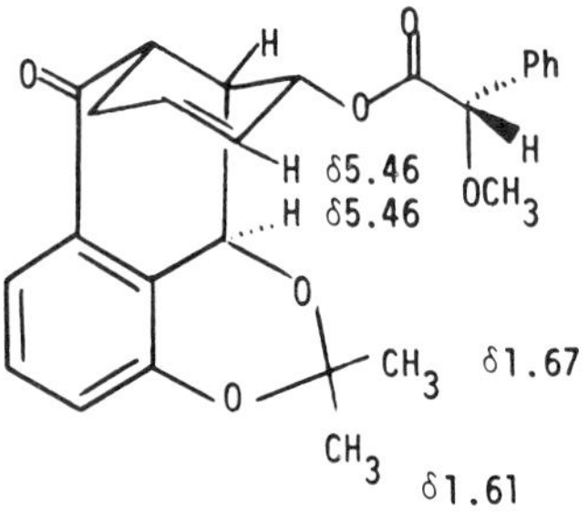

Chart 1. Continued. *Determination of absolute configuration using mandelate esters. Configurations denoted refer only to carbinol carbon atom. Key:* see *page 11.*

SCHEME 2. A Synthesis of Optically Active Ibogamine

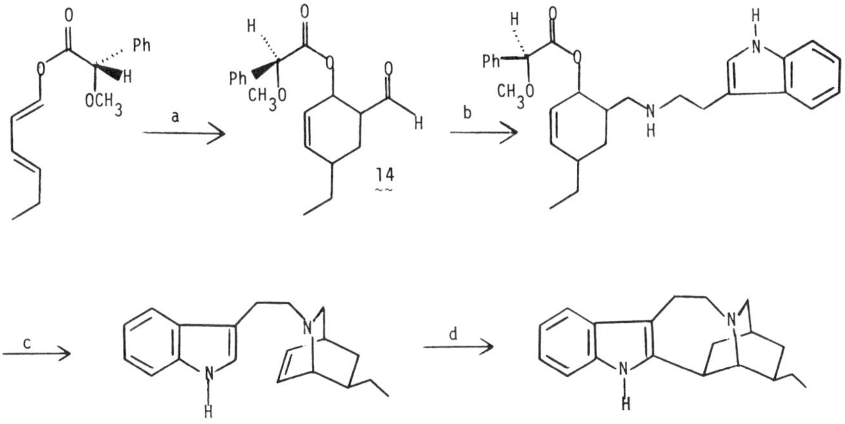

(a) acrolein, BF_3·ether. (b) tryptamine, $MgSO_4$, $PhCH_3$ then add $NaBH_4$, CH_3OH.
(c) 3%$(Ph_3P)_4Pd$, CH_3CN.
(d) $(CH_3CN)_2PdCl_2$, $AgBF_4$, $(C_2H_5)_3N$, CH_3CN then add $NaBH_4$.

The ee and absolute configuration of the initial adduct **14** assigned by the NMR method (see Chart 1) was verified by comparison of the final product with an authentic sample. A similar strategy was employed for the synthesis of catharanthine **15** (11).

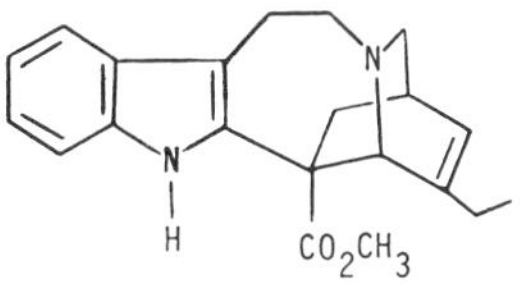

15

An approach toward anthracyclinone antitumor agents emerges from the juglone adduct **13** (12). Pillaromycinone **16** has all of its chiral centers in ring D, and they can emanate from a cyclohexenone **17** in which the *cis* ring juncture would direct the stereochemical introduction of the remaining substituents (see equation 5). The *cis* ring juncture of **17** can readily derive from a Diels-Alder reaction starting from a cyclohexenone **18**. This requires partial saturation of the B ring but in such a manner as to facilitate rearomatization, thus utilizing the presence of the two oxygen substituents. Furthermore, the *cis* B/C ring juncture of **18** will direct an incoming diene so as to create the correct

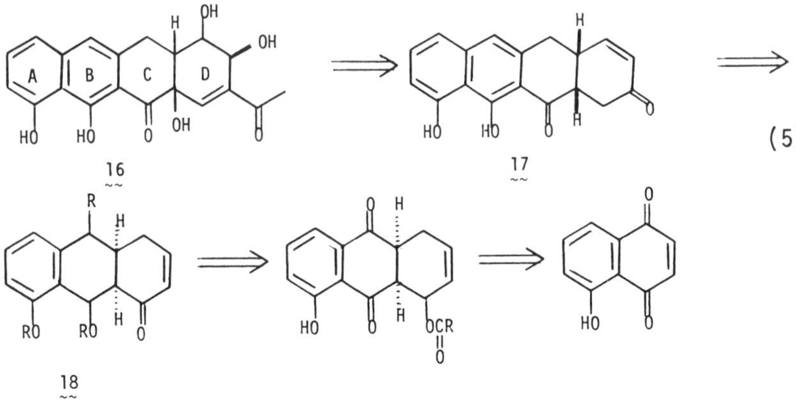
A
B
C
D
H
OH
OH
HO
HO
O
OH
O
16
H
HO
HO
O
H
O
17
R
H
RO
RO
H
O
18
O
H
HO
O
H
OCR
O
O
HO
O

(5)

stereochemistry of 17. The stereochemistry of 18 and, by extrapolation, that of 16 is established in the first step of synthesis, the Diels-Alder reaction, even though the pertinent chiral centers (in ring B) disappear in the final product. They represent stereochemical relay centers which are to be discharged once they served their function. Scheme 3 outlines the approach. Preliminary experiments suggest that treatment of 19 with $BF_3\cdot$ ether in methanol does produce the acetonide of deoxypillaromycinone.

Resolution Strategy

The mandelate esters serve yet an additional role in chiral synthesis - facilitation of resolution by hplc. The diastereomers in entries 1,2, and 4 of Chart 1 are all readily resolved on a Waters Prep 500 column utilizing 2-10% ethyl acetate in hexane (7,8,10,13). The diastereomer labelled S of entry 2, Chart 1, may be used for a synthesis of verrucarinic acid 21 as outlined in equation 6; the acid is produced as its bis t-butyldimethylsilyl ether for incorporation into verrucarin A (8).

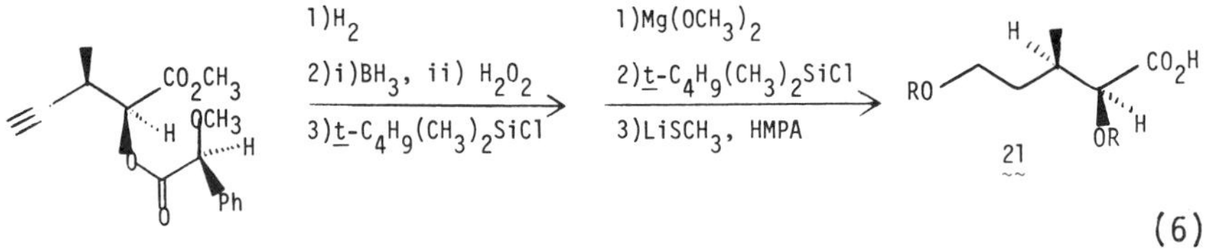
CO2CH3
H
OCH3
H
Ph
O
1)H2
2)i)BH3, ii) H2O2
3)t-C4H9(CH3)2SiCl
1)Mg(OCH3)2
2)t-C4H9(CH3)2SiCl
3)LiSCH3, HMPA
RO
H
CO2H
OR
H
21

(6)

The macrolide erythrynolide B, 22, offers a particularly striking strategy utilizing this type of resolution (equation 7). In particular utilizing 23 as a key intermediate which focuses on an allylic alkylation for the crucial ring formation, the alcohol 24 and acid 25 become simple precursors containing all the critical centers. Straightforward analysis rapidly converts

SCHEME 3. An Approach to Pillaromycinone

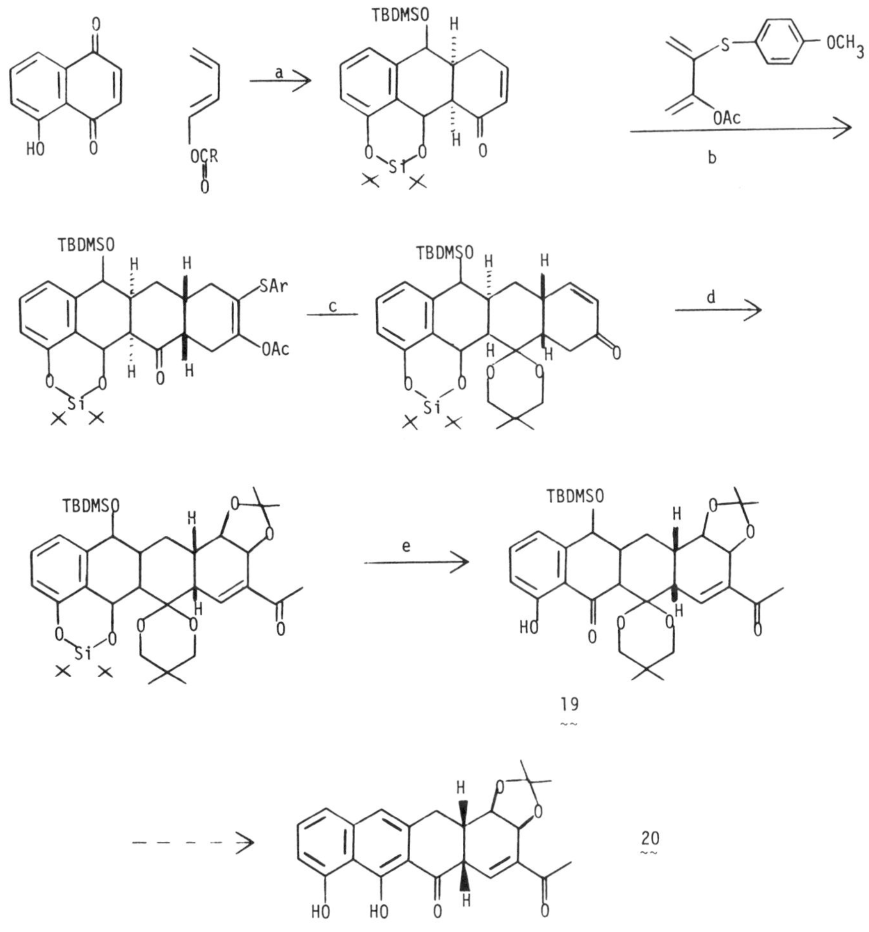

O
HO O
OCR
O
a
TBDMSO H
O Si O H O
OCH3
S
OAc
b

TBDMSO H H
SAr
OAc
H O H
O Si O
c
TBDMSO H H
H H O
O Si O O O
d

TBDMSO H
O O
O Si O O O H O
e
TBDMSO H
O O
HO O O O H O

19

H
O O
HO HO O H O
20

(a) i) $BF_3 \cdot$ether, $CHCl_3$, 25°; ii) $NaBH_4$, CH_3OH, $PhCH_3$; iii) $(\underline{t}\text{-}C_4H_9)_2SiCl_2$, 1-hydroxybenztriazole, $(C_2H_5)_3N$, CH_3CN; iv) $LiBH_4$, THF, 0°; v) $\underline{t}\text{-}C_4H_9(CH_3)_2SiCl$, imidazole, DMF, 50°; vi) DIBAL-H, $PhCH_3$, -78°; vii) Ac_2O, DMSO, $PhCH_3$.

(b) BCl_3, CH_2Cl_2, 0°. (c) i) $(CH_3)_2C(CH_2OH)_2$, TsOH, PhH; ii) MCPBA, CH_2Cl_2, -78°; iii) CH_3Li, ether; iv) $PhCH_3$, reflux.

(d) i) $Ph_2P(O)CH(CH_3)OCH_3$, $\underline{n}\text{-}C_4H_9Li$, THF, -78°; ii) O_2, $h\nu$, sensitizer, CH_2Cl_2.

(e) i) $C_5H_5N(HF)_x$, C_5H_5N, THF; ii) MnO_2, $(CH_3)_2CO$.

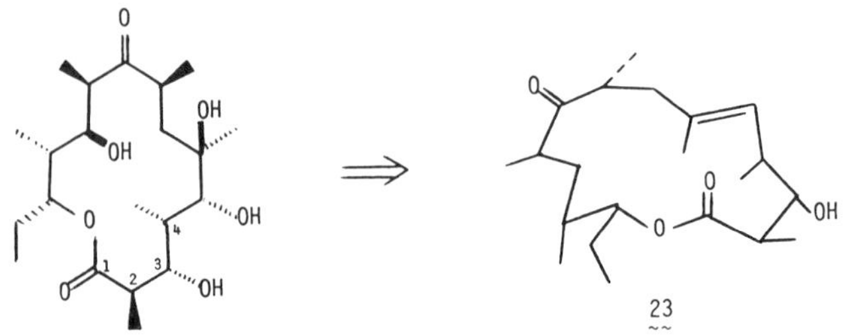

(7)

hydroxyketone **24** to alcohol **26** and acid **25** to alcohol **27**. However, **26** and **27** are simply mirror images. Thus, in this strategy, both enantiomers are needed - the 5*R*,6*R* isomer **26** for the northwestern half and the 5*S*,6*S* isomer **27** for the southeastern half. Preparative hplc of the mandelate ester of the racemate readily leads to their epimerically pure diastereomers whose optical purities and configurations were assigned by NMR spectroscopy (see Chart 1, entry 4) and confirmed by potassium carbonate hydrolysis to enantiomerically pure **26** and **27** ; the configurations so deduced were in agreement with the literature assignment (10). The recovered *O*-methylmandelic acid was suitable for recycling. Equations 8 and 9 show the approaches to **24** (10) and **25** (14) respectively.

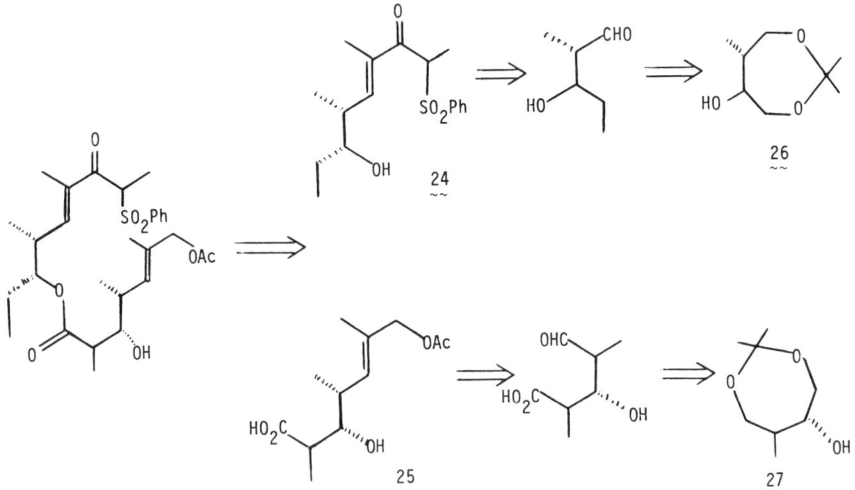

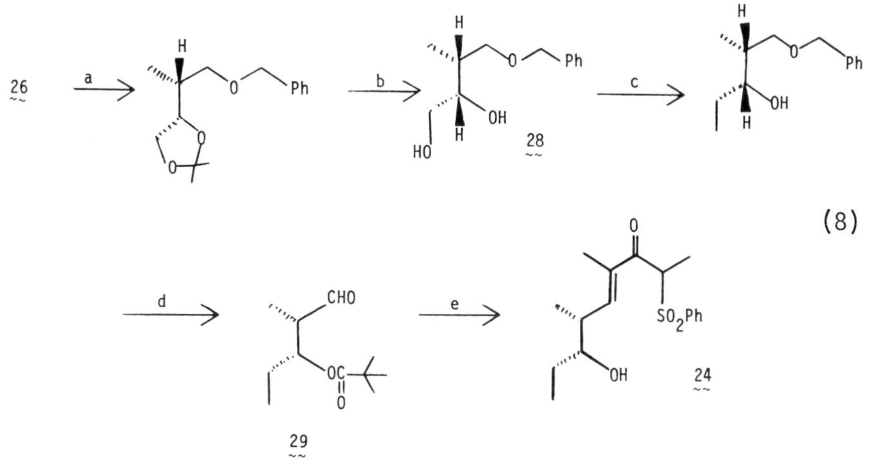

26 —a→ H, O, Ph, O, O —b→ H, O, Ph, OH, H, HO, 28 —c→ H, O, Ph, OH, H

(8)

—d→ CHO, OC, O, 29 —e→ O, SO_2Ph, OH, 24

(a) i) conc HCl, Δ; ii) NaH, $PhCH_2Br$, DMF, rt. (b) HCl, H_2O, CH_3CN, rt. (c) i) 2,4,6-($\underline{i}$-C_3H_7)$_3C_6H_2SO_2Cl$, C_5H_5N, rt; iii) $Li(CH_3)_2Cu$, ether, -78°. (d) i) $\underline{t}$-C_4H_9COCl, C_5H_5N; ii) 10% Pd/C, C_2H_5OAc, rt; iii) $(COCl)_2$, DMSO, $(C_2H_5)_3N$, CH_2Cl, -60°. (e) i) LDA, $CH_3CH_2C(O)CH(CH_3)SO_2Ph$, THF, -78°; ii) $SOCl_2$, C_5H_5N, -50°.

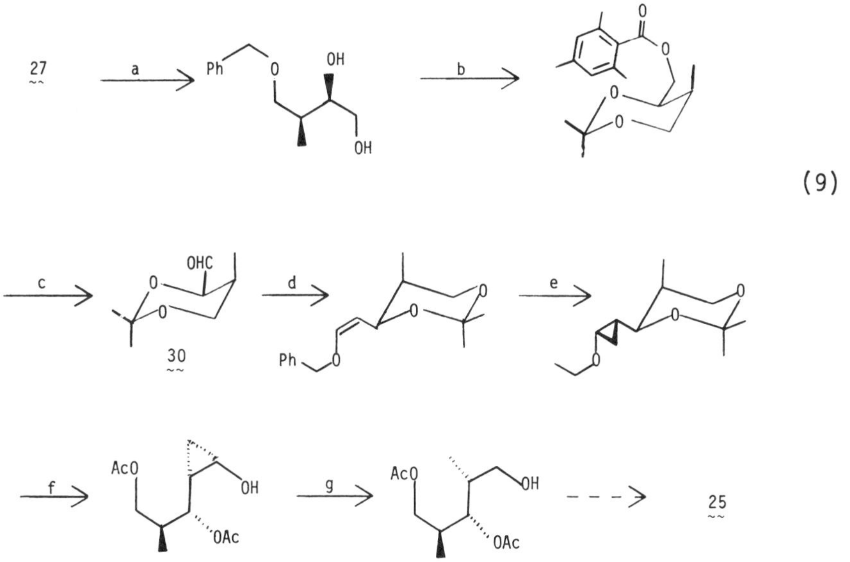

27 —a→ Ph, O, OH, OH —b→ O, O, O, O

(9)

—c→ OHC, O, O, 30 —d→ O, O, Ph, O —e→ O, O, O

—f→ AcO, OH, OAc —g→ AcO, OH, OAc - - -→ 25

(a) Steps a and b in equation 8. (b) i) 2,4,6$(CH_3)_3C_6H_2COCl$, C_5H_5N, -40°; ii) H_2, 10% Pd/C, CH_3OH, HOAc, rt; iii) $(CH_3)_2C(OCH_3)_2$ TsOH, rt. (c) i) $NaOCH_3$, CH_3OH, 65°; ii) DMSO, $(COCl)_2$, CH_2Cl_2,

Continued on p. 18.

Optically Active Building Blocks

An even more efficient strategy might be based upon the utilization of the same chiral building block for all the asymmetric centers provided such a building block is readily available in enantiomerically pure form. The benzyl ether of **31** (see equation 10) is the intermediate in the cuprate coupling of **28**

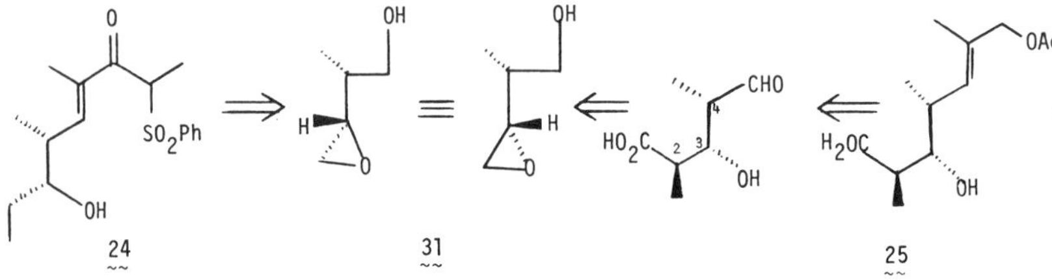

(10)

in equation 8, step c. Thus, the relationship of **31** to **24** is apparent. On the other hand, whereas the two original chiral centers of **27** (equation 7) correspond to C(2) and C(3) of erythrynolide B, the two chiral centers of **31** correspond to C(3) and C(4) (equation 10). In this approach, these two centers assume responsibility for creating C(2) of an appropriate configuration. Thus, **31** represents a single enantiomer from which all the chiral centers of erythrynolide B will spring.

The fact that **31** is a four carbon chain in which every carbon bears a substituent, three of them oxygen and one a methyl group, suggests *R*,*R*-tartaric acid as a logical precursor (15,16). Scheme 4 outlines an extraordinarily efficient route from *R*,*R*-tartaric acid to **31** in an overall yield of 64%. Correlation of **31** via cuprate coupling and selective formation of the pivalate at the secondary alcohol gives **32** which was previously derived from **29**.

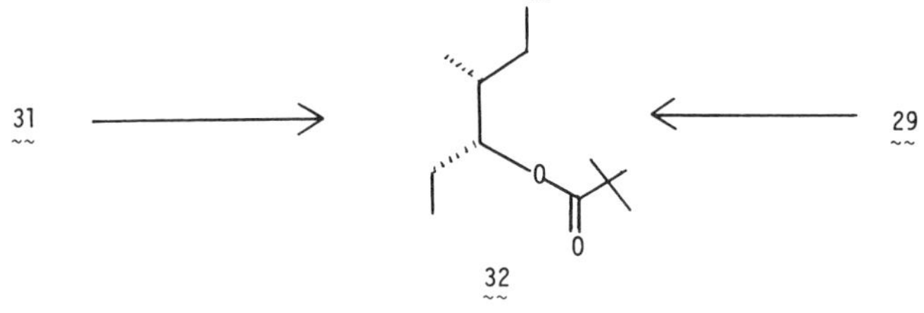

---

footnotes to equation 9 contd...
$(C_2H_5)_3N$, -60°. (d) $PhCH_2OCH_2\overset{+}{P}Ph_3Cl^-$, *t*-$C_4H_9OK$, THF, -78°. (e) CH_2I_2, Zn(Ag), DME, rt. (f) i) HOAc, CH_3OH, rt; ii) Ac_2O, C_5H_5N, rt; iii) H_2, Pd, CH_3OH. (g) i) $Hg(OAc)_2$, rt; ii) $NaBH_4$, CH_3OH.

SCHEME 4. Synthesis of a Key Synthon for Erythrynolide B

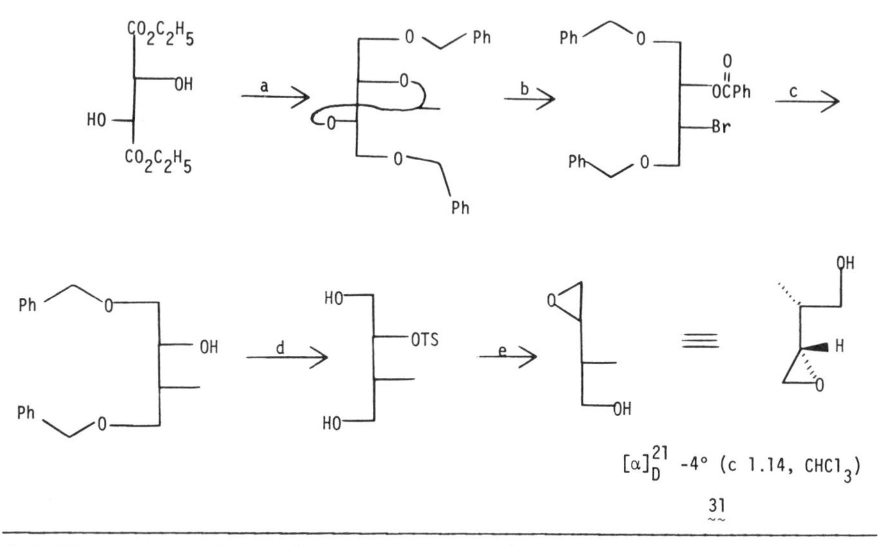

(a) i) PhCHO, $HC(OC_2H_5)_3$, TsOH, rt; ii) LAH, ether, Δ; iii) NaH, $PhCH_2Br$, THF. (b) NBS, CCl_4, Δ. (c) $Li(CH_3)_2Cu$, ether, 0°. (d) i) TsCl, C_5H_5N, rt; ii) H , 10%Pd/C, CH_3OH, HOAc. (e) NaOH, CH_3OH.

This correlation also provides unambiguous confirmation of the absolute stereochemistry of 29. Furthermore, 31 can be converted to aldehydes 33 and 34 which are related to 30 (equation 9) as shown in equation 11.

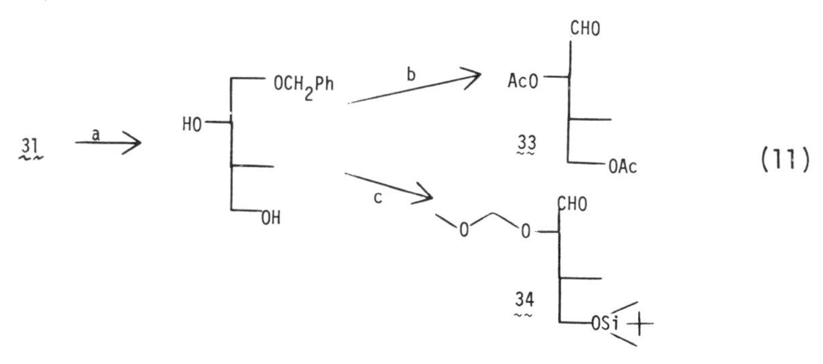

(11)

(a) $PhCH_2ONa$, $PhCH_2OH$, rt. (b) i) Ac_2O, C_5H_5N, rt; ii) H_2, 10% Pd/C, CH_3OH, HOAc; iii) PCC, CH_2Cl_2, 0°.
(c) i) $\underline{t}$-$C_4H_9(CH_3)_2SiCl$, imidazole, DMF, rt; ii) MEM-Cl, ($\underline{i}$-$C_3H_7)_2NC_2H_5$, CH_2Cl_2; iii) step ii and iii of b.

The creation of the final chiral center can be envisioned analogous to equation 9 but remains yet to be accomplished.

While the completion of the synthesis awaits the outcome of the above studies, the critical macrocyclization step has been demonstrated in a model. Thus, this approach provides much promise of efficiently creating the macrolide antibiotics.

Acknowledgment

I am indebted to a talented group of collaborators who transformed these ideas from dreams to reality. They are individually recognized in the references. Financial support was generously provided by the National Science Foundation, the General Medical Sciences Institute and the National Cancer Institute of the National Institutes of Health, and the University of Wisconsin.

Literature Cited

1. Trost, B.M.; Timko, J.M.; Stanton, J.L. Chem. Commun. 1978, 436.
2. Cohen, N.; Lopresti, R.J.; Neukom, C.; Saucy, G. J. Org. Chem. 1980, 45, 482 and references therein.
3. Saddler, J.C.; Donaldson, R.E.; Fuchs, P.L. J. Am. Chem. Soc. 1981, 103, 2110.
4. Mukaiyama, T.; Iwasara, N. Chem. Lett. 1981, 29. David, S.; Eustache, J.; Lubineau, A. J. Chem. Soc. Perkin I, 1971, 1795. Corey, E.J.; Ensley, H.E. J. Am. Chem. Soc. 1975, 97, 6908.
5. Trost, B.M.; O'Krongly, D.; Belletire, J.L. J. Am. Chem. Soc. 1980, 102, 7595.
6. Dale, J.A.; Mosher, H.S. J. Am. Chem. Soc. 1973, 95, 512.
7. Schmuff, N., unpublished work in these laboratories.
8. McDougal, P.G., unpublished work in these laboratories.
9. Trost, B.M.; Godleski, S.A.; Genet, J.P. J. Am. Chem. Soc. 1978, 100, 3930.
10. Belletire, J., unpublished work in these laboratories.
11. Trost, B.M.; Godleski, S.A.; Belletire, J. J. Org. Chem. 1979, 44, 2052.
12. Caldwell, C., unpublished work in these laboratories.
13. Cf. Bandi, P.C.; Schmid, H.H.O. Chem. Phys. Lipids, 1976, 17, 267.
14. Nishimura, Y., unpublished work in these laboratories.
15. Lubineau, A., unpublished work in these laboratories.
16. For an independent investigation on the use of tartaric acid derivatives as chiral building blocks, see Hüngerbühler, E.; Seebach, D.; Helv. Chim. Acta, 1981, 64, 687.

RECEIVED December 14, 1981.

2

Synthetic Control Leading to Natural Products

TERUAKI MUKAIYAMA

University of Tokyo, Department of Chemistry, Faculty of Science, Tokyo, Japan 113

New and useful asymmetric reactions have been developed based on the concept of "Synthetic Control". The concept of "Synthetic Control" is characterized by the utilization of common metal chelates for inter-or intramolecular interactions leading to highly stereospecific or entropically advantageous reactions. A variety of optically active compounds are obtained in much higher enantiomeric purities, compared with conventional methods, by utilizing chiral heterocyclic compounds such as chiral pyrrolidine or oxazepine derivatives, which have strong interactions with organometallic compounds to form tight complexes as intermediates. Similarly, asymmetric intramolecular Diels-Alder reactions are realized by the utilization of effective intramolecular metal chelation. Various natural products are successfully synthesized by the application of these reactions.

In this article, we summarize a variety of asymmetric syntheses guided by the principle of "Synthetic Control" described in the abstract.

0097-6156/82/0185-0021$05.00/0

Asymmetric syntheses based on chiral diamines. Optically active secondary alcohols are obtained by reduction of prochiral ketones with the chiral hydride reagent 1 prepared from lithium aluminium hydride and (S)-2-(N-substituted aminomethyl)-pyrrolidines, derived easily in four steps from commercially available (S)-proline.[1]

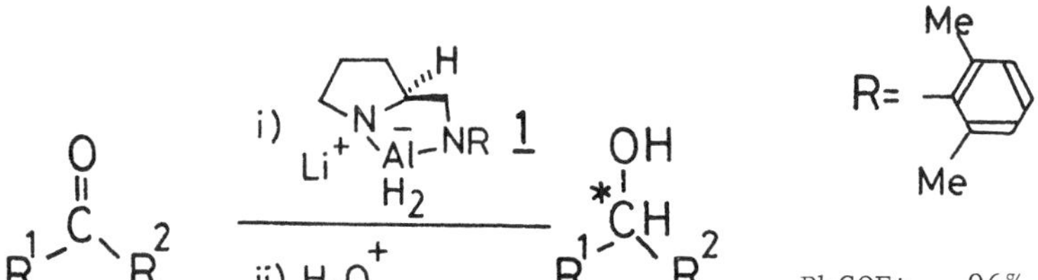

PhCOEt 96% ee

α-Tetralone 86% ee

The diamine 2 (R=Ph) was also applied to the synthesis of optically active α-hydroxyaldehydes. Treatment of the aminal, prepared from the chiral diamine and phenylglyoxal, with Grignard reagents affords hydroxyanimals, which, in turn, are hydrolyzed to yield α-alkyl-α-hydroxyphenylacetaldehydes.[2a]

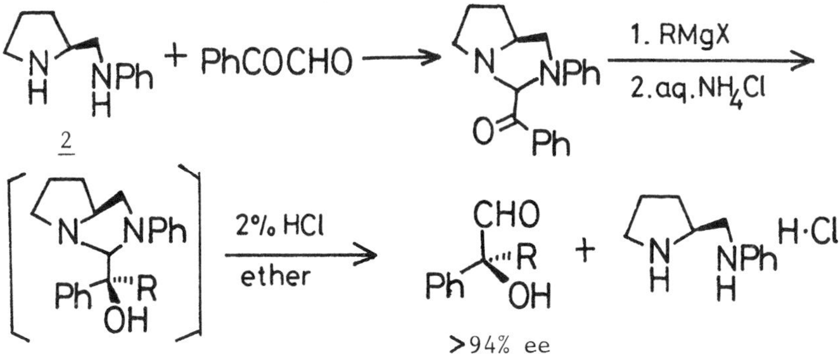

>94% ee

A more general and versatile method for the preparation of α-hydroxyaldehydes was also developed.[2b] Such aldehydes, in the desired configuration, are obtained by the following reaction sequence: i) reaction of one kind of Grignard reagent (RMgX) with the aminal of methyl glyoxylate, ii) diastereoselective addition of a second kind of Grignard reagent (R'MgX) to the ketoaminal, iii) hydrolysis of the resulting α-hydroxyaminal.

The two enantiomers of frontalin, a pheromone of several species of beetles belonging to the genus Dendroctonus, were separately synthesized by applying this asymmetric reaction.[2c]

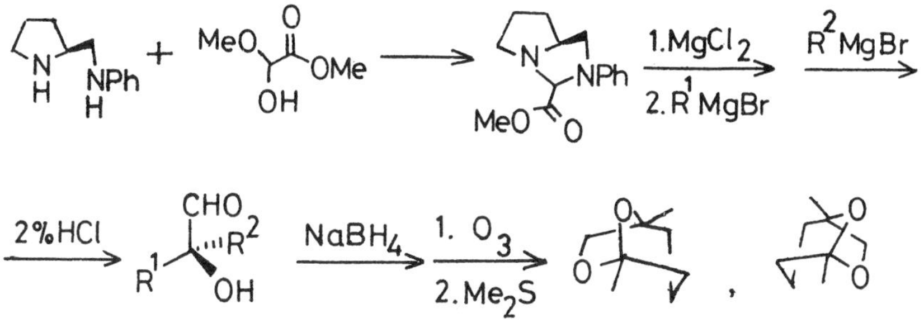

(−) 84% ee (+) 100% ee

Frontalin

(−): R^1=−$(CH_2)_3C(CH_3)=CH_2$, R^2= Me

(+): R^1= Me, R^2=−$(CH_2)_3C(CH_3)=CH_2$

Furthermore a new marine antibiotic, (−)-Malyngolide, discovered in the marine blue green alga Lyngbya majuscula Gomont, was synthesized in high optical yield by way of another application of this asymmetric reaction.[2d] The sequence is outlined in Scheme 1.

Optically active β-formyl-β-hydroxycarboxylic esters are obtained by employing either the lithium or zinc enolate of ethyl acetate in place of Grignard reagents in the above mentioned reaction.[2e]

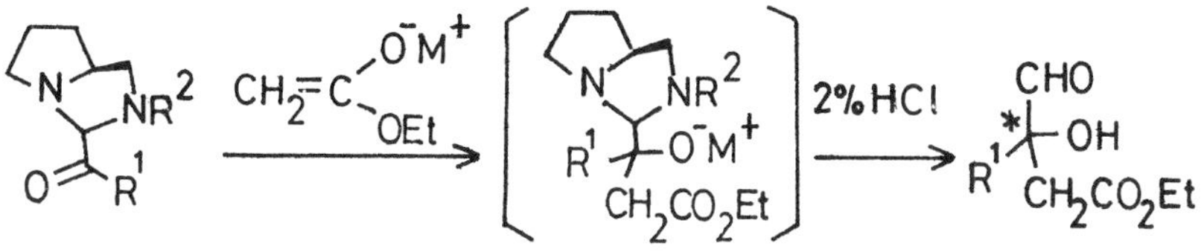

84–92% ee

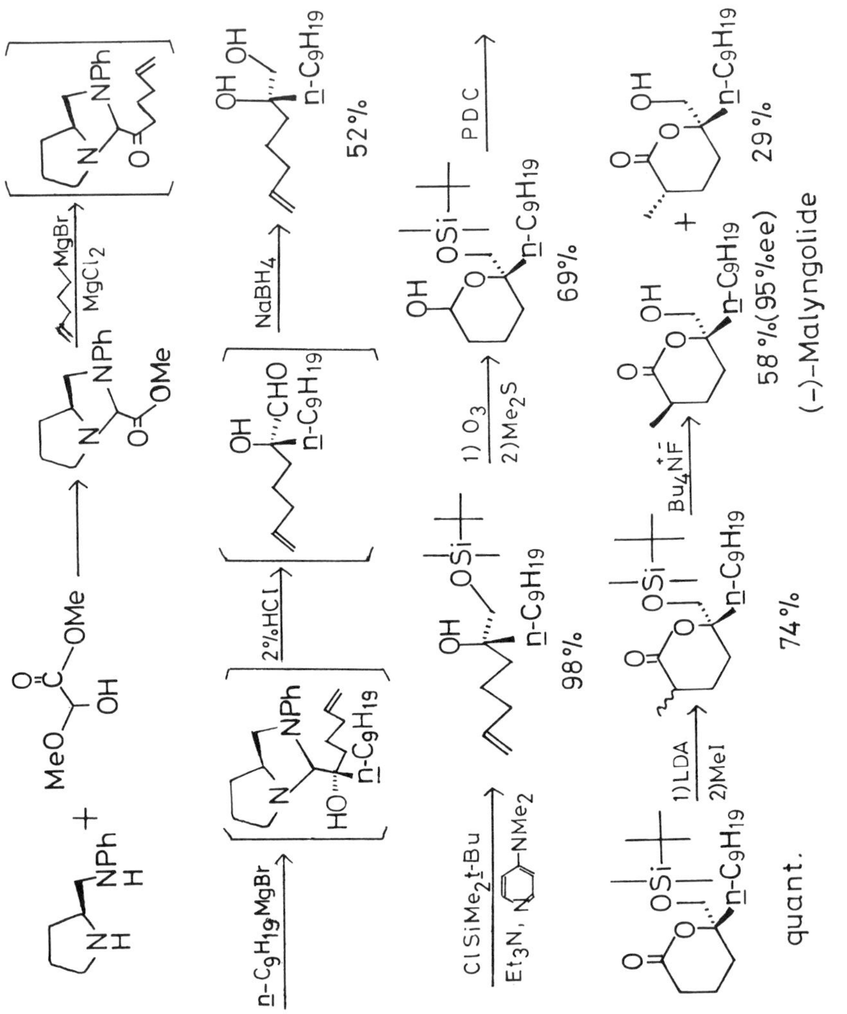

NPh
MeO
OMe
OH
MgCl2
NaBH4
52%
n-C9H19MgBr
2%HCl
CHO
ClSiMe2t-Bu
Et3N
NMe2
98%
1) O3
2)Me2S
69%
PDC
quant.
1)LDA
2)MeI
74%
Bu4NF
58 %(95%ee)
29%
(−)-Malyngolide

Scheme 1.

Asymmetric syntheses based on chiral aminoalcohols. Various chiral aminoalcohols 3, 4, 5, 6 were synthesized starting from (S)-proline, and the enantioselective addition of organometallic compounds to aldehydes in the presence of these aminoalcohols was investigated.

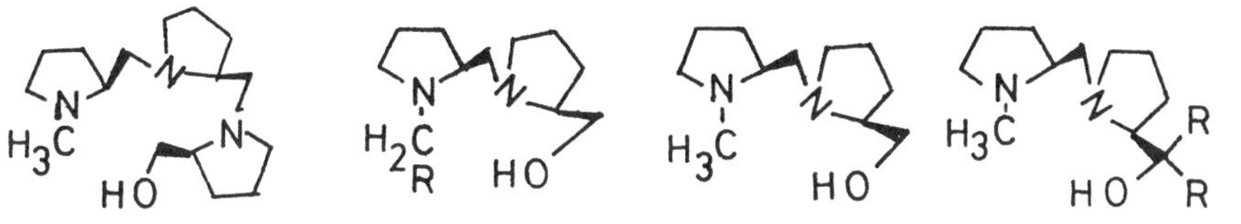

3 4 5 6

The enantioselective addition of alkyllithium reagents to aldehydes in the presence of the lithium salt of aminoalcohol 5 yields optically active secondary alcohols. High optical yields are achieved when the reaction is carried out in dimethyl ether and dimethoxymethane at low temperature.[3]

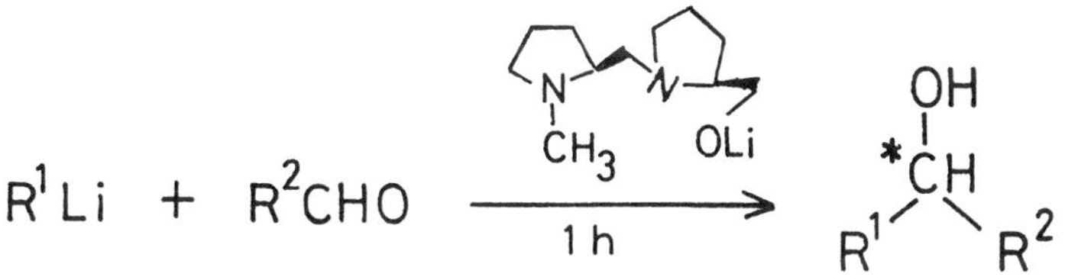

$$R^1Li + R^2CHO \xrightarrow[1\,h]{} R^1\overset{*}{C}H(OH)R^2$$

54-94% ee

It is of interest that the alcohols which possess the R configuration are produced by the reaction of dialkylmagnesiums with aldehydes, whereas the alcohols obtained by the similar reaction of alkyllithiums possess the S or R configuration, depending on the size of the alkyllithium.[3]

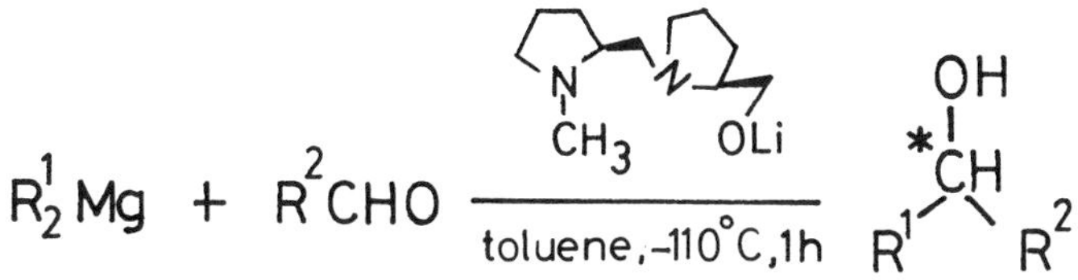

$$R^1_2Mg + R^2CHO \xrightarrow[toluene, -110^{\circ}C, 1h]{} R^1\overset{*}{C}H(OH)R^2$$

(R)-(+) 22-92% ee

By extending the above mentioned asymmetric addition of alkyllithium to other organolithium reagents such as lithium salts of methyl phenyl sulfide, 2-methylthiothiazoline, acetonitrile, N-nitrosodimethylamine, and trialkylsilylacetylenes, optically active oxiranes, thiiranes, aminoalcohols, and acetylenic alcohols are readily obtained.[4,5]

RCHO → (R'C≡CLi; chiral lithium alkoxide ligand) → R-*CH(OH)-C≡C-R'   R' = -H, -SiMe$_3$ etc

40–80% ee

Some of the optically active acetylenic alcohols were successfully converted to, e.g., γ-ethyl-γ-butyrolactone, the insect pheromone of Trogoderma, and important intermediates for the synthesis of substances with antibacterial activities.[6]

RCHO → ($Me_3SiC{\equiv}CLi$; chiral lithium alkoxide ligand) → R-*CH(OH)-C≡C-SiMe$_3$ → R-*CH(OH)-C≡CH → R-*CH(OH)-C≡C-COOH → lactones

Et (Trogoderma insect pheromone)   n-C_4H_9 → canadensolide   n-C_8H_{17} → avenaciolide

A chiral aminoalcohol 7, derived from ℓ-4-hydroxyproline, is found to be a superior catalyst for the enantioselective 1,4-addition of arylthiols to 2-cyclohexen-1-one to yield 3-arylthiocyclohexanones in high optical purities.[7]

RSH + 2-cyclohexen-1-one —(2 mol% 7)→ 3-(RS)-cyclohexanone, R = t-Bu-C$_6$H$_4$-

88% ee

Asymmetric synthesis based on a chiral oxazepine. (2R,3S)-3,4-Dimethyl-2-phenylperhydro-1,4-oxazepine-5,7-dione (8) was prepared from the half ester of malonic acid and ℓ-ephedrine and syntheses of various optically active carboxylic acids starting from this chiral oxazepine 8 were investigated.

(2-chloro-1-methylpyridinium TsO$^-$, Et$_3$N; CH_2Cl_2, 64%); LiOH, THF-H_2O; H_3O^+; (2-chloro-1-methylpyridinium TsO$^-$, Et$_3$N; CH_2Cl_2, 64%) → 8

i) Synthesis of β-substituted carboxylic acids.

Optically active β-substituted alkanoic acids are obtained by the reaction of the 6-alkylidene derivatives of 8 (9), which are easily prepared from 8 and aldehydes, with Grignard reagents in the presence of a catalytic amount of nickel chloride, followed by hydrolysis.[8]

RCHO + 8 →($TiCl_4$, PyH-THF) 9 (76-99%, E:Z=100:0-85:15)

→($R'MgX$, $NiCl_2$, THF, -78°C) →(H_3O^+) R–C*H(R')–CH₂–COOH (82- >99% ee)

The antibiotic indolmycin was synthesized in high optical purity as one application of this chiral reagent (Scheme 2).[9]

ii) Synthesis of optically active cyclopropanedicarboxylic acids.

The reaction of dimethylsulfoxonium methylide with 6-alkylideneoxazepine 9 followed by hydrolysis gives almost optically pure cyclopropanedicarboxylic acids in good yields.[10]

9 →($CH_2S(O)(CH_3)_2$, THF, rt) → (i) OH^-, ii) CH_2N_2) → dimethyl cyclopropanedicarboxylate (>90% ee)

iii) Synthesis of optically active 3-substituted γ-butyrolactones.

Almost optically pure 3-substituted γ-butyrolactones were obtained by the following sequence; i) the reaction of 6-alkylideneoxazepine 9 with phenylthiomethyllithium in the presence of a catalytic amount of nickel chloride, ii) the transformation of the adduct 10 to the dihydrofuran derivatives 11 by trimethyloxonium tetrafluoroborate, iii) acid hydrolysis of the dihydrofuran derivative 11.[11]

Scheme 2.

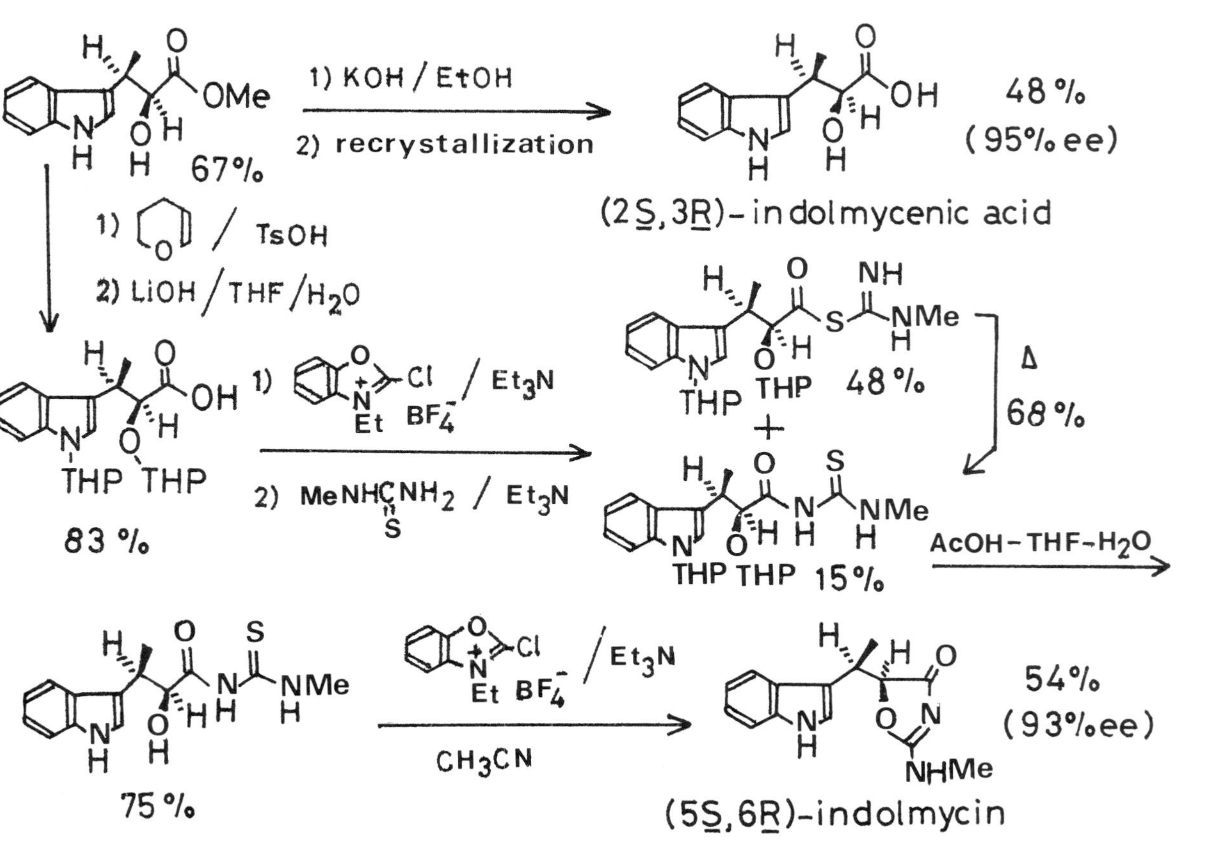

Scheme 2. (continued)

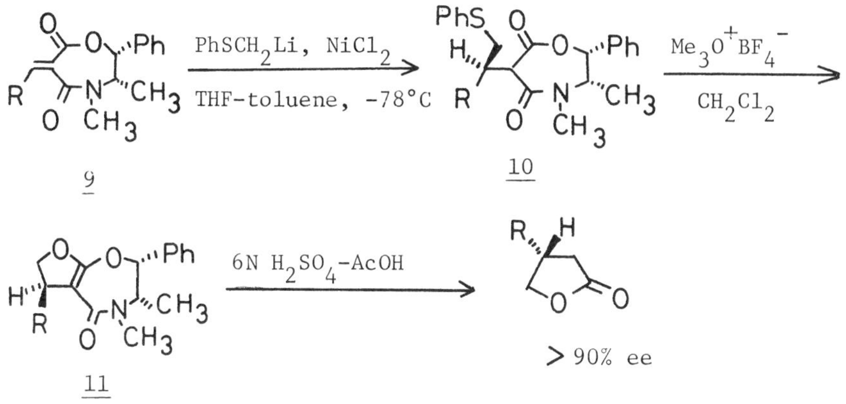

PhSCH$_2$Li, NiCl$_2$ / THF-toluene, -78°C; $Me_3O^+BF_4^-$ / CH_2Cl_2; 6N H_2SO_4-AcOH

9 10 11

> 90% ee

Diels-Alder reaction assisted by intramolecular interactions. The Diels-Alder reaction is one of the most important reactions in organic synthesis and has been applied to the synthesis of various natural products which possess a six-membered ring system. Unfortunately, however, there are some limitations to the structures of dienes and dienophiles that can be used successfully. For example, the Diels-Alder adducts between furan derivative and β,β-dimethylacrylic acid derivatives have not yet been isolated.

The adducts of some sterically hindered dienophile and furan derivatives are successfully obtained in good yields by the intramolecular Diels-Alder reaction of the diene and dienophile activated by an alkoxymagnesium salt coordinated to the same molecule. The acceleration of the reaction is apparently due to the coordination of the dienophile and a proximity effect which makes the reaction entropically advantageous.[12]

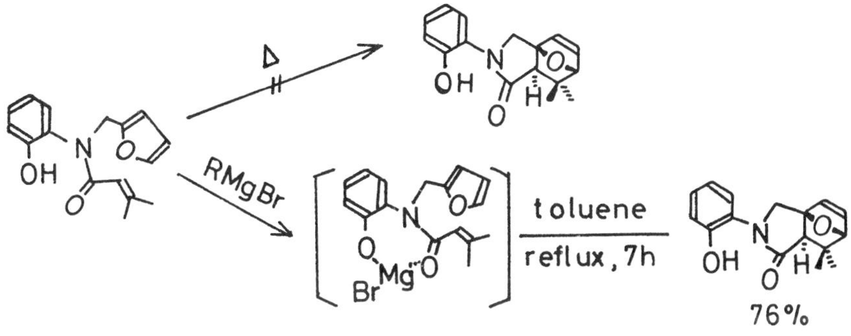

Δ; RMgBr; toluene / reflux, 7h

76%

This process was applied to the synthesis of the Karahana ether.[13]

60% HNO_3, AcOH, 80°C, 1.5h; 82%; i) KOH / EtOH, ii) H_3O^+; $LiAlH_4$, Et_2O; 68%; n-BuLi; $TiCl_4$, $LiAlH_4$; 64%; CH_2Cl_2; TsO⁻

Karahana ether

The above sequence was adapted to an asymmetric Diels-Alder reaction. The reaction of β,β-dimethylacrylic acid derivative 12, derived from (R)-2-amino-2-phenylethanol, afforded preferentially one diastereomer 13 in good yield, as did the crotonic acid derivative.

12 —RM→ —i) Δ, ii) H_3O^+→ 13 (88) + (12)

The adduct 13 is transformed into (+)-Farneciferol C.[14]

13 —H_2, Pd/C→ —Na, liq.NH_3-THF→ —i) $NaNO_2$ AcOH, ii) KOH EtOH, iii) H_3O^+→

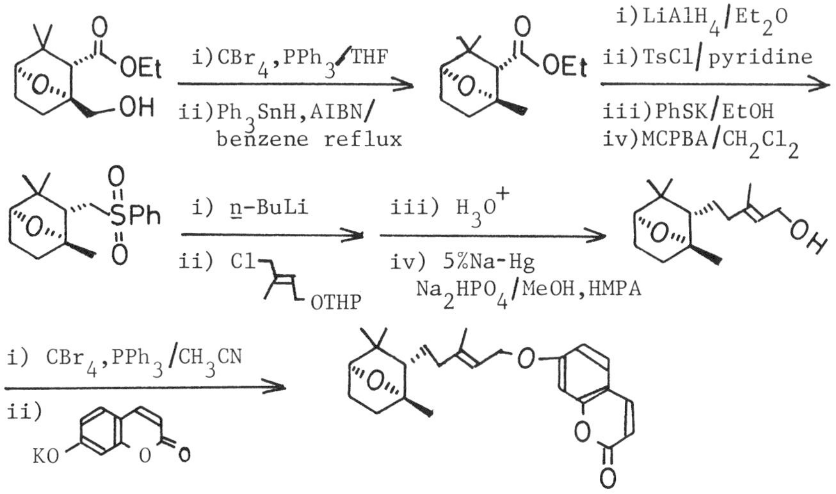

i) CBr_4, PPh_3/THF
ii) Ph_3SnH, AIBN/ benzene reflux

i) $LiAlH_4$/Et_2O
ii) TsCl/pyridine
iii) PhSK/EtOH
iv) MCPBA/CH_2Cl_2

i) n-BuLi
ii) Cl–(CH=C)–OTHP

iii) H_3O^+
iv) 5%Na-Hg Na_2HPO_4/MeOH, HMPA

i) CBr_4, PPh_3/CH_3CN
ii) KO–(coumarin)

(+)-Farneciferol C

This methodology, i.e., the introduction of an intramolecular chelate effect for highly selective reaction, has been further extended to the asymmetric Michael reaction. The reaction of N-crotylephedrine 14 with Grignard reagents, followed by acid hydrolysis, constitutes a simple procedure for obtaining highly optically pure carboxylic acids.[15]

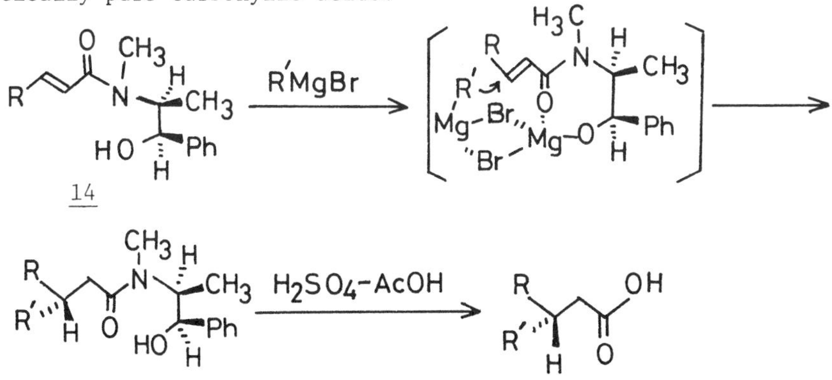

R'MgBr

14

H_2SO_4-AcOH

85- >99% ee

Stereospecific reactions leading to optically active sugar derivatives. The cadmium salt of the 2-allyloxybenzimidazole derivative 15 reacts with various aldehydes to afford adduct 16 in high regio- and stereoselectivity. Adducts 16 are subsequently transformed into trans-vinyloxirane 17.[16]

D- or L-Ribose is synthesized starting from 2,3-O-isopropylidene-D- or L-glyceraldehyde, respectively, by the application of this reaction (Scheme 3).[17]

15 → (n-BuLi, CdI_2; THF, -100°C) → [CdI^+ intermediate] → (R^1CHO)

16 57-98% (α/γ=75/25-90/10) → (NaH; THF, reflux) → 17 60-91%

$R^2=(CH_2CH_2O)_3Me$

Further, the mild allylation of carbonyl compounds with allyltin dihaloiodide, formed in situ by the oxidative addition of stannous fluoride to allyl iodide, has been applied to the synthesis of 2-deoxy-D-ribose starting from 2,3-O-isopropylidene-D-glyceraldehyde.[18]

allyl iodide + 2,3-O-isopropylidene-D-glyceraldehyde → (1) SnF_2 2) $PhOCH_2COCl$; DMI/DMF) → 18 74% ($OCOCH_2OPh$) (erythro:threo=81:19)

erythro-18 → (NH_4OH) → 71% → ($AcOH/H_2O$) → 95% → (1) O_3 2) Me_2S) → 2-deoxy-D-ribose 75%

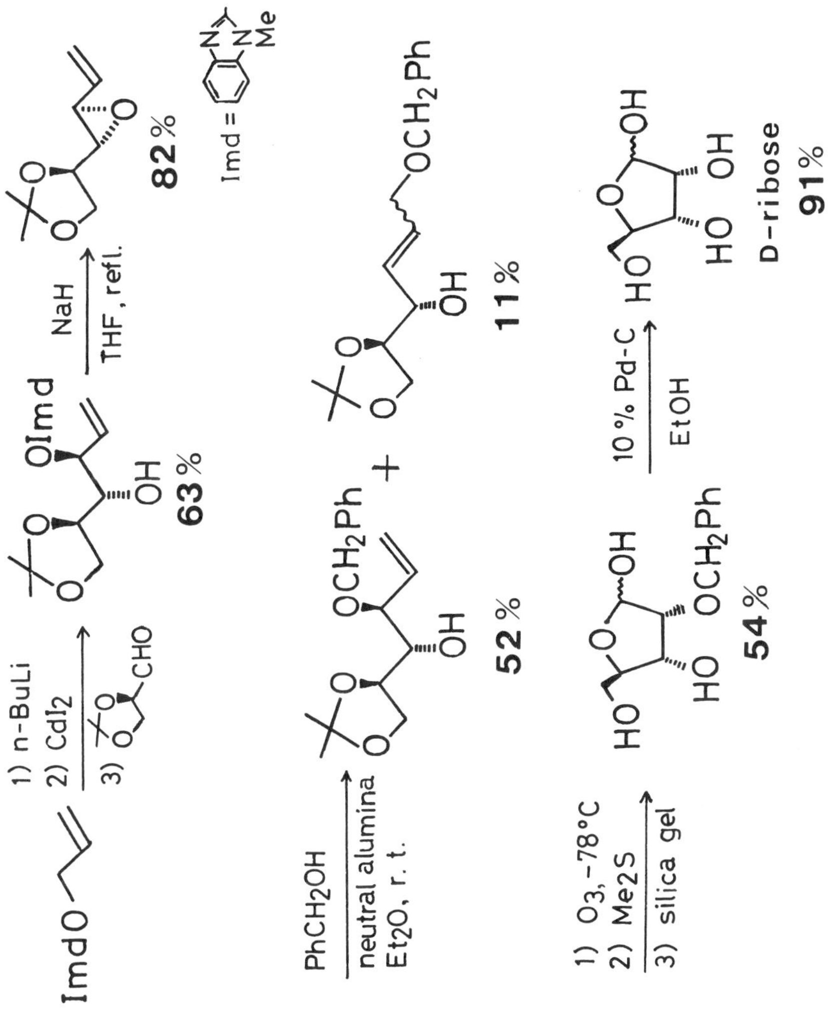

ImdO

1) n-BuLi
2) CdI_2
3) CHO

OImd
OH
63 %

NaH
THF, refl.

82 %

Imd = N N Me

$PhCH_2OH$
neutral alumina
Et_2O, r. t.

OCH_2Ph
OH
52 %

+

OH
OCH_2Ph
11 %

1) O_3, −78 °C
2) Me_2S
3) silica gel

HO
OH
HO OCH_2Ph
54 %

10 % Pd-C
EtOH

HO
OH
HO OH
D-ribose
91 %

Scheme 3.

Literature Cited

1) a) Mukaiyama, T.; Asami, M.; Hanna, J.; Kobayashi, S. Chem. Lett. 1977, 783. b) Asami, M.; Ohno, H.; Kobayashi, S.; Mukaiyama, T. Bull. Chem. Soc. Jpn. 1978, 51, 1869. c) Asami, M.; Mukaiyama, T. Heterocycles 1979, 12, 499.
2) a) Mukaiyama, T.; Sakito, Y.; Asami, M. Chem. Lett. 1978, 1253. b) Mukaiyama, T.; Sakito, Y.; Asami, M. ibid. 1979, 705. c) Sakito, Y.; Mukaiyama, T. ibid. 1979, 1027. d) Sakito, Y.; Tanaka, S.; Asami, M.; Mukaiyama, T. ibid. 1980, 1223. e) Sakito, Y.; Asami, M.; Mukaiyama, T. ibid. 1980, 455.
3) a) Mukaiyama, T.; Soai, K.; Kobayashi, S. ibid. 1978, 219. b) Soai, K.; Mukaiyama, T. ibid. 1978, 491. c) Mukaiyama, T.; Soai, K.; Sato, T.; Shimizu, H.; Suzuki, K. J. Am. Chem. Soc. 1979, 101, 1455. d) Sato, T.; Soai, K.; Suzuki, K.; Mukaiyama, T. Chem. Lett. 1978, 601.
4) Mukaiayama, T.; Suzuki, K.; Soai, K.; Sato, T. ibid. 1979, 447.
5) Soai, K.; Mukaiyama, T. Bull. Chem. Soc. Jpn. 1979, 52, 3371.
6) Mukaiyama, T.; Suzuki, K. Chem. Lett. 1980, 255.
7) Mukaiyama, T.; Ikegawa, A.; Suzuki, K. ibid. 1981, 165.
8) a) Mukaiyama, T.; Takeda, T.: Osaki, M. ibid. 1977, 1165. b) Mukaiyama, T.; Takeda, T.; Fujimoto, K. Bull. Chem. Soc. Jpn. 1978, 51, 3368.
9) Takeda, T.; Mukaiyama, T. Chem. Lett. 1980, 163.
10) Mukaiyama, T.; Fujimoto, K.; Takeda, T. ibid. 1979, 1207.
11) Mukaiyama, T.; Fujimoto, K.; Hirose, T.; Takeda, T. ibid, 1980, 635.
12) Mukaiyama, T.; Tsuji, T.; Iwasawa, N. ibid. 1979, 697.
13) Mukaiyama, T.; Iwasawa, N.; Tsuji, T.; Narasaka, K. ibid. 1979, 1117.
14) Mukaiyama, T.; Iwasawa, N. ibid. 1981, 29.
15) Mukaiyama, T.; Iwasawa, N. ibid. 1981, 913.
16) Yamaguchi, M.; Mukaiyama, T. ibid. 1979, 1279.
17) Yamaguchi, M.; Mukaiyama, T. ibid. 1981, 1005.
18) Harada, T.; Mukaiyama, T. ibid. 1981, 1109.

RECEIVED December 21, 1981.

3

Asymmetric Synthesis of Chiral Tertiary Alcohols in High Enantiomeric Excess

ERNEST L. ELIEL, JORMA K. KOSKIMIES, BRUNO LOHRI, W. JACK FRAZEE, SUSAN MORRIS-NATSCHKE, JOSEPH E. LYNCH, and KENSO SOAI

University of North Carolina, Department of Chemistry, Chapel Hill, NC 27514

Chiral 1,3-oxathianes have been used as adjuvants for highly stereoselective asymmetric syntheses. Acylation (direct or via an intermediate carbinol) proceeds to give exclusively equatorial products in which the chirality has been transferred to C(2) of the 1,3-oxathiane. Reaction of the acyl compounds with Grignard reagents gives predominantly one diastereomer of a 2-oxathianylcarbinol bearing two different alkyl groups on the carbinol carbon, following Cram's rule. Conditions for maximum stereoselectivity have been worked out. Cleavage of the oxathiane ($NCS/AgNO_3$) leads to α-hydroxy-aldehydes, RR'C(OH)CHO, from which glycols, RR'C-(OH)CH_2OH, tertiary alcohols, RR'C(OH)CH_3 and other derivatives can be prepared, generally in enantiomeric purity exceeding 90%. Suitable chiral 1,3-oxathianes can be conveniently derived from camphor (either enantiomer) or (+)-pulegone.

In 1971 we discovered (1) that the reaction of conformationally locked 2-dithianyllithium compounds with electrophiles (Corey-Seebach reaction) proceeds with remarkable stereoselectivity, giving virtually exclusively the equatorial substitution products, as exemplified in Scheme 1 (R=H). The preference for the 1,3-dithianyl-2-carbanion to undergo electrophilic substitution from the equatorial side amounts to over 6 kcal/mol (2), corresponding to a selectivity factor in excess of 10,000. This high preference was subsequently shown (3) to be due to stereoelectronic factors, in accord with theoretical predictions (4, 5, 6).

Stereoselectivities of such magnitudes resemble those found in enzymatic reactions and we resolved to try to apply the high

0097-6156/82/0185-0037$05.00/0

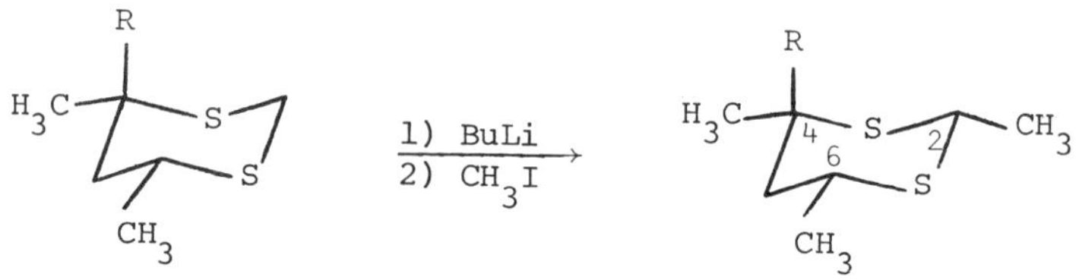

Scheme 1

selectivity to the design of a very efficient asymmetric synthesis. Before entering into details, we wish to state here the conditions of an effective asymmetric synthesis in general terms (7):

1) The synthesis must be highly stereoselective.

2) If a chiral adjuvant (chiral auxiliary reagent) is built into the starting material, the chiral center (or other chiral element) created in the asymmetric synthesis must be readily separable from the chiral adjuvant without racemization.

3) The chiral adjuvant itself must be recoverable in good yield and without loss of enantiomeric purity.

4) The chiral adjuvant should be readily (cheaply) available in enantiomerically pure form.

In addition, of course, the synthesis must proceed in acceptable overall chemical yield. Statement 1) is qualitative and its translation into quantitative terms is a matter of taste. Syntheses producing 85-90% enantiomeric excess (e.e.) usually allow purification of the chiral product (or an intermediate on the way) to enantiomeric purity by simple recrystallization. However, higher demands may be made in cases of syntheses of chiral liquids when crystalline intermediates are not accessible.

With respect to statements 2) and 3), these conditions are particularly easy to fulfill in catalytic asymmetric synthesis where 2) simply demands separation of the chiral product from the chiral catalyst and 3) is superseded by a requirement of reasonable turnover. (A turnover number of 100, modest by the standards of many catalytic reactions, is equivalent to a 99% recovery of chiral auxiliary reagent, which is rarely achieved!) Catalytic asymmetric syntheses are therefore particularly attractive, but, of course, they are often not available. Statement 3) may not apply when the chiral auxiliary reagent is very cheap (e.g. sucrose).

Contemplation of Scheme 1 suggests that its application to asymmetric synthesis should be facile if R = CH_3 and the starting dithiol is resolved. However, separation of the new chiral center (C-2) from the original one (C-6) would appear to be unachievable. We felt that this dilemma could be overcome by the use of 1,3-oxathianes instead of 1,3-dithianes and, indeed, it was found that the electrophilic reactions of 2-lithio-1,3-oxathianes are of the same order of stereoselectivity as those of the corresponding dithianes (8). However, an asymmetric synthesis based on the stereoselective cleavage of the C(2)-S bond followed by scission of the C(6)-O linkage in 2-alkyl-4,6,6-trimethyl-1,3-oxathianes gave disappointingly low optical yields (9).

A chance discovery pointed us toward the successful synthetic approach. 2-Acyl-1,3-oxathianes can be obtained with exclusively equatorial acyl groups by acylation of conformationally biassed 2-lithio-1,3-oxathianes or (in better yield) by reaction with aldehydes followed by Swern oxidation (10). We discovered (8) that reaction of these ketones with Grignard reagents once again proceeds highly stereoselectively giving one of the two possible tertiary alcohols in large excess over the other. The sequence of the two highly stereoselective steps is shown in Scheme 2. It is clear that starting from a chiral 1,3-oxathiane such as **1**,

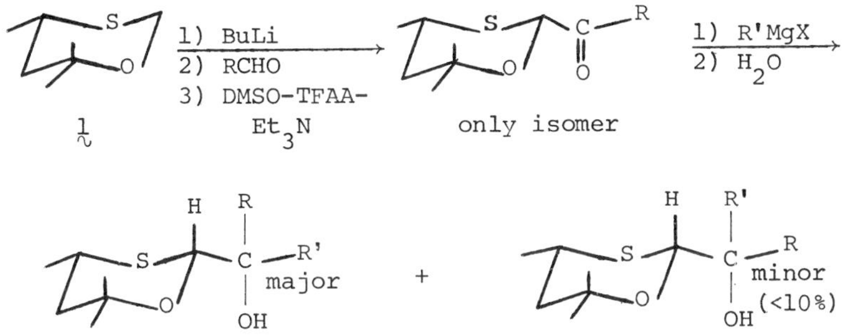

1) BuLi 2) RCHO 3) DMSO-TFAA-Et_3N — only isomer — 1) R'MgX 2) H_2O

major + minor (<10%)

Scheme 2

proceeding as shown in Scheme 2, and hydrolyzing the resulting product, one should be able to synthesize chiral α-hydroxyaldehydes in high optical yields. The process should also allow for recovery of the hydroxythiol from which the original 1,3-oxathiane was synthesized. The hydroxyaldehyde products, in turn, might be converted to chiral α-hydroxyacids, primary-tertiary glycols, primary-tertiary epoxides and other products bearing a chiral tertiary carbinol function of the type RR"C(OH)C-. Tertiary alcohol derivatives of the latter type are not easily available in high enantiomeric purity by conventional methods. If, in addition, the chiral oxathiane is readily available in high enantiomeric purity, all four conditions for a viable asymmetric synthesis previously discussed may be fulfilled.

Before discussing the experimental details and applications of the asymmetric synthesis outlined in Scheme 2, we shall take up its rationale and discuss briefly optimization of experimental conditions to obtain high optical yields.

The two-step synthesis summarized in Scheme 2 proceeds with high stereoselectivity in each step. The stereoelectronic rationale for the selectivity of the first step has already been discussed (2, 3). The second step is based on Cram's rule and, in particular, on the rule involving the rigid model where the metal of the organometallic reagent is complexed to an oxygen or nitrogen atom linked to the chiral center adjacent to the ketone function of the substrate (11). As shown in Scheme 3, application of this rule predicts the stereoselective outcome of the Grignard additions to 2-acyl-1,3-oxathianes if one also considers the fact

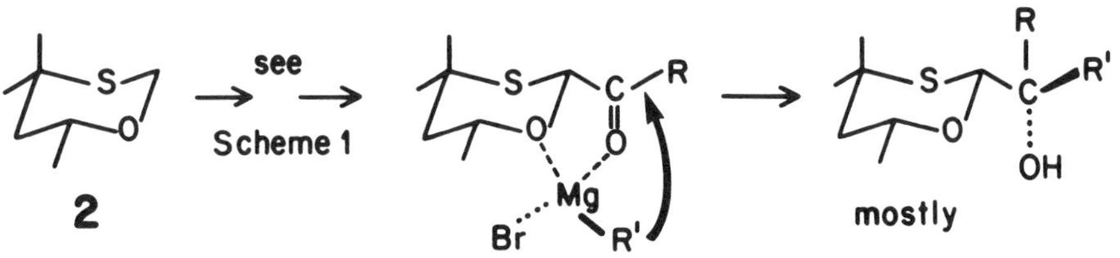

Scheme 3

that magnesium is a hard acid center and will therefore complex to the hard oxygen rather than to the soft sulfur in the oxathiane (12). Indeed, highly stereoselective additions of Grignard reagents involving operation of the rigid-model Cram's rule have been repeatedly observed in the literature (13).

For the success of the projected asymmetric synthesis it is crucial that the stereoselectivity of the Grignard addition be extremely high, with one diastereomer predominating over the other by 95% or more. (A 95% predominance of one diastereomer will translate into a 95% predominance of one enantiomer after hydrolysis, i.e. into a 90% e.e.) We therefore studied (14) variations in the structure of the oxathiane (**1**, Scheme 2 or **2**, Scheme 3), the organometallic reagent (R'MgI, R'MgBr, R'MgCl, R'$_2$Mg or R'Li), solvent (diethyl ether, tetrahydrofuran or mixtures thereof) and temperature (reflux temperature down to -78°C). We also studied the effect of varying R and R' (Scheme 2). The results of these experiments are summarized in Table 1. It was not necessary to employ optically active starting materials for this preliminary study, since, starting with racemic 2-acyl-1,3-oxathiane, one obtains two diastereomeric carbinols in the Grignard synthesis (Scheme 2) whose ratio is equal (or nearly so) to the ratio of enantiomers attainable when one starts with completely resolved **1** or **2** and subsequently hydrolyzes the tertiary carbinol product to a chiral α-hydroxyaldehyde. Analysis of the mixture of diastereomeric tertiary alcohols (Scheme 2) was generally performed by NMR spectroscopy: either the ratio of the areas of the C(2) protons (distinct in the two isomers) or the ratio of C-13 signals of carbon nuclei adjacent to the exocyclic chiral center [notably C(2), C(OH) or the carbon nuclei in R and R' adjacent to the carbinol function] could be employed to this end. Authentic mixtures of the diastereomers expected in each reaction were synthesized (for NMR comparison) from the lithio derivative of **1** or **2** and a ketone RR'C=O.

Perusal of Table 1 indicates the following features: a) Oxathiane **2** (Scheme 3) is somewhat superior to oxathiane **1** (Scheme 2), notably with alkyl (as distinct from phenyl) ketones. We ascribe this to the more ready accessibility of the oxygen atom in **2** (as compared to **1**) for complexation. b) High stereoselectivity is easier to achieve with phenyl ketones than with alkyl ketones. This is a common feature in asymmetric synthesis; we ascribe it to the lesser reactivity of phenyl compared to alkyl ketones caused by conjugation of the C=O double bond. (Other things being equal, reactions with higher activation energies have a better chance of being selective than those with lower activation energies; when the activation energies for reaction of both enantiomers are low, their difference must of necessity be small - the difference between two small numbers cannot be a large number - and so selectivity must also of necessity be low. Unfortunately the converse does not follow: If the activation energies for both enantiomers are high, their difference may or may not be large and selectivity may or may not be satisfactory.) c) Stereoselectivity is increased by lowering the temperature; this is particularly important for aliphatic ketones whose stereoselectivity is not satisfactory at room temperature but often becomes so at -78°C. The enhancement of stereoselectivity with temperature is a fairly general phenomenon based on the kinetic equation $k_R/k_S = e^{\Delta\Delta S^{\ddagger}/R}\, e^{-\Delta\Delta H^{\ddagger}/RT}$; as T decreases (and assuming that $\Delta\Delta H^{\ddagger}$ is negative, i.e. that predominance of the R-isomer is activation enthalpy controlled) k_R/k_S (i.e. selectivity) increases. d) Grignard reagents are superior to alkyllithium or dialkylmagnesium reagents. This may be due to better complexation of the metal in the former (cf. Scheme 3). e) Alkylmagnesium iodides are marginally superior to alkylmagnesium bromides which are appreciably better than alkylmagnesium chlorides. This order came as somewhat of a surprise since we had expected the chlorides to be better coordinating species. Earlier literature on this particularly point (13) is not clear-cut. f) In some instances, a mixture of ether and tetrahydrofuran is superior to diethyl ether itself, but the effect is marginal. Again this order was unexpected, since we had expected the better coordinating THF to interfere more with the formation of the complex postulated (Scheme 3) to be responsible for the stereoselectivity. Clearly

Table 1

Reactions of 2-Acyl-1,3-Oxathianes with Grignard and Other Organometallic Reagents (cf. Scheme 2)

Oxathiane	Acyl Group	Organometallic	Solvent	Temp.°C	% e.e.
1	C_6H_5CO	CH_3MgI	ether-THF (5:1)	reflux	98
1	C_6H_5CO	CH_3Li	ether	R.T.	50
1	CH_3CO	C_6H_5MgBr	THF	reflux	70
1	CH_3CO	C_6H_5MgBr	ether	reflux	28
1	CH_3CO	C_6H_5Li	ether-benzene (30:1)	0	44
2	C_6H_5CO	CH_3MgI	ether-THF (3:1)	reflux	96
2	C_6H_5CO	CH_3Li	ether	R.T.	72
2	CH_3CO	C_6H_5MgBr	ether	reflux	78
2	CH_3CO	C_6H_5MgBr	ether	-40	86
2	CH_3CO	C_6H_5Li	ether-benzene (30:1)	0	76
2	CH_3CO	$(C_6H_5)_2Mg$	dioxane-ether (1:10)	reflux	56
2	C_6H_5CO	C_2H_5MgI	ether	reflux	>99
2	C_2H_5CO	C_6H_5MgBr	ether	reflux	60
2	C_2H_5CO	C_6H_5MgBr	ether	-78	92
1	C_6H_5CO	C_2H_5MgI	ether-THF (4:1)	reflux	>99

2	C_6H_5CO	$(CH_3)_2CHMgI$	ether	reflux	98
2	$(CH_3)_2CHCO$	C_6H_5MgBr	ether	reflux	70
2	$(CH_3)_2CHCO$	C_6H_5MgBr	ether	-78	99
2	CH_3CO	C_2H_5MgI	ether	-78	94
2	C_2H_5CO	CH_3MgI	ether	reflux	60
2	CH_3CO	n-C_3H_7MgI	ether	reflux	80
2	CH_3CO	n-C_3H_7MgBr	ether	reflux	68
2	CH_3CO	n-C_3H_7MgCl	ether	reflux	54
2	CH_3CO	n-C_4H_9MgBr	ether	reflux	74
2	CH_3CO	(n-$C_4H_9)_2Mg$	ether-hexane	reflux	52
2	$(CH_3)_2CHCO$	CH_3MgI	ether	-78	54[a]
2	$(CH_3)_3CO$	CH_3MgI	ether	-78	92
2	CH_3CO	$H_2C=CHMgBr$	ether-THF (3:1)	-78	80
2	$CH_2=CHCO$	CH_3MgI	ether	-78	90
2	CH_3CO	$HC \equiv CMgBr$	THF	R.T.	88

[a] The reverse procedure (addition of isopropylmagnesium iodide to the methyl ketone at -78°) gave no addition product.

this is not the only effect of THF in the transition state. Similar observations had been made earlier (13).

In general, under appropriate conditions, the stereoselectivity of the reaction is remarkably high with ratios of 95:5 or more leading, eventually, to enantiomeric excesses above 90%. Similarly high stereoselectivity had been observed in a number of earlier examples involving Cram's rule in the rigid model (13).

With the information given in Table 1 as background, we proceeded to design an asymmetric synthesis involving a highly stereoselective 1,3-oxathiane alkylation (Scheme 2) followed by an almost equally selective Grignard step (Schemes 2, 3). The target compound chosen was atrolactic acid methyl ether and the chiral adjuvant was the resolved oxathiane **1**, obtained as shown in Scheme 4. Complete resolution of 3-benzylthiobutyric acid

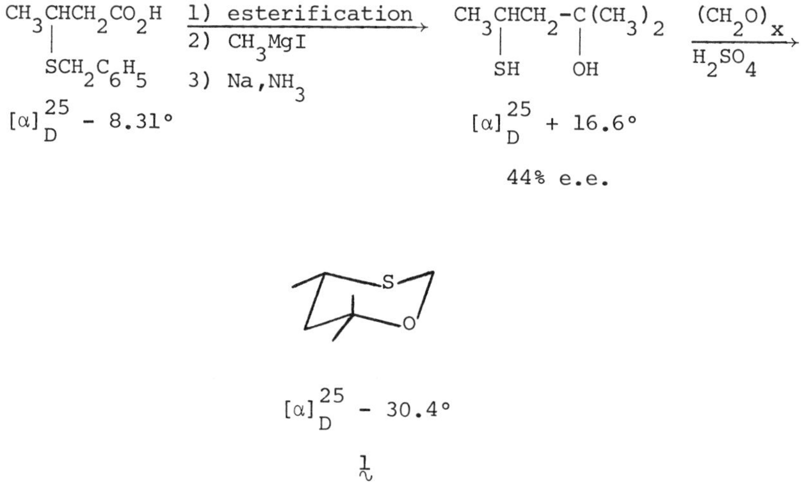

$CH_3CHCH_2CO_2H$ (with $SCH_2C_6H_5$ on C-3), $[\alpha]^{25}_D$ - 8.31° → [1) esterification; 2) CH_3MgI; 3) Na, NH_3] → $CH_3CH(SH)CH_2\text{-}C(CH_3)_2OH$, $[\alpha]^{25}_D$ + 16.6°, 44% e.e. → [$(CH_2O)_x$, H_2SO_4] → oxathiane, $[\alpha]^{25}_D$ - 30.4°, **1**

62% overall yield

Scheme 4

(from crotonic acid and benzyl mercaptan (15)) with brucine has been described in the literature (16) but is tedious, requiring seven recrystallizations. We employed cinchonidine for this task and contented ourselves with using starting material of 44% enantiomeric purity. (Actually, according to optical rotation values of our acid and of that in the literature, ours should have been 48% enantiomerically pure; however, the hydroxythiol obtained (Scheme 4) showed only 44% e.e. by a chiral shift reagent determination. Either the rotation of the "completely resolved" acid reported in the literature (16) is too low - i.e. the acid was not, in fact, completely resolved - or some racemization - by elimination-addition of benzyl mercaptan - occurred at the ester stage of the synthesis. If the former explanation is correct, this represents one of many failures we have encountered in trying to derive accurate enantiomeric excess data from optical rotations.) This may have been fortunate, since it is much easier to determine, with accuracy, e.e.'s of 44% than e.e.'s of 95% or more! Since reaction of lithio-oxathiane 1 (Scheme 4) with ethyl benzoate proceeded poorly, we carried out, instead, reaction with benzaldehyde followed by selective oxidation (Scheme 5) employing dimethyl sulfoxide - trifluoroacetic anhydride (or oxalyl chloride) - triethylamine (10). (Other oxidants tend to oxidize the sulfur as well as the carbinol.) Reaction of the ketone so obtained with methylmagnesium iodide gave the corresponding carbinol (Scheme 5) in virtually complete diastereomeric purity.

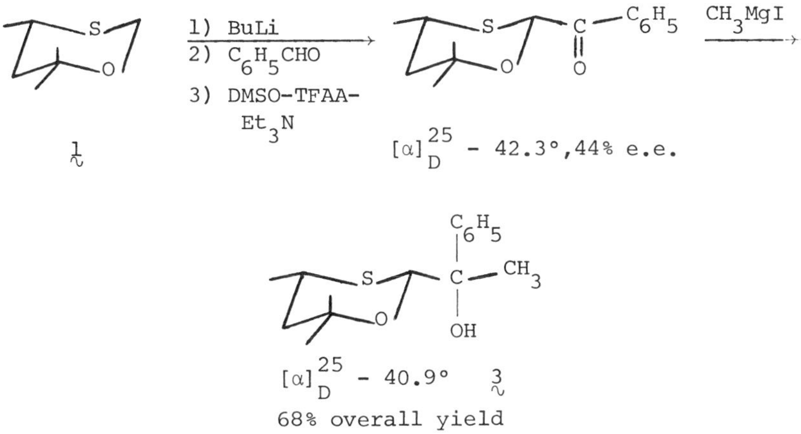

1 —— 1) BuLi; 2) C_6H_5CHO; 3) DMSO-TFAA-Et_3N ——→ ketone ($C(=O)C_6H_5$), $[\alpha]_D^{25}$ - 42.3°,44% e.e. —— CH_3MgI ——→

carbinol ($C(OH)(CH_3)C_6H_5$), $[\alpha]_D^{25}$ - 40.9° 3

68% overall yield

Scheme 5

The hydrolysis of oxathianes (17) is not as simple a reaction as one might surmise, especially in the present case where the molecule bears an acid-sensitive tertiary alcohol group. We finally carried out this step in good yield using methyl iodide/water/acetonitrile/calcium carbonate (18). Since the resulting α-hydroxyaldehyde could not readily be characterized and suffered extensive cleavage upon oxidation, we preceded the cleavage step by O-methylation and followed it by oxidation of the α-methoxyaldehyde to atrolactic acid methyl ether (4, Scheme 6). The optical purity of this acid was inferred from its rotation (19) to be 44% but for further confirmation the acid was converted (with diazomethane) to the methyl ester whose enantiomeric purity was found to be 44% by NMR analysis employing

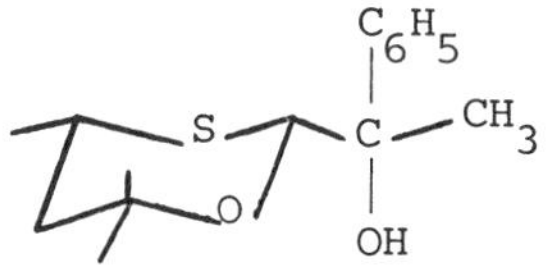

C_6H_5, S, C, CH_3, O, OH

3

$H-C(=O)-C(C_6H_5)(OCH_3)-CH_3$ $\xrightarrow[\text{reagent}]{\text{Jones}}$ $HO-C(=O)-C(C_6H_5)(OCH_3)-CH_3$

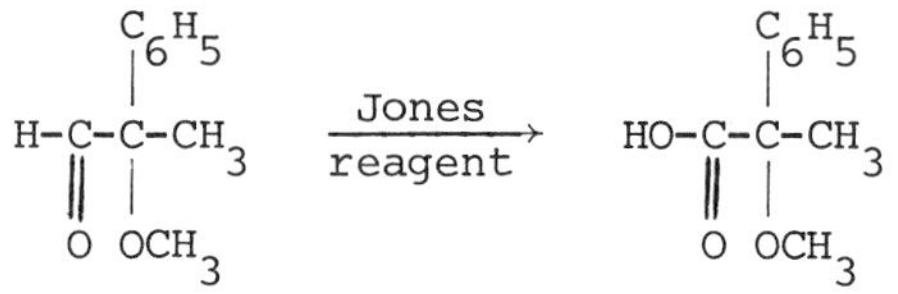

$[\alpha]_D^{25}$ -44.5° 4, $[\alpha]_D^{25}$ + 13.9° (MeOH)

59% overall yield

Scheme 6

a chiral shift reagent. The "optical yield" (= 100 x e.e. of product/e.e. of starting material) was thus quantitative. Using reasonable confidence limits for the chiral shift reagent determination, the enantiomeric purity of the product is 44±1% (and likewise for the starting material) giving an optical yield of 97-100%. Configurational correlation (9) of starting material 1 and product indicates the stereochemistry of the reaction to be compatible with the operation of Cram's (rigid model) rule. A similar synthesis based on 2 gave 92% optical yield (9).

Although both optical and chemical yield in these syntheses (reported (9) in 1978) are satisfactory, they are subject to criticism on two grounds, related to the conditions for a viable asymmetric synthesis laid down above: 1) The methyl iodide cleavage of the oxathianylcarbinol 3, while producing the α-hydroxyaldehyde in good yield, does not permit recovery of the chiral oxathiane (or hydroxythiol) moiety and thus falls afoul of condition 3. 2) The chiral oxathiane adjuvant is not enantiomerically pure and while, in principle, it can be obtained in enantiomerically pure form, this would be far from facile. Thus condition 4, relating to easy availability of the chiral adjuvant in optically pure form, is not fulfilled either.

It would appear that the easiest way of preparing an optically pure chiral 1,3-oxathiane might be from a chiral (and enantiomerically pure) natural product. In principle, all that is required is the existence of an S-C-C*-C-O moiety, with the linking C-C*-C unit being chiral. We have synthesized two pertinent oxathianes, one (**5**) derived from (+)- or (-)-camphor-10-sulfonic acid (20) (Scheme 7) and the other (**6**) derived from (+)-pulegone (21) (Scheme 8).

Both (+)- and (-)-camphorsulfonic acids are available commercially, the latter in form of its ammonium salt. While the free acid (a hemihydrate) is difficult to characterize, the ammonium salt is easily recrystallized to enantiomeric purity and characterized by specific rotation. Treatment of either acid or salt with thionyl chloride followed by lithium aluminum hydride reduction of the resulting sulfonyl chloride gives the exo alcohol 10-mercaptoisoborneol. Unfortunately the yield is only 50-55%, partly because the reduction of the sulfonyl chloride function to mercaptan is not clean, and partly because the reduction of the ketone function is not entirely stereoselective: about 10% of 10-mercaptoborneol (endo) is also formed and the isomer must be purified by column chromatography. Once pure, the hydroxythiol is converted to the corresponding oxathiane (**5**, Scheme 7) and the rest of the synthesis (20) proceeds in analogous fashion as that captioned in Schemes 5 and 6, except that conversion of oxathiane **5** to the lithium derivative required sec-butyllithium in this case and that the methyl iodide cleavage did not proceed as well as for the trimethyloxathiane case summarized in Schemes 5 and 6. Using (+)-camphorsulfonic acid as chiral adjuvant, we were able to obtain (+)-atrolactic acid methyl ether (2-phenyl-2-methoxypropionic acid, cf. Scheme 6) in 97±2% enantiomeric purity (20), as determined by NMR spectroscopy of the ester in presence of a chiral shift reagent.

Although this synthesis produces an essentially enantiomerically pure product, it has several drawbacks: the difficulty of purifying the starting material, the need for sec-butyllithium to make the lithio derivative of **5** and difficulties we encountered (22) at times in the methyl iodide cleavage. We decided therefore to look for yet another chiral oxathiane adjuvant - preferably one which would closely resemble the 4,4,6-trimethyloxathiane used in most of the preliminary experiments shown in Table 1 - and for a better method of cleavage.

The oxathiane which we now prefer is the one derived from pulegone as shown in Scheme 8 (21). Enantiomerically pure (+)-pulegone is a cheaply available perfume chemical. 1,4-Addition of benzyl mercaptan in the presence of base gives predominantly the more stable trans isomer 7-benzylthiomenthone which, upon reduction with sodium in liquid ammonia in presence of a proton donor, yields the more stable equatorial alcohol with simultaneous cleavage of the benzyl group, thus leading to 7-mercaptomenthol as the predominant isomer. This material can readily be freed of diastereoisomers by high pressure liquid chromatography. Alternatively, the crude material may be converted to the corresponding oxathiane (**6**, Scheme 8) which crystallizes and can thus be freed of diastereomeric impurities. (The major side product 7, derived from 7-mercaptoneomenthol, has found use as a chiral adjuvant in its own right; it can be isolated in pure form from the mother liquor by hplc.) The overall yield of the pure oxathiane from (+)-pulegone is about 30% (Scheme 8).

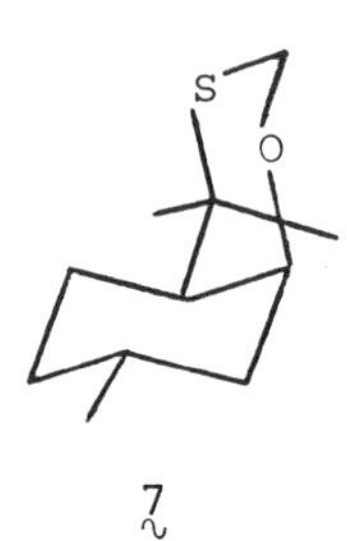

7

The oxathiane 6 shown in Scheme 8 can be used as discussed earlier (Schemes 5, 6). We shall describe here three syntheses undertaken with oxathiane 6: that of (+)-ethylmethylpropylcarbinol (21) (93% e.e., Scheme 9), that of (-)-ethylmethylphenylcarbinol (23) (100% e.e., Scheme 9) and that of mevalolactone (24) (87-97% e.e., Scheme 10).

Synthesis of carbinols 7a and 7b (Scheme 9) is entirely analogous to the syntheses described earlier. However, cleavage in the case of 7a and 7b was carried out by means of N-chlorosuccinimide (NCS) - silver nitrate (25). This cleavage produces not only the chiral α-hydroxyaldehydes in good yields but also returns the hydroxythiol chiral adjuvant in form of the pair of diastereomeric sultines 8. (Lithium aluminum hydride reduction of 8 yields the hydroxythiol which, upon treatment with paraformaldehyde and acid, regenerates oxathiane 6.) In this way condition 3 for a viable asymmetric synthesis (vide supra), namely recovery of the chiral adjuvant, is finally fulfilled. Reduction of the rather sensitive hydroxyaldehydes with sodium borohydride gives glycols 9a and 9b whose enantiomeric purity was determined by either Mosher's acid or chiral shift reagents and NMR spectroscopy to be 94% in the case of 9a and 99% in the case of 9b (26). Conversion of the glycols to monotosylates followed by hydride reduction gives the tertiary carbinols 10a and 10b in 93% and 100% enantiomeric purity, respectively (27). (The increase from 99% to 100% e.e. in going from 9b to 10b if indeed significant, is probably due to the fact that the intermediate tosylate was crystallized.) The enantiomer of 9b was synthesized similarly using oxathiane 7 (Scheme 9) as a chiral template; the enantiomeric excess of the product was 97%. Tertiary alcohols such as 10 have in the past been accessible in enantiomerically pure form only with difficulty, though recently alternative means of asymmetric synthesis have been disclosed (27, 28).

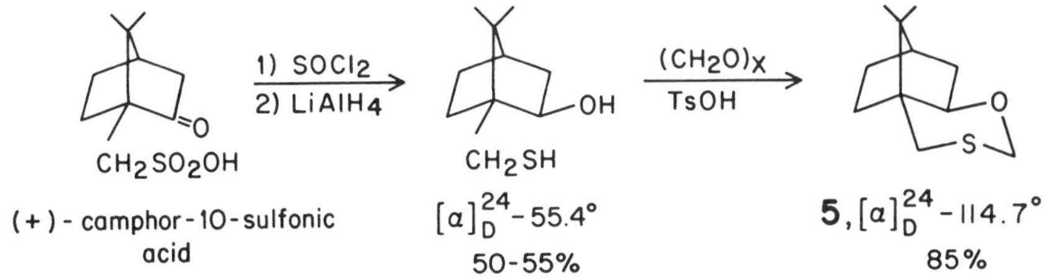

(+)-camphor-10-sulfonic acid (CH_2SO_2OH) → 1) $SOCl_2$ 2) $LiAlH_4$ → (CH_2SH, OH) $[\alpha]_D^{24}$ −55.4° 50-55% → $(CH_2O)_x$, TsOH → **5**, $[\alpha]_D^{24}$ −114.7° 85%

Scheme 7

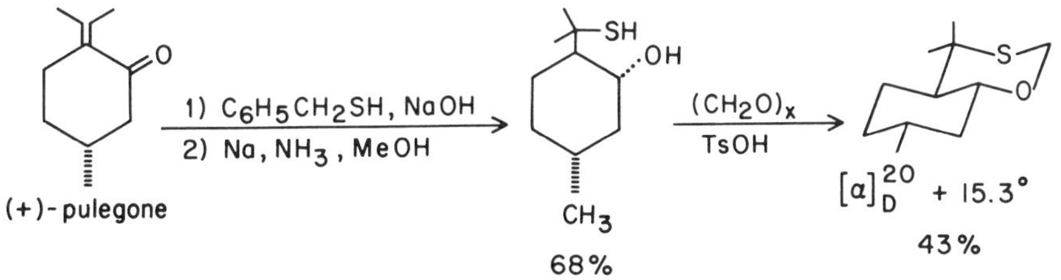

(+)-pulegone → 1) $C_6H_5CH_2SH$, NaOH 2) Na, NH_3, MeOH → (SH, OH, CH_3) 68% → $(CH_2O)_x$, TsOH → $[\alpha]_D^{20}$ +15.3° 43%

Scheme 8

6 → 1) BuLi, RCHO 2) DMSO-TFAA-Et_3N → (S, O, C(=O)R) → R'MgX → (S, O, C(R)(R')OH) → NCS, $AgNO_3$ → **8** (S~O, O) + $HC(=O)-C(R)(OH)-R'$ → $NaBH_4$ → $HOH_2C-C(R)(OH)-R'$ **9**

8 → LAH → (SH, OH) → $(CH_2O)_x$, TsOH → **6**

9a, 94%, e.e.
9b, 99%, e.e.

1) TsCl, C_5H_5N 2) $LiBHEt_3$ or LAH → $H_3C-C(R)(OH)-R'$ **10**

Series a: R = C_2H_5, R' = n-C_3H_7
Series b: R = C_6H_5, R' = C_2H_5

10a, 93% e.e., 17% overall yield
10b, 100% e.e., 29% overall yield

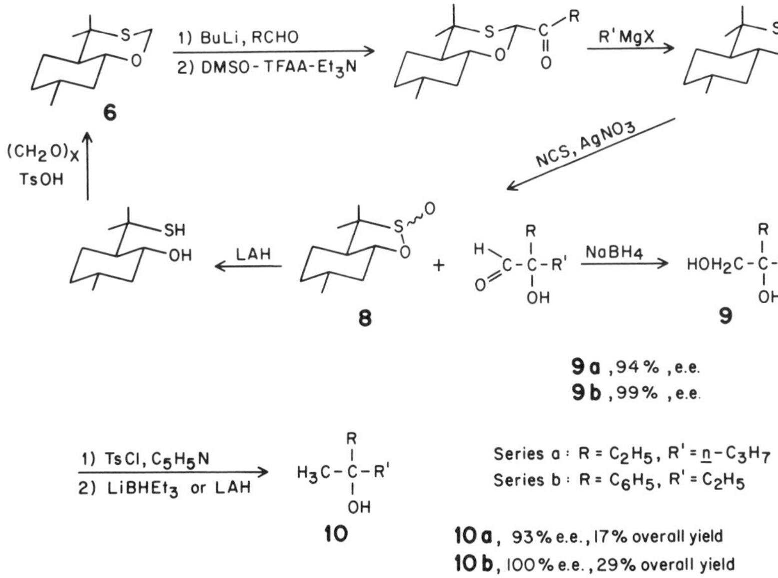

Scheme 9

Experiments presently in progress suggest that the monotosylate precursors of the tertiary alcohols can alternatively be converted, by base, to primary-tertiary epoxides which in turn may be opened at the primary site by a variety of nucleophiles, thus greatly amplifying the opportunities for synthesis of compounds bearing chiral tertiary carbinol moieties. It has also proved possible (23, 30) to oxidize the α-hydroxyaldehydes to α-hydroxyesters which, in turn, may be saponified to α-hydroxyacids.

Our synthesis of (R)-(-)-mevalolactone (24) is shown in Scheme 10. The crucial intermediate, synthesized by the earlier described method, is (+)-2-methyl-3-butene-1,2-diol, **11**. Its oxathiane precursor was obtained in 90% diastereomer excess, increased to 97% by one recrystallization. Diol **11** was then converted to the monotosylate which yielded chiral 3-hydroxy-3-methyl-4-pentenonitrile, **12**, when heated with KCN/EtOH. Hydroboration-oxidation was attended with *in situ* hydrolysis and lactonization to give directly (R)-(-)-mevalolactone. Its optical purity was estimated to be at least 87% as judged from optical rotation; the upper limit, of course, is 97%, the diastereomer excess of the precursor. Unfortunately an attempted determination of enantiomeric purity by NMR using a chiral shift reagent with the benzhydrylamide gave only a single signal for the benzhydrylic hydrogen, though this proton was clearly doubled in an analogous experiment with racemic mevalolactone. This failure is but a demonstration of the general difficulty of determining enantiomeric purities in excess of 90% by NMR with chiral shift reagents, given that the peak of the minor enantiomer, amounting to 5% or less, becomes hard to quantify and even hard to detect. Better methods (e.g. chromatographic ones (31)) are urgently needed for the determination of such high enantiomeric purities.

It should be mentioned here that the highest previously attained enantiomeric excess in the synthesis of mevalolactone was 17% (32). Although the reported (24) overall chemical yield in the synthesis summarized in Scheme 10 is low, recent improvements (use of $NaBH_4$ instead of $LiAlH_4$ in reduction to glycol **11** followed by efficient continuous extraction) have increased this yield by about a factor of five and the possibility of further improvement is under investigation (33).

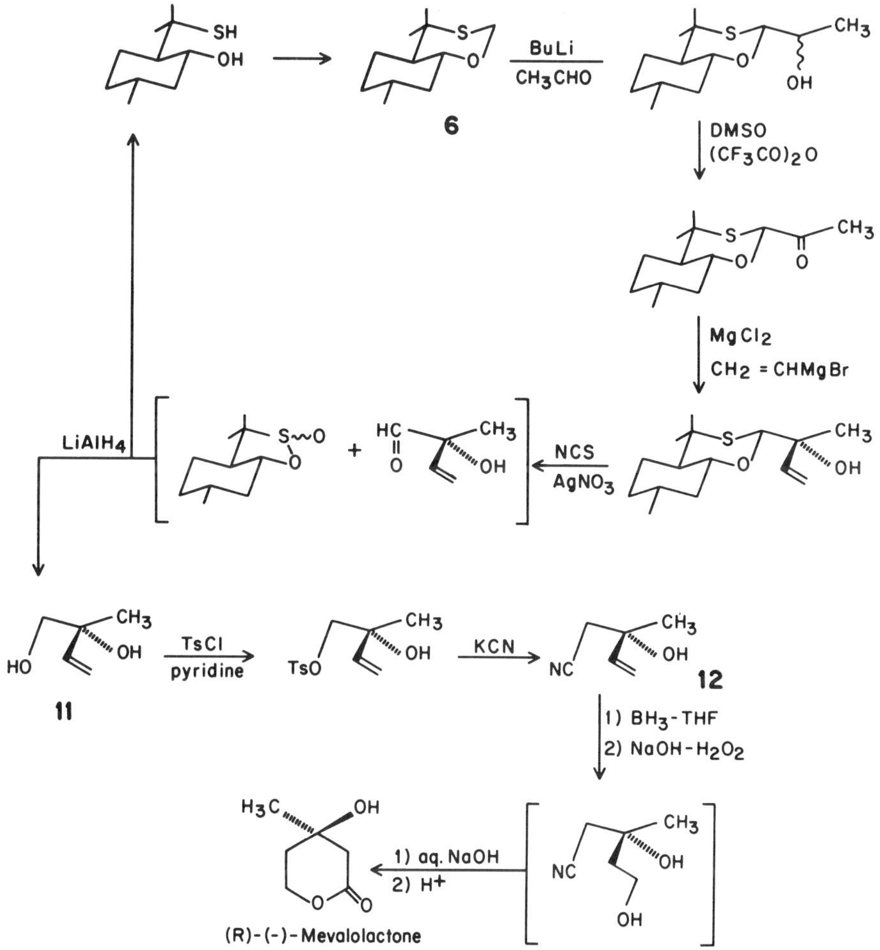
SH
OH
6
BuLi
CH3CHO
CH3
OH
DMSO
(CF3CO)2O
CH3
O
MgCl2
CH2 = CHMgBr
LiAlH4
S~O
O
HC
O
CH3
OH
NCS
AgNO3
CH3
OH
HO
CH3
OH
11
TsCl
pyridine
TsO
CH3
OH
KCN
NC
CH3
OH
12
1) BH3-THF
2) NaOH-H2O2
H3C
OH
1) aq. NaOH
2) H+
NC
CH3
OH
OH
(R)-(-)-Mevalolactone

Scheme 10

Acknowledgement

This work was supported, in part, under NSF grants CHE75-20052 and CHE78-28118. Acknowledgement is also made to the Donors of the Petroleum Research Fund, administered by the American Chemical Society, for partial support of this research.

Literature Cited

1. Eliel, E.L.; Hartmann, A.A. J.Am.Chem.Soc. 1971, 93, 2572.
2. Eliel, E.L.; Hartmann, A.A.; Abatjoglou, A.G. J.Am.Chem.Soc. 1974, 96, 1807.
3. Abatjoglou, A.G.; Eliel, E.L.; Kuyper, L.F. J.Am.Chem.Soc. 1977, 99, 8262.
4. Bernardi, F.; Csizmadia, I.G.; Mangini, A.; Schlegel, H.B.; Whanbo, M.-H.; Wolfe, S. J.Am.Chem.Soc. 1975, 97, 2209.
5. Lehn, J.-M.; Wipff, G. J.Am.Chem.Soc. 1976, 98, 7498.
6. Epiotis, N.D.; Yates, R.L.; Bernardi, F.; Wolfe, S. J.Am.Chem.Soc. 1976, 98, 5435.
7. Eliel, E.L. Tetrahedron 1974, 30, 1503.
8. Koskimies, J.K. Ph.D. Dissertation, University of North Carolina, Chapel Hill, NC, 1976.
9. Eliel, E.L.; Koskimies, J.K.; Lohri, B. J.Am.Chem.Soc. 1978, 100, 1614.
10. Omura, K.; Sharma, A.K.; Swern, D. J.Org.Chem. 1976, 41, 957.
11. Cram, D.J.; Kopecky, K.R. J.Am.Chem.Soc. 1959, 81, 2748.
12. cf. Ho, T.-L. "Hard and Soft Acids and Bases"; Academic Press: New York, 1977; Ho, T.-L. Chem.Rev. 1975, 75, 1.
13. For example: Cram, D.J.; Wilson, D.R. J.Am.Chem.Soc. 1963, 85, 1245; Stocker, J.H.; Sidisunthorn, P.; Benjamin, B.M.; Collins, C.J. J.Am.Chem.Soc. 1960, 82, 3913; Stocker, J.H. J.Org.Chem. 1964, 29, 3593; Angiolini, L.; Costa-Bizzari, P.; Tramontini, M. Tetrahedron 1969, 25, 4211; Méric, R.; Vigneron, J.-P. Bull.Soc.Chim.France 1973, 327; Nicolaou, K.C.; Charemon, D.A.; Barnette, W.E. J.Am.Chem.Soc. 1980, 102, 6611; Still, W.C.; McDonald, III, J.H. Tetrahedron Lett. 1980, 1031 and refs. there cited.
14. Morris-Natschke, Susan Ph.D.Dissertation, University of North Carolina, Chapel Hill, NC, 1981.
15. Owen, L.N.; Sultanbawa, M.U.S. J.Chem.Soc. 1949, 3098.
16. Hagberg, C.E.; Allenmark, S. Chem.Scripta 1974, 5, 13.
17. cf. Greene, T.W. "Protective Groups in Organic Synthesis"; Wiley-Interscience: New York, 1981; pp. 133-140.
18. Takano, S.; Hatekeyama, S.; Ogasawara, K. J.Chem.Soc.Chem. Commun. 1977, 68; Fetizon, F.; Jurion, M. ibid. 1972, 382.
19. Mizurro, H.; Yamada, S.-i. Chem.Pharm.Bull. 1975, 23, 527.
20. Eliel, E.L.; Frazee, W.J. J.Org.Chem. 1979, 44, 3598.
21. Eliel, E.L.; Lynch, J.E. Tetrahedron Lett. 1981, 22, 2855.
22. Müller, N.; Eliel, E.L. unpublished observations.

23. Lynch, J.E.; Eliel, E.L. unpublished observations.
24. Eliel, E.L.; Soai, K. Tetrahedron Lett. 1981, 22, 2859.
25. Corey, E.J.; Erickson, B.W. J.Org.Chem. 1971, 36, 3553.
26. Regarding the rotation and configuration of **9b**, see Mitsui, S.; Imaizumi, S.; Senda, Y.; Konno, K. Chem.Ind.(London) 1964, 233.
27. Regarding previous reports on chiral **10a** and **10b**, cf. Richter, W.J. Justus Leibigs Ann.Chem. 1975, 401 and Zeiss, H.H. J.Am.Chem.Soc. 1951, 73, 2391.
28. Mukaiyama, T.; Sakito, Y.; Asami, M. Chem.Lett. 1978, 1253; 1979, 705.
29. Inch, T.D.; Lewis, G.J.; Sainsbury, G.L.; Sellers, D.J. Tetrahedron Lett. 1969, 3657; Inch, T.D. Synthesis 1970, 466.
30. Inch, T.; Ley, R.V.; Rich, P. J.Chem.Soc.(C) 1968, 1693.
31. cf. Krull, I.S. Advances in Chromatography 1978, 16, 175; see also the article by Pirkle, W.H.; Finn, J.M.; Hamper, B.C.; Schreiner, J.; Pribish, J.R. in this volume.
32. Abushanab, E.; Reed, D.; Suzuki, F.; Sih, C.J. Tetrahedron Lett. 1978, 3415.
33. Kogure, T.; Eliel, E.L. unpublished observations.

RECEIVED December 14, 1981.

4

Acyclic Stereoselection via the Aldol Condensation

CLAYTON H. HEATHCOCK

University of California, Department of Chemistry, Berkeley, CA 94720

The aldol condensation, one of the oldest organic reactions, is emerging as a powerful method for control of relative and absolute stereochemistry in the synthesis of conformationally flexible compounds. Some of the research which has been carried out at Berkeley over the past five years is reviewed in this article. Points discussed are the factors that control simple *erythro*, *threo* diastereoselection, the use of double stereodifferentiation to influence the "Cram's rule" preference shown by chiral aldehydes, and some recent experiments that shed light on the role that the solvent and other nucleophilic ligands play in determining the stereochemistry of the reaction.

It has been quite apparent for some time that control of stereostructure in the synthesis of acyclic and other conformationally mobile compounds is a problem for which synthetic chemists have few solutions. For example, in his 1956 article in "Perspectives in Organic Chemistry," the late Professor R.B. Woodward characterized the macrolide antibiotic erythromycin as a synthetic challenge which is "...quite hopelessly complex, especially in view of its plethora of asymmetric centers." (1,2) In fact, it was this

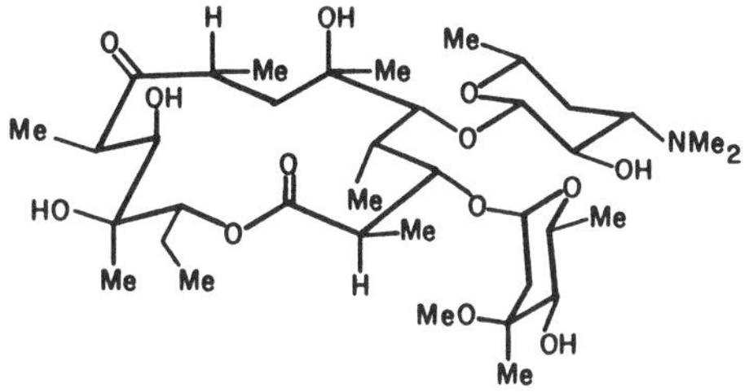

erythromycin-A

0097-6156/82/0185-0055$05.00/0

molecule and the above mentioned quote from the master of organic synthesis which planted the seed of the project which is discussed in this article. In casting about for a method which might be used to attack the synthetic problem posed by erythromycin, I came across a statement by another master -- of the field of biosynthesis. In fact, it was also in "Perspectives in Organic Chemistry" that J.W. Cornforth wrote: "Nature, it seems, is an organic chemist having some predilection for the aldol and related condensations..." (3) Thus, the erythromycin aglycone is presumably constructed by Nature by a series of aldol-type condensations of propionate units:

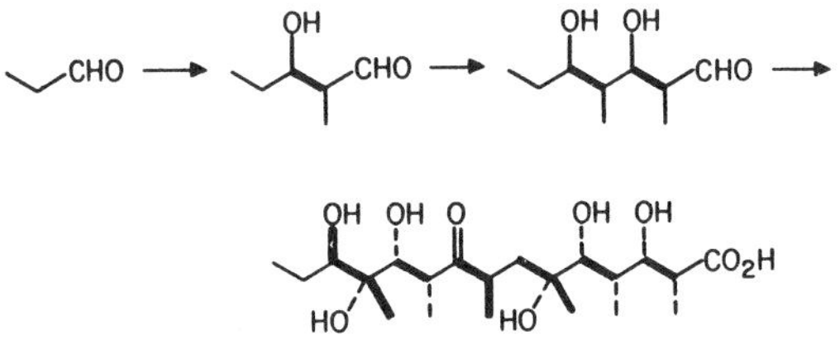

CHO → CHO → CHO →

CO_2H

In 1976 we began an investigation of the aldol condensation with the ultimate goal of learning enough about its stereochemistry to use it as the sole method for constructing the aglycone of erythromycin-A and, at the same time, controlling the relative stereochemistry of its ten centers of chirality. Here, following a brief review of our early work, I shall discuss some recent unpublished experiments that pertain to the coordination chemistry of the enolate counterion and how that can influence the stereochemistry of the aldol condensation.

In evaluating the aldol condensation as a method for building acyclic molecules containing many stereocenters, such as erythronolide-A, there are two types of diastereoselection which must be considered. The first is referred to as <u>simple diastereoselection</u> and arises from the fact that two newly-created chiral

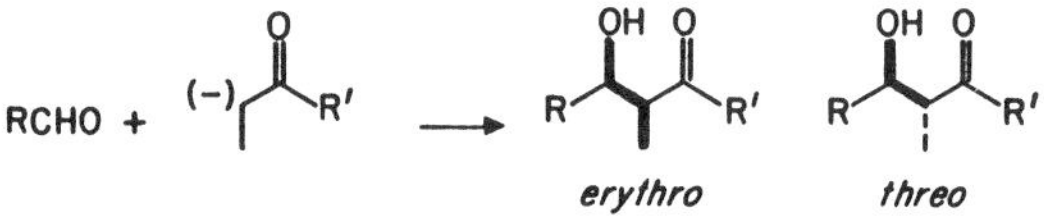

RCHO + (-) R' → R R' (*erythro*) R R' (*threo*)

centers may be formed with either the *erythro* or *threo* relative configuration.* It is now clear that simple diastereoselection is controlled by two factors -- the configuration (*E* or *Z*) of the enolate and the orientation of the enolate and aldehyde in the transition state of the aldol reaction itself. Deprotonation of an ethyl carbonyl compound can give either a *trans*-or a *cis*-enolate:*

Since the aldehyde and enolate both have enantiotopic faces, there are two *relative* ways they can approach one another. That is, the *Si* face of the aldehyde can become bonded to the *Re* face of the enolate (giving the *erythro* diastereomer) or the *Re* face of the aldehyde can be attacked by the *Re* face of the enolate (giving the *threo* diastereomer):

It has now been fairly well established what factors control enolate configuration. Deprotonation of ethyl carbonyl compounds in which the other group attached to the carbonyl is small gives predominately the *trans*- enolate. As the R group becomes larger, more *cis*-enolate is produced. With very large R groups, the *cis*-enolate is the overwhelming diastereomer produced (Table 1).**

It has also been established what factors control the orientation of the two reactants in the transition state for the aldol

---

*The stereochemical descriptors *erythro* and *threo* are used in the following sense. When the aldol is written so that its main chain is in an extended (zig-zag) arrangement, an *erythro* isomer is one in which the bonds to the α-alkyl group and the β-hydroxy group both project toward (bold bonds) or away from (dashed bonds) the viewer.

*I shall use the configurational descriptors *cis* and *trans* with the O^- or O^-M^+ group always being the point of reference. In the *E-Z* system, depending on the nature of R, either OM (O^-) or R may be fiducial, which is confusing.

**"Large" and "small" require definition at this point. What is important is the size of the group from the standpoint of the substituent attached to the α position relative to the carbonyl group. Thus, in propionate esters, all alkoxy groups are "small," since the methyl senses only the oxygen part of the group.

Table I

R	*trans*	*cis*	
$-OCH_3$	97%	3%	small
$-OC(CH_3)_3$	97%	3%	small
$-CH_2CH_3$	70%	30%	medium
$-CH(CH_3)_2$	44%	56%	medium
$-C_6H_5$	0%	100%	large
$-C(CH_3)_3$	0%	100%	large

reaction. When the R group is small, neither cis- nor trans- enolates show any simple diastereoselection. Thus, the cis-enolate from propionaldehyde and the trans-enolate from methyl propionate both react with aldehydes to afford essentially 1:1 mixtures of erythro and threo isomers.

However, when the R group is large, both cis- and trans- enolates show substantial stereoselectivity, giving rise to opposite diastereomers. Thus, the cis- enolate from ethyl t-butyl ketone gives almost entirely erythro aldols while the trans enolate from 2,6-dimethyl propionate gives threo-aldols with good stereoselectivity. (4,5) This behavior is readily interpreted in terms of a transition state model first put forth by Zimmerman and Traxler for the Ivanov condensation in 1957. (6) In the Zimmerman model (Figure 1) it is assumed that there is carbon-carbon bonding in the transition state of the reaction, and that the two oxygens of the reacting array are both disposed in the general direction of the cation. This leads to six-center arrangements in which the R group of the aldehyde and the R' group of the enolate are either close to one another or remote from one another. The argument goes that, if the interaction of R with R' is large, the top transition state (R's remote) will be favored. Thus, a cis-enolate will lead to an erythro-aldol and, similarly, a trans-enolate to a threo-aldol.

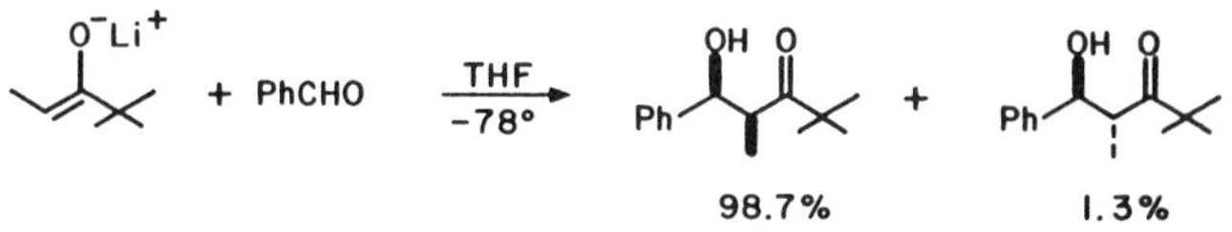

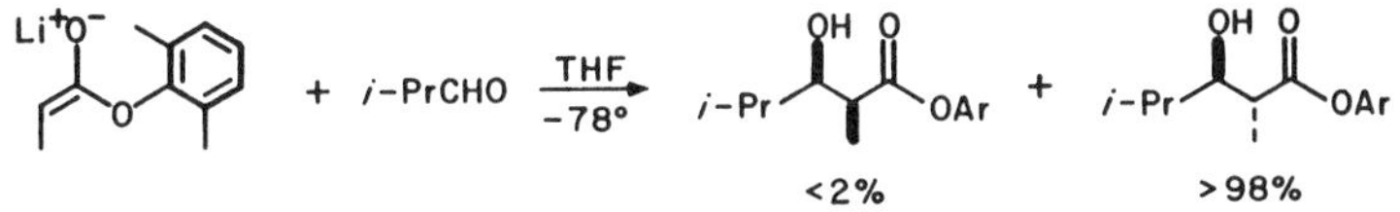

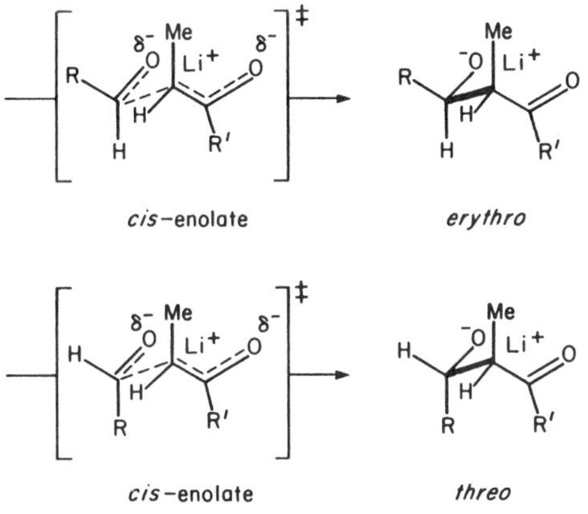

Figure 1. Zimmerman-Traxler transition state (6) showing the transformation of a cis-*enolate to* erythro- *and* threo-*aldols.*

The other aspect of the aldol condensation to be considered in using this reaction for the construction of compounds such as erythronolide-A is diastereoface selection. That is, in many cases one will want to carry out aldol condensations on aldehydes already having one or more chiral centers. The carbonyl faces in these molecules are diastereotopic, rather than enantiotopic, and there are four relative ways such aldehydes can react with achiral enolates:

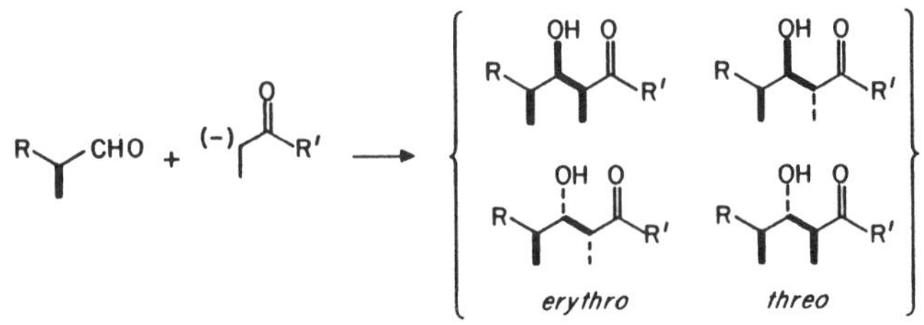

Of course, this is no more than the problem of relative asymmetric induction which was first examined systematically by Cram (7) and Prelog (8) nearly thirty years ago. Using various erythro or threo selective reagents, even though one can reasonably control simple diastereoselection, one still obtains mixtures of "Cram" and "anti-Cram" diastereomers:

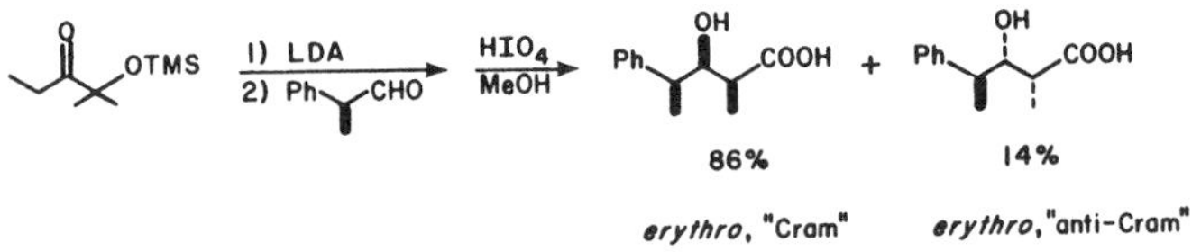

Although a diastereomer ratio of 6:1 is not bad, it is not nearly high enough for the purpose at hand. That is, if one is to use the aldol condensation as a method for the repetitive addition of propionaldehyde synthons to build up large molecules having many centers of chirality, one will soon be defeated by the effects of a geometric progression unless each condensation proceeds with very high stereoselectivity. Thus, five condensations, each proceeding with 80% stereoselectivity, will lead to an overall stereochemical yield of only 33%.

We have examined a purely logical way in which the "Cram's rule problem" can be attacked -- double stereodifferentiation.* For example, either reactant in an aldol condensation can be chiral and exhibit diastereoface selectivity. Suppose we have an aldehyde which reacts with achiral enolates to give the two possible erythro adducts in a 10:1 ratio:

*The phenomenon has also been referred to as "double asymmetric induction." (9) We have used the term double stereodifferentiation, first introduced by Izumi and Tai (10) in order to avoid confusion in cases involving racemates.

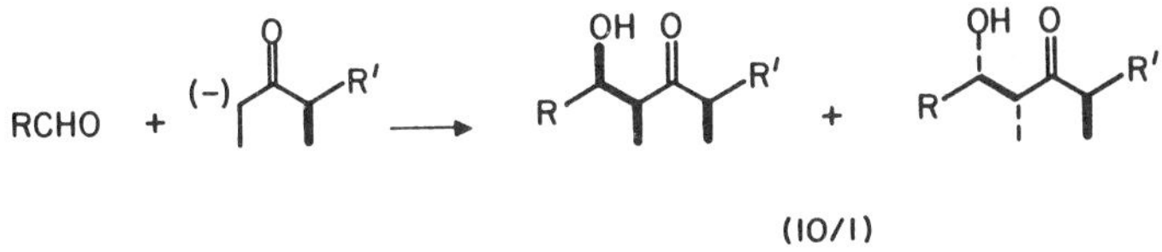
RCHO
(-)
R'
OH O
R
(10/1)

Also suppose that we have a chiral enolate which reacts with achiral aldehydes to give the two erythro-aldols in a 10:1 ratio:

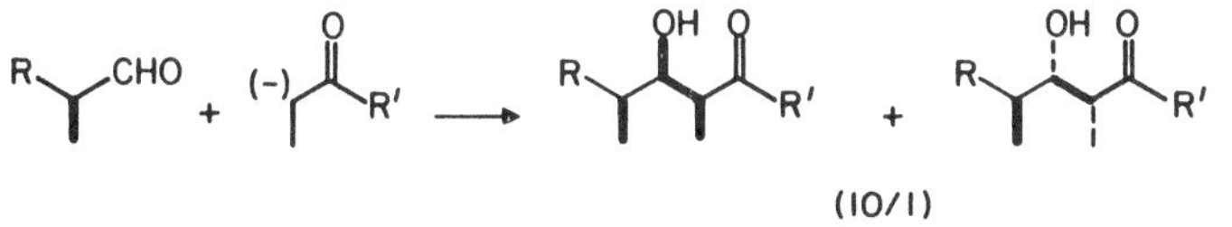
R
CHO
(-)
R'
OH O
(10/1)

Now suppose that we allow one enantiomer of the chiral aldehyde to react in turn with the two enantiomers of the chiral enolate. In one case the two reactants will both promote the same absolute configuration (chirality) at the two new chiral centers. In this case, the effective "Cram's rule selectivity" shown by the aldehyde will be greater than in its reactions with representative achiral enolates. For the selectivities chosen in this example, the "Cram:anti-Cram ratio" should be on the order of 100:1.

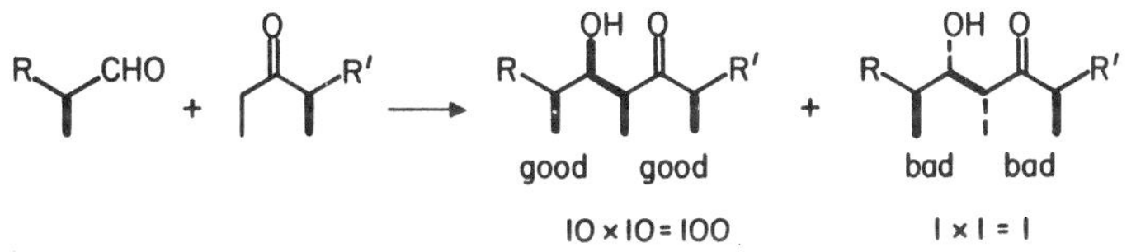
R
CHO
R'
OH O
good good
bad bad
10 × 10 = 100
1 × 1 = 1

Of course, in the other combination, neither reactant gets its way. In this case, the effective diastereoface selectivity shown by the aldehyde should be poorer than is seen in reactions of the same aldehyde with representative achiral enolates.

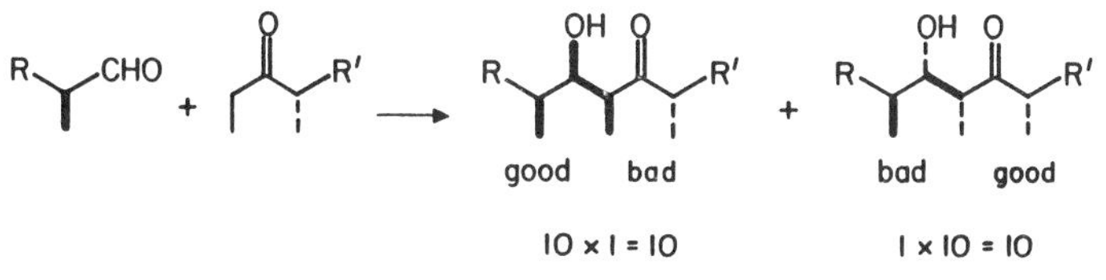
R
CHO
R'
OH O
good bad
bad good
10 × 1 = 10
1 × 10 = 10

In order to test this concept as a way of controlling the problem of diastereoface selectivity in aldol condensations involving chiral aldehydes, we prepared the chiral ethyl ketone shown below, which is available in four straightforward steps from D-fructose. This compound shows modest inherent diastereoface selectivity, reacting with benzaldehyde to give the two *erythro*-aldols in a ratio of 4:1, with the *R,R* diastereomer predominating. As a reaction partner, we chose the acetonide of glyceraldehyde, since both enantiomers are readily available. This aldehyde also shows modest inherent diastereoface selectivity in its reactions with achiral enolates -- on the order of 4.5:1. The sense of the stereoselectivity is such that *R*-glyceraldehyde acetonide gives predominantly the aldol with the *S,S* configuration at the two new

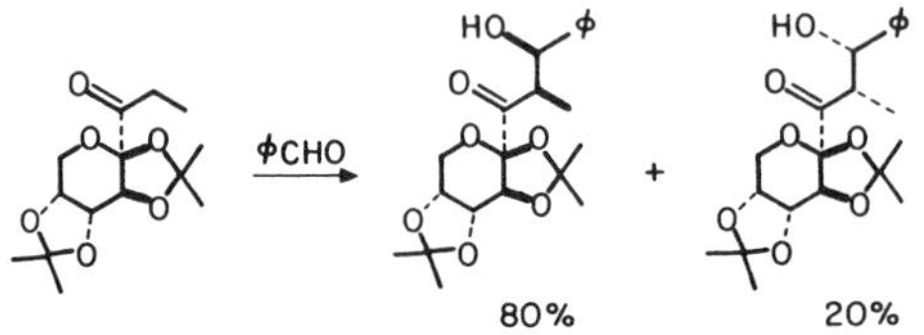

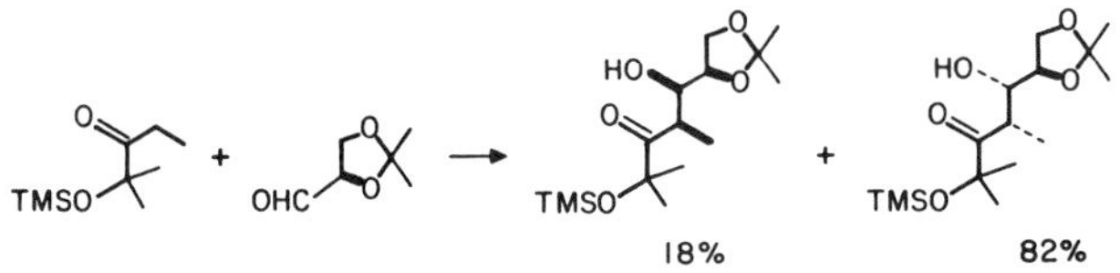

ϕCHO
80% 20%

TMSO OHC TMSO TMSO
18% 82%

centers. Thus, the fructose-derived ketone and the acetonide of *R*-glyceraldehyde show unproductive double stereodifferentiation and give an almost equal mixture of the two *erythro*-aldols. However, the other combination is reinforcing, and only a single *erythro*-aldol results. (11)

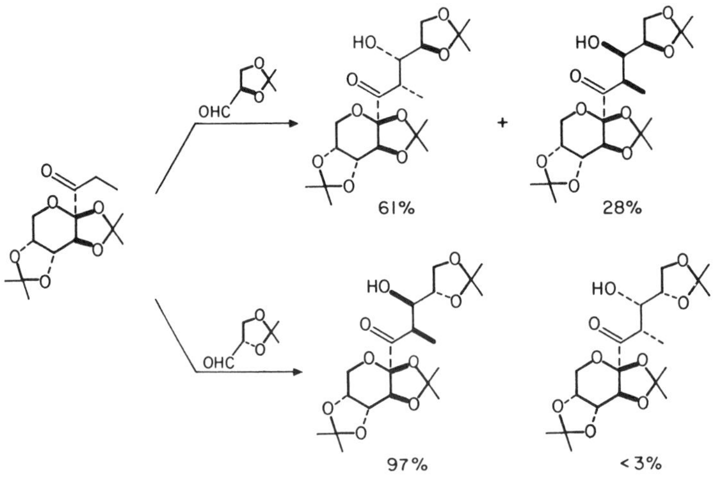

OHC
61% 28%

OHC
97% <3%

We have also observed double stereodifferentiation when one of the chiral elements is a reactant and the other is the solvent. One example of this phenomenon is shown below for the condensation of the indicated achiral ethyl ketone with the acetonide of *R*-glyceraldehyde. Although the magnitude of the effect is small, it is significant. (11)

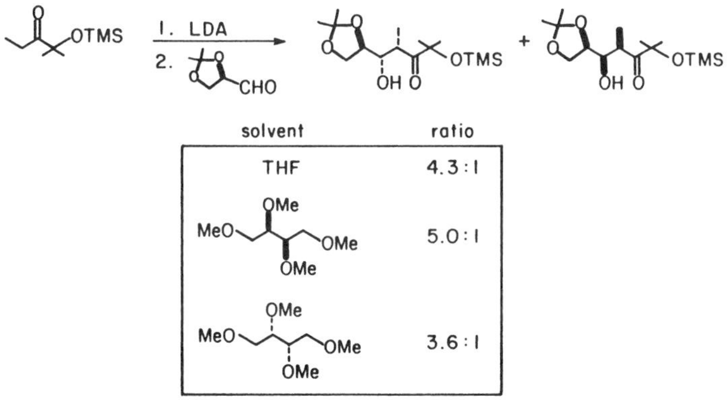

1. LDA
2. CHO

solvent	ratio
THF	4.3:1
MeO, OMe, OMe, OMe	5.0:1
MeO, OMe, OMe, OMe	3.6:1

To capitalize on the concept of double stereodifferentiation as a method for enhancing mediocre diastereoface selectivity in aldol condensations of chiral aldehydes, we synthesized the chiral ethyl ketone shown below. This compound shows good to excellent inherent diastereoface selectivity with achiral aldehydes. The selectivity appears to increase dramatically with the steric demand of the group attached to the aldehyde carbonyl. Thus, with pivaldehyde and diacetaldehyde, only one aldol is produced. The diastereoface selectivity in these two cases is at least 19:1 and 10:1, respectively. (12)

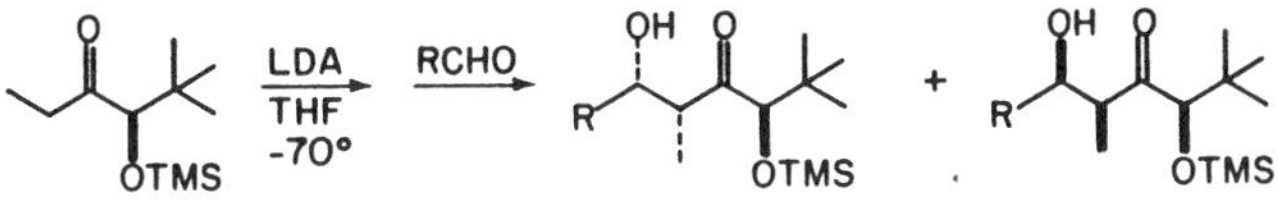

LDA, THF, -70° ; RCHO

R		
R = C_6H_5–	75%	25%
= *i*-C_3H_7–	75%	25%
= *t*-C_4H_9–	>95%	<5%
= $C_6H_5CH_2$–	87%	13%
= $(C_6H_5)_2CH$–	>90%	<10%

What was surprising was the discovery that condensation of the racemic ketone with racemic α-phenylpropionaldehyde gives a single racemic aldol! In order for this to be observed, the *R* enantiomer of the aldehyde must react selectively with the *R* enantiomer of the ketone, and not with the *S* enantiomer. That is, the reaction must show kinetic resolution. Since the racemates of both reactants are

C_6H_5 CHO + OTMS OLi → (85%) C_6H_5 OTMS OH O

Racemic Racemic

HIO_4, MeOH (80%)

(No more than 2% C_6H_5 CO_2H OH) C_6H_5 CO_2H OH

involved in the reaction, we refer to the phenomenon as *mutual kinetic resolution*, so as to avoid confusion. We may understand the origin of this kinetic resolution with the aid of the same multiplicative model that we used to understand double stereodifferentiation. Thus, let us assume that the erythro:threo diastereoselectivity is 80:1, that the inherent diastereoface selectivity of the aldehyde is 10:1, and that the inherent diastereoface selectivity of the enolate is 20:1. For each of the sixteen possible stereoisomers that can result from the reaction of one of the enantiomers of the aldehyde with one of the enantiomers of the enolate, we can compute the probability of formation by multiplying together three numbers. By summing all of the products, and taking the appropriate ratios, we can therefore estimate the relative product distribution expected from the reaction. The results of this exercise are summarized in Figure 2 (only one-half of the products are shown -- the other eight are the enantiomeric set and will be produced in precisely the same ratio in this double racemic reaction).

One thing which is seen immediately upon examination of Figure 2 is that the *threo* products are not expected to be formed in any significant amount, because of the high *erythro:threo* ratio we assumed. We also see that, for the *R* + *R* combination (top row), there is a productive and a nonproductive combination. That is, the two *erythro*-aldols resulting from the reaction of *R* aldehyde with *R* enolate are expected to be produced in a ratio of 200:1. On the other hand, the *erythro*-aldols resulting from reaction of *R* aldehyde with *S* enolate will be produced in a ratio of about 1:2, since the two reactants are working at cross-purposes in this

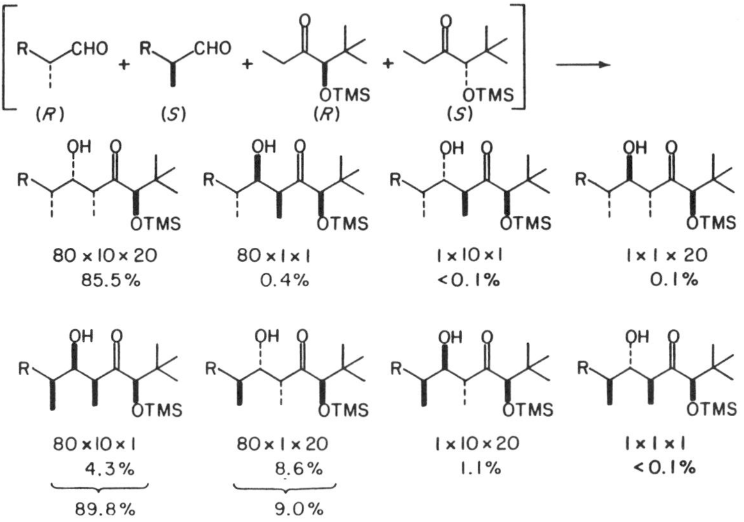

Figure 2. One half of the stereoisomers possible from aldol condensation of racemic aldehyde with racemic ketone.

case.* Of course, cleavage of the aldol products to the respective β-hydroxy acids would still lead to a 10:1 ratio of the two *erythro* products corresponding to "Cram" and "anti-Cram" addition to the aldehyde -- the multiplicative approach does not change the net diastereoface selectivity of either reactant.

However, the use of this approach points up the origin of the observed mutual kinetic resolution. In part, at least, it is a consequence of the inherent diastereoface selectivity exhibited by the two reactants. Thus, if we ignore the *threo* isomers, the approximation used in Figure 2 leads to a prediction that the rate of the *R* + *R* reaction will be approximately seven times the rate of the *R* + *S* reaction.

The question remains of why our double racemic reaction shows net diastereoface selectivity of >49:1. It might be that we have been too conservative in selecting stereoselectivity factors. For example, there is a trend that the enolate used shows higher

*Although the *threo*-aldols are produced in such small amount that they are not observed it is interesting to note that the situation is just reversed in this set of stereoisomers -- the *R* + *S* combination shows productive double stereodifferentiation and leads to a high ratio while the *R* + *R* combination shows nonproductive double stereodifferentiation.

inherent diastereoface selectivity as the R group attached to the aldehyde carbonyl with which it reacts becomes larger. In a similar manner, there is evidence that the diastereoface selectivity of α-phenylpropionaldehyde increases as the steric bulk of the nucleophile increases. It may also be that there is some independent factor which favors the R + R reaction over the R + S reaction. Thus, if there is some other reason favoring kinetic resolution, the net diastereoface selectivity of the aldehyde will benefit dramatically since this combination results in a very disparate ratio of the two *erythro* products. In any event, the reaction just discussed is but one example of this phenomenon; we have observed a total of five other cases at this point.

I believe that the phenomenon of double stereodifferentiation with mutual kinetic resolution may have general ramifications beyond the specific aldol condensations being discussed here. We can generalize as shown below:

Principle of Mutual Kinetic Resolution

$$A^* + B \longrightarrow \text{diastereomers } C + D$$

$$\text{inherent diastereoselectivity of } A^* = \frac{C}{D} = X$$

$$E^* + F \longrightarrow \text{diastereomers } G + H$$

$$\text{inherent diastereoselectivity of } E^* = \frac{G}{H} = Y$$

$$(R)\text{-A} + (R)\text{-E} \xrightarrow{k_1}$$

$$(R)\text{-A} + (S)\text{-E} \xrightarrow{k_2}$$

$$(S)\text{-A} + (S)\text{-E} \xrightarrow{k_1}$$

$$(S)\text{-A} + (R)\text{-E} \xrightarrow{k_2}$$

$$\text{mutual kinetic resolution} = \frac{k_1}{k_2} = \frac{XY+1}{X+Y}$$

That is, in order for the phenomenon to be observed, both reactants must show inherent diastereoface selectivity in their reactions with achiral partners. If one of the reactants shows no inherent diastereoface selectivity in its reactions with achiral reactants, then mutual kinetic resolution will not be observed regardless of the stereoselectivity of the other reactant. For an example, consider the case of an enzyme which mediates some reaction, say reduction of the carbonyl group. We can let the enzyme be A* and assume that, because of its uniquely-evolved molecular structure, it shows very high inherent diastereoface selectivity (thus, it will reduce prochiral carbonyl compounds to chiral alcohols with very high enantiomeric excess). For B*, let us take a chiral aldehyde that shows no inherent diastereoface selectivity in its reaction with achiral reducing agents -- for example, 4-methylhexanal. The principle being expounded here predicts that the enzyme will not discriminate between R- and S-4-methylhexanal. On the other hand, consider an aldehyde which does show inherent diastereoface selectivity, such as α-phenylpropionaldehyde. In this case, double stereodifferentiation will result and the enzymatic reduction of one enantiomer of the aldehyde will be faster than for the other enantiomer; kinetic resolution will result.

I would now like to briefly discuss some experiments we have recently carried out that have a bearing on the role of the solvent and the other nucleophilic ligands on these enolate reactions. The experiment shown below clearly indicates that the species present in solution immediately following the aldol condensation (Y) is different from the species produced by treatment of the isolated aldol with LDA in the same solvent.

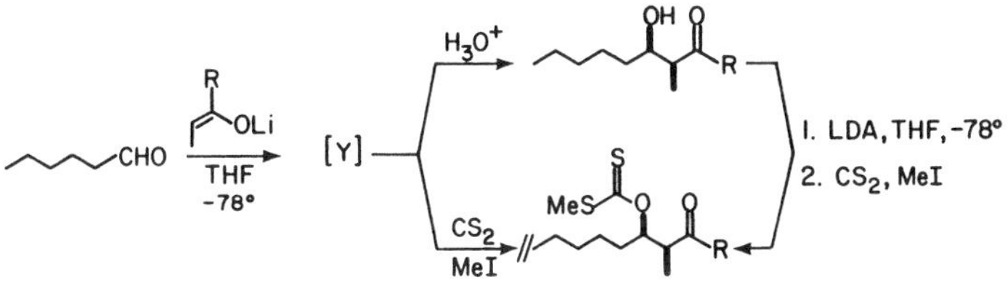

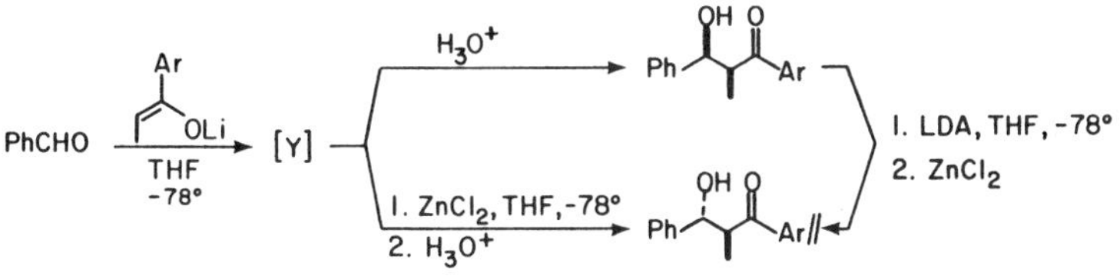

A similar dichotomy is seen in the following reaction, except that in this case it is the species produced from deprotonation of the aldol which is unreactive. At this point we do not have definite proof as to the nature of the two species. However, our working hypothesis is that the aldol condensation leads directly to an aldolate that has the lithium cation chelated between the two oxygens. Since the lithium is no doubt tetra-coordinate, the other two ligands are probably a THF molecule and the molecule of diisoproylamine originally bonded to the lithium in the form of LDA. We believe that this species reacts slowly to form the xanthate and rapidly undergoes the $ZnCl_2$-catalyzed isomerification to the *threo*- aldolate. We further believe that the product produced by deprotonation of the aldol is not chelated. In this case the lithium is presumably bonded to the aldolate oxygen, the diisopropylamine, and two solvent molecules. This species must react rapidly to give the xanthate and be unreactive with respect to *erythro*:*threo* equilibration.

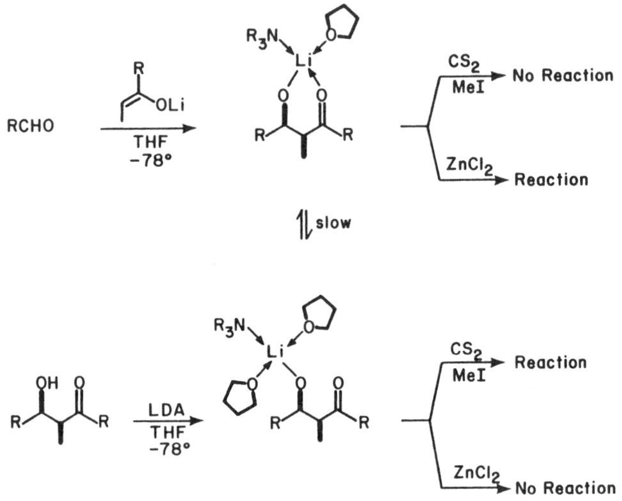

RCHO
THF
-78°
R_3N
Li
CS_2
MeI
No Reaction
$ZnCl_2$
Reaction
slow
LDA
THF
-78°
R_3N
Li
CS_2
MeI
Reaction
$ZnCl_2$
No Reaction

The foregoing hypothesis requires that ligand exchange on the lithium cation be slow under the conditions of the experiments described. In fact, there is little information available in the literature on this point. Just recently, however, there has been a convincing demonstration that lithium exchange on a lithium cation can be *very slow*. Dr. Patricia Watson, of the Central Research Department at the Dupont company, prepared an organometallic complex by addition of bis(pentamethylcyclopentadienyl)methylytterbium to a THF solution of methyllithium. (13) The product is a crystalline 1:1 adduct, which was recrystallized *from diethyl ether* and subjected to single crystal x-ray analysis. The structure of the recrystallized material is as follows:

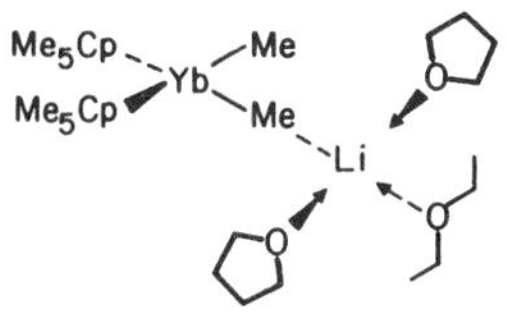

Thus, even though the material has been recrystallized from ether, two of the original THF ligands still remain bonded to the lithium. Upon repeated recrystallization from ether, or upon being boiled in ether, the remaining THF ligands are exchanged and eventually one may obtain a complex having three ether molecules.

This is an important result, which causes us to reexamine our view of the nature of many of these compounds. If oxygen ligands can exchange so slowly on the lithium cation, what about nitrogen ligands? It may be true that the lithium never becomes unbonded from the nitrogen when LDA is used to form a lithium enolate. Thus, the real structure of a monomeric lithium enolate might be:

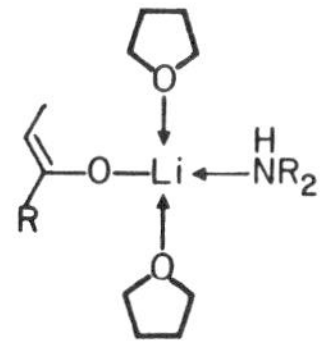

With this new view of lithium amides and lithium alkoxides in mind, we have begun to look more carefully at the effect of the ligands upon the stereochemistry of the aldol condensation. For example, instead of using LDA as the base to form an enolate, we have used the lithium amide derived from *O*-methylprolinol. Condensation of the derived enolate with benzaldehyde gives the (*S*,*S*)-aldol with 8% enantiomeric excess. This is slightly higher than the asymmetric induction we had previously seen with 1,2,3,4-tetramethoxybutane, even though the latter material was used as solvent (11). Similar results have been obtained with the lithium amides derived from *O*-methylephedrine and *O*-methylpseudoephedrine. At this point, the asymmetric induction observed with these chiral amides is only marginal. However, we think that with some careful design we can find chiral bases which will allow us to realize useful asymmetric induction in condensations of achiral enolates with achiral aldehydes.

One further insight into the coordination chemistry of these lithium enolates is obtained from the stereoselectivity observed in the reactions of chiral α-alkoxyketone enolates. For example, the

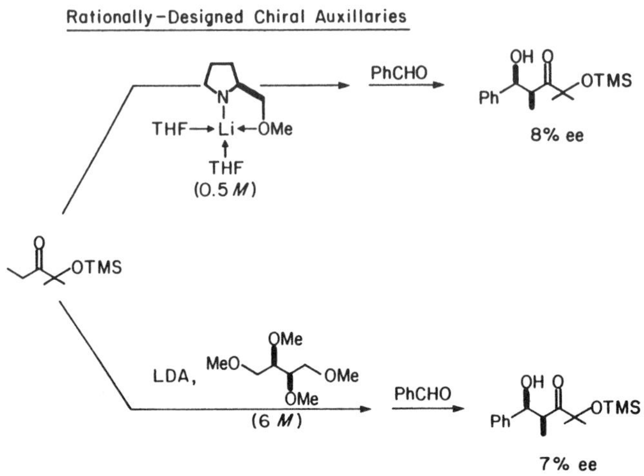
Rationally-Designed Chiral Auxillaries
THF→Li←OMe
THF
(0.5 M)
PhCHO
OTMS
Ph
OH O
8% ee
OTMS
LDA,
MeO
OMe
OMe
OMe
(6 M)
PhCHO
OH O
Ph
OTMS
7% ee

lithium enolate of the compound discussed earlier in this article shows selectivity in the following sense:

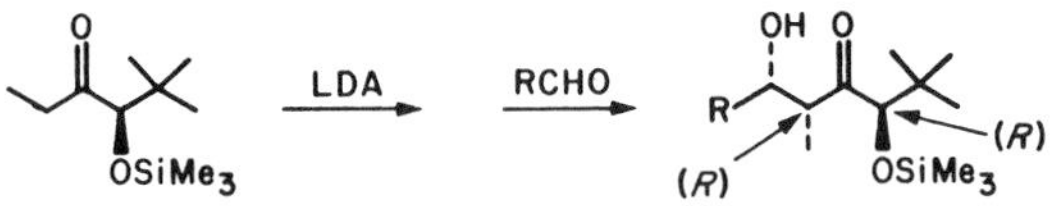
O
OSiMe3
LDA
RCHO
OH O
R
(R)
(R)
OSiMe3

However, Masamune has examined the reactions of the boron enolates of this type of compound and observes exactly the opposite sense of diastereoface selectivity: (14)

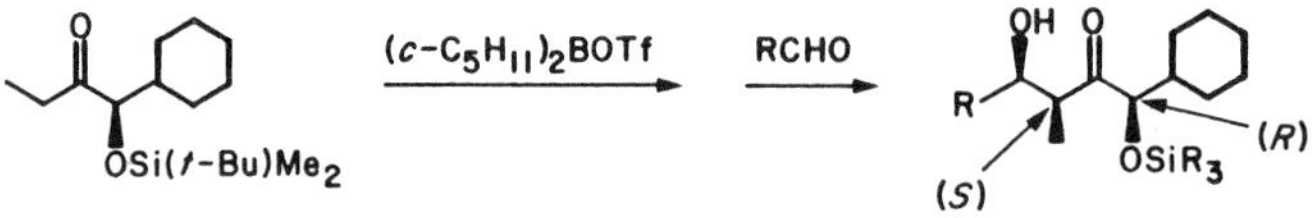
O
OSi(t-Bu)Me2
(c-C5H11)2BOTf
RCHO
OH O
R
(S)
OSiR3
(R)

These apparently divergent results can be readily understood in terms of the coordination chemistry of the two enolates. With lithium, coordination can occur to the aldehyde and both sites in the enolate. This imposes a structure on the enolate such that the *Re* face of the (*R*)-enolate is shielded by the *t*-butyl group. Thus, reaction occurs on the *Si* face. However, with the enolbronoate, two of the boron ligands are alkyl groups. If the boron is to be ligated to the enolate oxygen and the aldehyde carbonyl its tetravalency is saturated. Thus, it cannot also bond to the oxygen of the α-alkoxy group. In this case, it seems that the enolate prefers to react from a conformation having the α C–O bond *syn*-coplanar with the C=C. Thus, the cyclohexyl group shields the *Si* face of the (*R*)-enolate, and reaction occurs on the *Re* face.

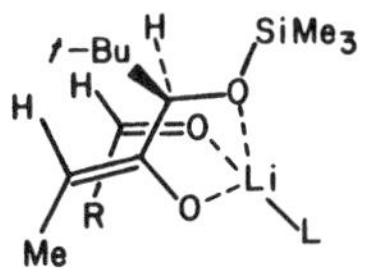

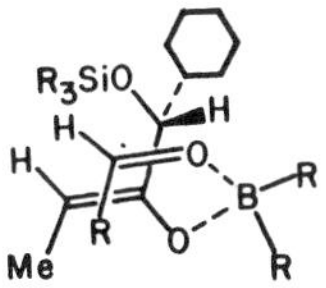

I believe that these observations presage a period during which we, and others, will look more and more closely at the details of important enolate reactions such as the aldol condensation and take account not only of the reactants themselves, but also of the solvent and other potential ligands which are present in the reaction medium.

Acknowledgements: The work described was carried out at Berkeley by a talented group of graduate students and postdoctoral associates. Special acknowledgement is due Charles Buse, James Hagen, Esa Jarvi, William Kleschick, John Lampe, Stephen Montgomery, Michael Pirrung, John Sohn, Charles White, and Steven Young. I am also indebted to the United States Public Health Service for support of the research.

Literature Cited

1. R.B. Woodward in "Perspectives in Organic Chemistry," edited by A. Todd, Interscience Publishers, New York, 1956.

2. It is fitting that Woodward's last major achievement was the total synthesis of erythromycin-A. the synthesis, which was carried out in classical Woodwardian fashion, was communicated posthumously: R.B. Woodward, *et. al*, *J. Am. Chem. Soc.*, **103**, 3210, 3213, 3215 (1981).

3. J.W. Cornforth in "Perspectives in Organic Chemistry," edited by A. Todd, Interscience Publishers, New York, 1956.

4. C.H. Heathcock, C.T. Buse, W.A. Kleschick, M.C. Pirrung, J.E. Sohn, and J.Lampe, J. Org. Chem., 45, 1066 (1980).

5. C.H. Heathcock, M.C. Pirrung, S.H. Montgomery and J. Lampe, Tetrahedron, in press.

6. H. Zimmerman and M. Traxler, J. Am. Chem. Soc., 79, 1920 (1957).

7. D.J. Cram and F.A. Abd. Elhafez, J. Am. Chem. Soc., 74, 5828 (1952).

8. V. Prelog, Helv. Chim. Acta., 38, 308 (1953).

9. A. Horeau, H.-B. Kagan, and J.-P. Vigneron, Bull. Soc. Chim. Fr., 3795 (1968).

10. Y. Izumi and A. Tai, "Stereodifferentiating Reactions," Kodansha Ltd., Tokyo; Academic Press, New York, 1977.

11. C.H. Heathcock, C.T. White, J.J. Morrison, and D. VanDerveer, J. Org. Chem., 46, 1296 (1981).

12. C.H. Heathcock, M.C. Pirrung, J. Lampe, C.T. Buse, and S.D. Young, J. Org. Chem., 46, 2290 (1981).

13. P. Watson, private communication.

14. S. Masamune, M. Hirama, S. Mori, S.A. Ali, and D.S. Garvey, J. Am. Chem. Soc., 103, 1568 (1981).

RECEIVED December 14, 1981.

Asymmetric Carbon–Carbon Bond Forming Reactions via Chiral Chelated Intermediates

Diastereoselective Asymmetric Synthesis of 1,2-Disubstituted Cycloalkanecarboxaldehydes

KENJI KOGA

University of Tokyo, Faculty of Pharmaceutical Sciences, Hongo, Bunkyo-ku, Tokyo, Japan 113

1,4-Addition of Grignard reagents to Schiff bases of α,β-unsaturated aldehydes with optically active *tert*-leucine *tert*-butyl ester, followed by hydrolysis gives chiral β-substituted carboxaldehydes in 82-98% e.e. Similar reaction of α,β-unsaturated cycloalkanecarboxaldehydes followed by alkylation of either the corresponding magnesioenamine or the free aldehyde with alkyl halides leads to 1,2-disubstituted cycloalkanecarboxaldehydes of high diastereomeric as well as enantiomeric purity. The predominant diastereomer has the two alkyl groups *trans* to each other when the free aldehyde is alkylated but *cis* to each other when alkylation is performed on the *Z*-isomer of the magnesioenamine formed initially in the Grignard addition. Conversion of the *Z*-magnesioenamine to its *E*-isomer by heating followed by alkylation places the two alkyl groups *trans* to each other. Mechanistic explanations for this divergent behavior are provided.

The design of highly efficient asymmetric syntheses has been one of the most challenging and exciting fields in synthetic organic chemistry.[1] As part of our research program directed toward the development of new stereoselective reactions, we have reported an efficient method for asymmetric alkylations based on a strategy of fixing the intermediate conformation by chelation.[2] It appears that the conformation of the chiral moiety of Schiff bases (**3**), prepared from carbonyl compounds (**1**) and optically

0097-6156/82/0185-0073$05.00/0

$$\underset{\mathbf{1}}{\geq C=O} + \underset{\mathbf{2}}{R-\overset{*}{C}H(NH_2)-COOBu^t} \longrightarrow \underset{\mathbf{3}}{\geq C=N-\overset{*}{C}H(R)-COOBu^t} \xrightarrow{R'-M} \mathbf{4}$$

Chart 1.

active α-amino acid esters (2), can be effectively fixed as 4 in which both the imine nitrogen and ester oxygen coordinate to the metal (Chart 1). Optically active tert-leucine tert-butyl ester (2, R=Bu^t) was found to be an excellent chiral auxiliary reagent acting as a bidentate ligand. It was recovered without any racemization after the reaction. Some representative examples of its use in asymmetric synthesis are shown in Chart 2.

As an extension of this reaction, a diastereoselective asymmetric synthesis of 1,2-disubstituted cycloalkanecarboxaldehydes (19) was attempted[3] as shown in Chart 3. These compounds are useful chiral synthons having asymmetric tertiary and quaternary carbon atoms in vicinal positions. Three synthetic approaches were examined starting from cycloalkenecarboxaldehydes (8). In method A, optically active aldehydes (10), prepared by the above method via Grignard 1,4-addition to chiral α,β-unsaturated cyclic aldimines (9) followed by hydrolysis of the resulting magnesioenamines (18), were metalated and alkylated as usual. In method B, the reaction was performed by a one-pot procedure via Grignard 1,4-addition to 9 followed by alkylation of the resulting 18 with alkyl halides in the presence of HMPA. In method C, the reaction was performed as in method B, except that the reaction mixture was heated to reflux for several hours before alkylation. The

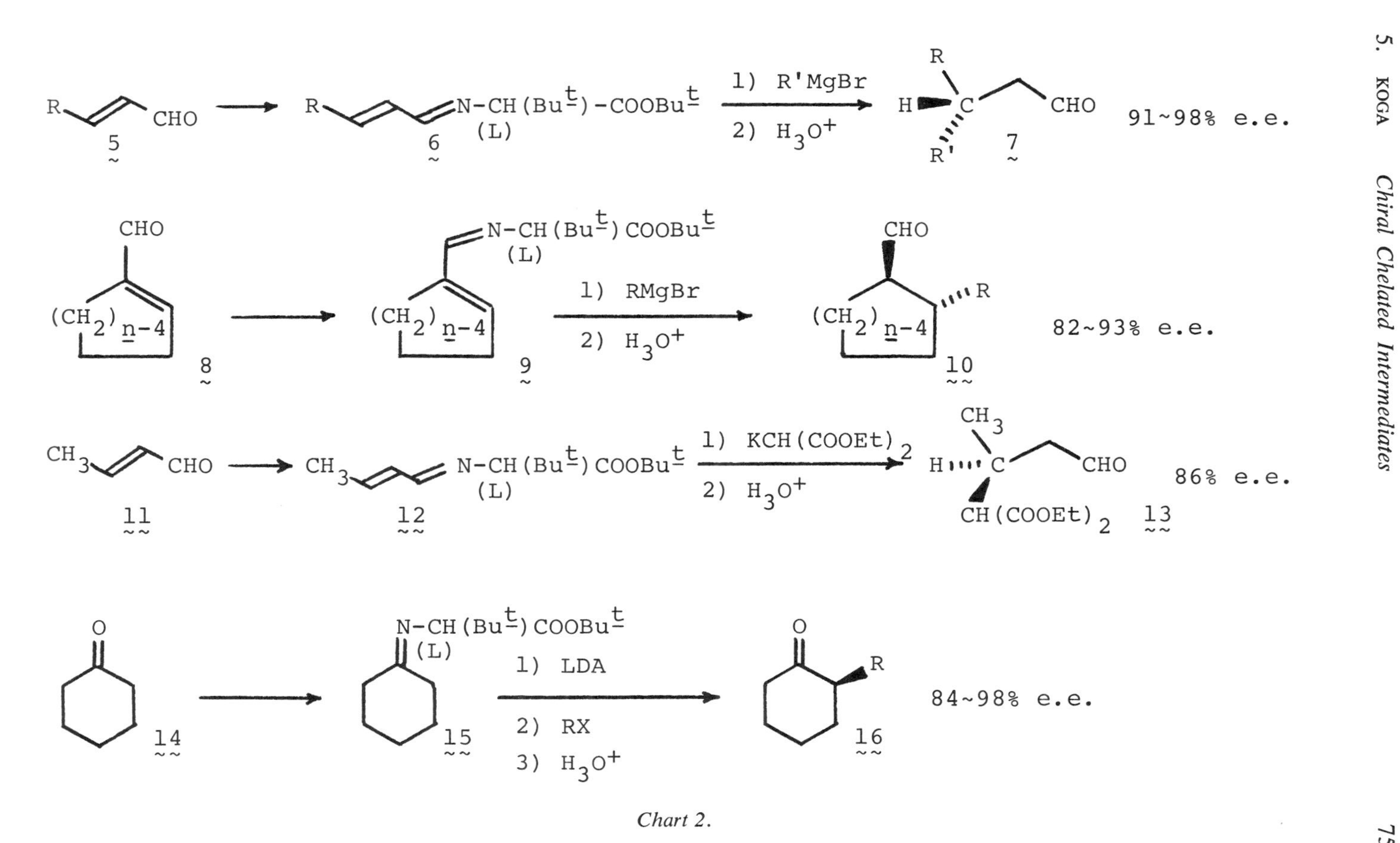
1) R'MgBr
2) H_3O^+
91~98% e.e.
1) RMgBr
2) H_3O^+
82~93% e.e.
1) $KCH(COOEt)_2$
2) H_3O^+
86% e.e.
1) LDA
2) RX
3) H_3O^+
84~98% e.e.

Chart 2.

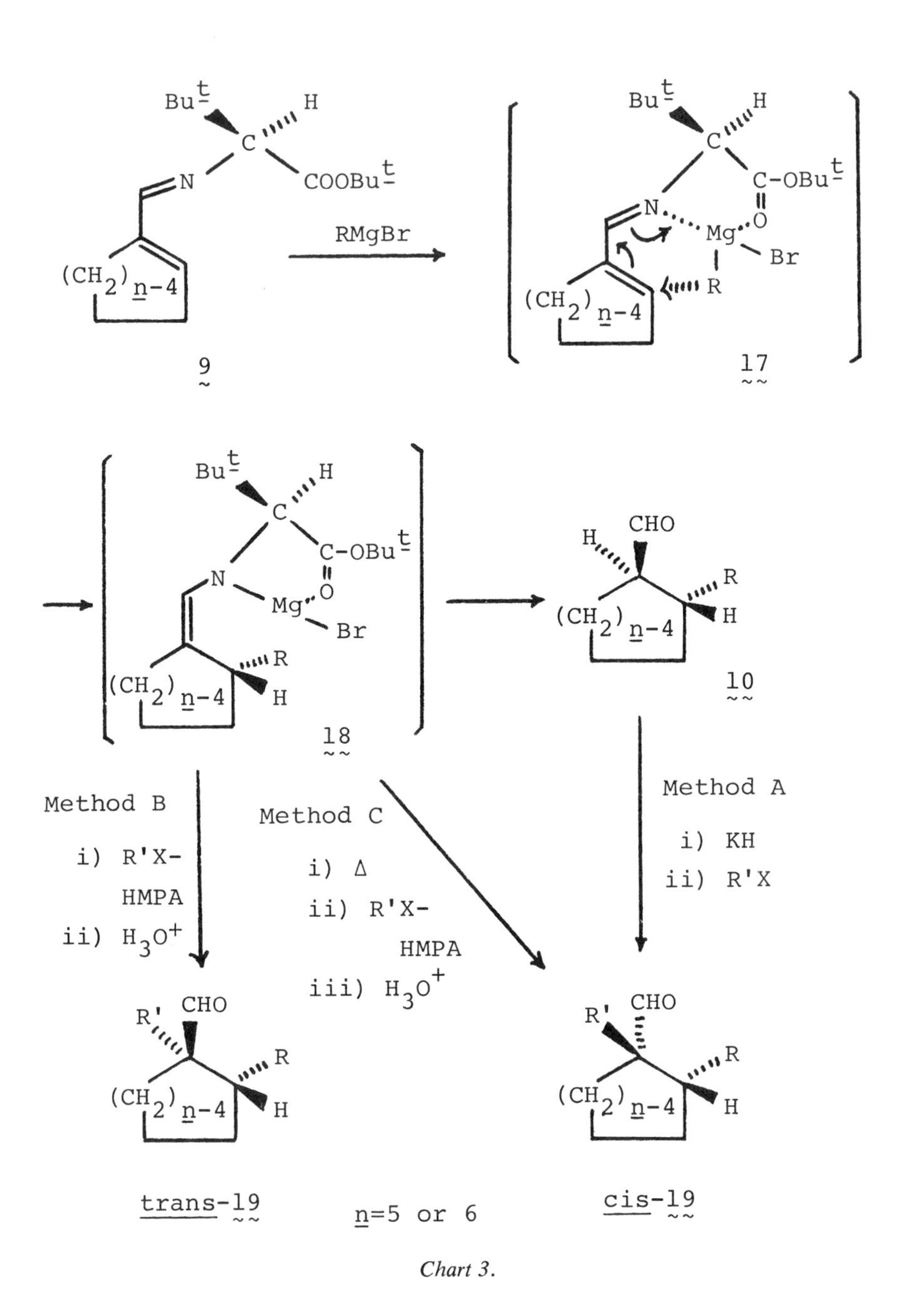
Bu^t
H
C
N
COOBu^t
(CH$_2$)$_{n-4}$
9
RMgBr
C-OBu^t
Mg
O
Br
R
17
18
CHO
10
Method B
i) R'X-HMPA
ii) H$_3$O$^+$
Method C
i) Δ
ii) R'X-HMPA
iii) H$_3$O$^+$
Method A
i) KH
ii) R'X
R'
trans-19
n=5 or 6
cis-19

Chart 3.

results are summarized in Table 1. It should be noted that the first step (conjugate addition) determines the enantioselectivity, while the second step (alkylation) determines the diastereoselectivity of the reaction.

The stereochemical course of the first conjugate addition step by Grignard reagents, common to methods A, B, and C, is already known to proceed via the *s-cis* conformation as shown in **17**.[2a,d,g] The observed alkylation of the anion of **10** by method A to give preferentially *cis*-**19** is quite reasonable, based on the premise that the reaction occurs from the side opposite to the R group. Therefore, it is striking to find that, except in run 5, the stereochemical course of the second alkylation step by method B is reversed to give *trans*-**19**, not readily accessible by the usual alkylation methods.

From mechanistic considerations of the first conjugate addition step proposed earlier,[2a,d,g] the double bond of the intermediate magnesioenamine should be in the *Z*-configuration as shown in **18**. Therefore, a possible cause for the reversal of the stereochemical course of the second alkylation step in method B is considered to be the steric effect of the original chiral center in the *Z*-configuration of the enamine moiety. The following experiments were undertaken to examine this point.

As shown in Table 2, the reaction of α,β-unsaturated aldimines (**20**), prepared from the corresponding cycloalkenecarboxaldehydes (**8**) and 2-methoxyethylamine, with Grignard reagents followed by treatment with methyl iodide using method B was found to give preferentially the corresponding *cis*-**19**. This means that the second methylation occurred from the side opposite to the initially introduced R group, as in the reaction of **10** by method A. The 2-methoxyethylamine moiety of **20** is expected to function as a bidentate ligand, similar to the α-amino acid ester moiety, during the Grignard 1,4-addition step. Thus, it is highly probable that magnesioenamine (**21**) with a structure similar to **18**, but

Table 1 Asymmetric Synthesis of 1,2-Disubstituted Cycloalkanecarboxaldehydes (**19**)

Run	Method	*n*	RMgBr	R'X	Isolated yield[a] trans-**19**	Isolated yield[a] cis-**19**	% e.e.[b]
1	A	5	C_6H_5MgBr	CH_3I	0	65	82
2	A	5	$CH_2=CHMgBr$	CH_3I	0	61	92
3	A	6	C_6H_5MgBr	CH_3I	1	62	91
4	A	6	$CH_2=CHMgBr$	CH_3I	11	61	93
5	B	5	C_6H_5MgBr	CH_3I	15	62	82
6	B	5	$CH_2=CHMgBr$	CH_3I	62	0	92
7	B	6	C_6H_5MgBr	CH_3I	55	0	91
8	B	6	$CH_2=CHMgBr$	CH_3I	67	0	93
9	B	6	$CH_2=CHMgBr$	$C_6H_5CH_2I$	67	0	93
10	B	6	$CH_2=CHMgBr$	$CH_2=CH-CH_2Br$	63	0	93
11	B	6	$CH_2=CHMgBr$	C_2H_5I	65	0	93
12	B	6	$CH_2=CHMgBr$	CH_3OCH_2Cl	52	0	93
13	C	5	C_6H_5MgBr	CH_3I	2	44	82
14	C	6	C_6H_5MgBr	CH_3I	0	49	91

a) After column chromatography. b) Corrected value for the optical purity of **2** (R=Bu^t) used.

Table 2 The Reaction of α,β-Unsaturated Aldimine (**20**) by Method B

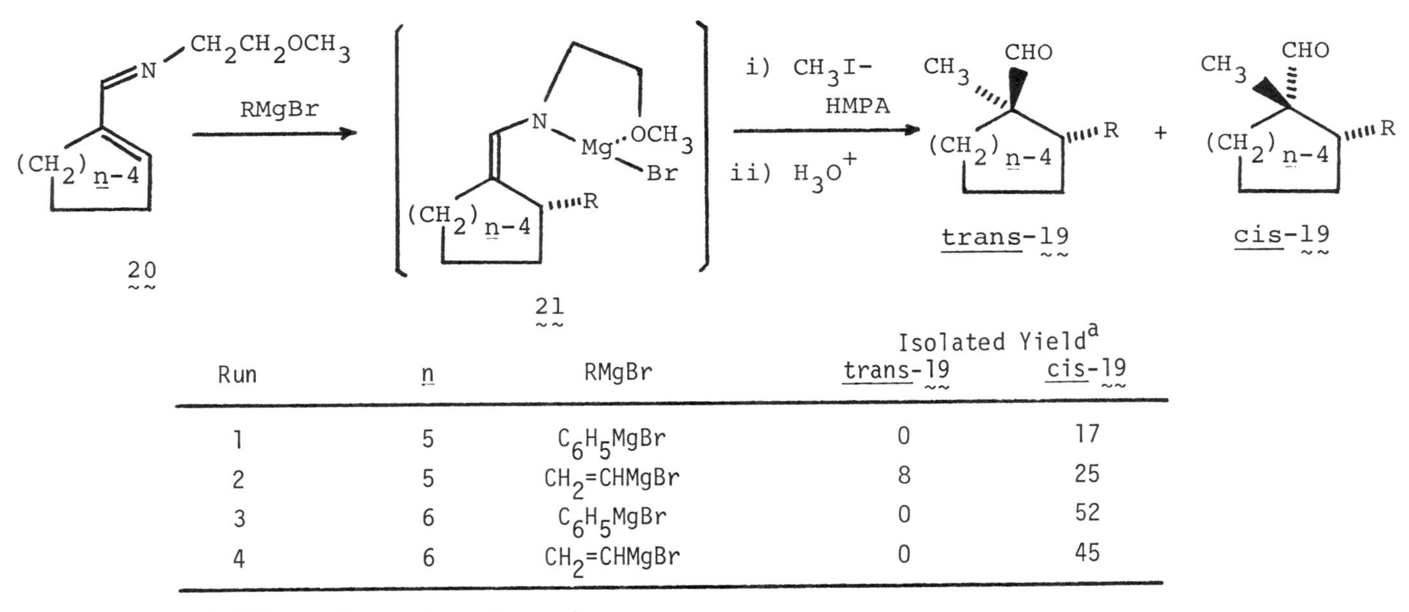

Run	n	RMgBr	Isolated Yield[a] trans-19	cis-19
1	5	C_6H_5MgBr	0	17
2	5	$CH_2=CHMgBr$	8	25
3	6	C_6H_5MgBr	0	52
4	6	$CH_2=CHMgBr$	0	45

a) After column chromatography.

without a chiral center in the chelated ring, is produced. The fact that the reaction of methyl iodide by method B gave preferentially *cis*-**19** from **21**, but preferentially (except in run 5 in Table 1) *trans*-**19** from **18** clearly suggests that the original chiral center in the chelated ring of **18** exerts a strong steric influence on the second alkylation step.

The importance of the *Z*-configuration for **18** to give *trans*-**19** by method B was also demonstrated. It is known that *E*-*Z* isomers of metalloenamines do not equilibrate under conditions similar to those employed in method B, but do equilibrate in THF under reflux.[4] It is thus probable that equilibration prior to alkylation will affect the diastereoselectivity of the reaction. Although the chemical yields were lower, probably due to the instability of magnesioenamines at higher temperatures, methylation of magnesioenamines after heating (method C) was found to give preferentially *cis*-**19** as shown in Table 1. Additional support for the occurrence of *E*-*Z* isomerization of magnesioenamines on heating was observed by ^{13}C NMR studies using an earlier reported method.[4] Thus, starting from cyclohexenecarboxaldehyde (**8**, *n*=6) enriched with ^{13}C at the aldehyde carbon, magnesioenamine (**18**, *n*=6, R=C_6H_5) was prepared with phenylmagnesium bromide in THF-d_8 at -23°C. The ^{13}C NMR spectrum of this magnesioenamine showed a single peak (146.0 ppm) of enriched carbon before heating, but a new peak (147.5 ppm) appeared after heating at 70°C. The intensity of the new peak relative to the initial peak increased gradually, and the signals became complex after 3 hr at 70°C, probably due to partial decomposition.

These data clearly show that the formation of magnesioenamine (**18**) having the *Z*-configuration and attack of alkyl halides under the steric influence of the original chiral center of the *tert*-leucine *tert*-butyl ester moiety are responsible for the predominant formation of *trans*-**19** by method B. However, the substituent R introduced is also expected to show a steric effect at the second alkylation stage. Therefore, assuming that the configuration of the magnesioenamine is retained in Z-stereochemistry throughout the reaction, the stereochemical course of the second alkylation step by method B is considered to be subject to the two opposing steric effects of the original chiral center and the newly created chiral center, of which the former apparently predominates in many cases, while the latter is predominant when a bulky R group is attached to the five-membered ring.

Literature Cited

1. (a) Morrison, J. D.; Mosher, H. S. "Asymmetric Organic Reactions"; Prentice-Hall: New Jersey, 1971. (b) Kagan, H. B.; Fiaud, J. C. Top. Stereochem. 1978, 10, 175. (c) Valentine, D., Jr.; Scott, J. W. Synthesis 1978, 329. (d) ApSimon, J. W.; Seguin, R. P. Tetrahedron 1979, 35, 2797.
2. (a) Hashimoto, S.; Yamada, S.; Koga, K. J. Am. Chem. Soc. 1976, 98, 7450. (b) Hashimoto, S.; Komeshima, N.; Yamada, S.; Koga, K. Tetrahedron Lett. 1977, 2907. (c) Hashimoto, S.; Koga, K. Ibid. 1978, 753. (d) Hashimoto, S.; Yamada, S.; Koga, K. Chem. Pharm. Bull. 1979, 27, 771. (e) Hashimoto, S.; Komeshima, N.; Yamada, S.; Koga, K. Ibid. 1979, 27, 2437. (f) Hashimoto, S.; Koga, K. Ibid. 1979, 27, 2760. (g) Hashimoto, S.; Kogen, H.; Tomioka, K.; Koga, K. Tetrahedron Lett. 1979, 3009.
3. (a) Kogen, H.; Tomioka, K.; Hashimoto, S.; Koga, K. Tetrahedron Lett. 1980, 21, 4005. (b) Idem Tetrahedron in press.
4. (a) Hoobler, M. A.; Bergbreiter, D. E.; Newcomb, M. J. Am. Chem. Soc. 1978, 100, 8182. (b) Meyers, A. I.; Snyder, E. S.; Ackerman, J. J. H. Ibid. 1978, 100, 8186.

RECEIVED December 14, 1981.

6

Asymmetric Carbon–Carbon Bond Forming Reactions via Chiral Oxazolines

ALBERT I. MEYERS

Colorado State University, Department of Chemistry, Fort Collins, CO 80523

The use of chiral oxazolines as auxiliaries in C-C bond forming reactions continues to provide chiral compounds in high enantiomeric excess. Conjugate addition-alkylation of α,β-unsaturated oxazolines leads to 2,3-disubstituted alkanoic acids in high ee's. Boron enolates of oxazolines, where the chirality resides in the oxazoline or boron substituents, react with aldehydes to give β-hydroxy esters in high erythro or threo selectivity with good ee's. Aromatic chiral oxazolines containing o-acyl groups react with organometallics furnishing, after hydrolysis, phthalides in high ee's. A further extension using aryl oxazolines leads to chiral binaphthyls in 70-100% ee's.

2,3-Disubstituted Carboxylic Acids

In our continuing program to utilize chiral oxazolines (1-7) as auxiliary reagents in asymmetric synthesis, several novel routes to chiral compounds have been developed. The previously reported (3) conjugate addition (Fig. 1) to vinyl oxazolines 1 (pure E enantiomer) by organolithium reagents furnishing the adduct 2 and subsequent hydrolysis gave 3,3-disubstituted carboxylic acids 3 in 95-99% ee. We have recently extended this methodology to provide an additional chiral center in carboxylic acids (Fig. 1). Thus, the intermediate lithio adduct 2 could be treated with an alkyl halide to give the alkylated oxazoline 5, which after hydrolysis affords the 2,3-disubstituted acids 6 in 77-82% diastereomeric purity. HPLC examination of the diastereomeric oxazolines 5 prior to hydrolysis indicates that the alkylation of 2 occurred with >99% stereoselectivity. Thus, virtually no presence of diastereomeric impurities was observed. It may, therefore, be concluded that the observed diastereomeric purity for 6 was the result of partial racemization during the hydrolysis

0097-6156/82/0185-0083$05.00/0

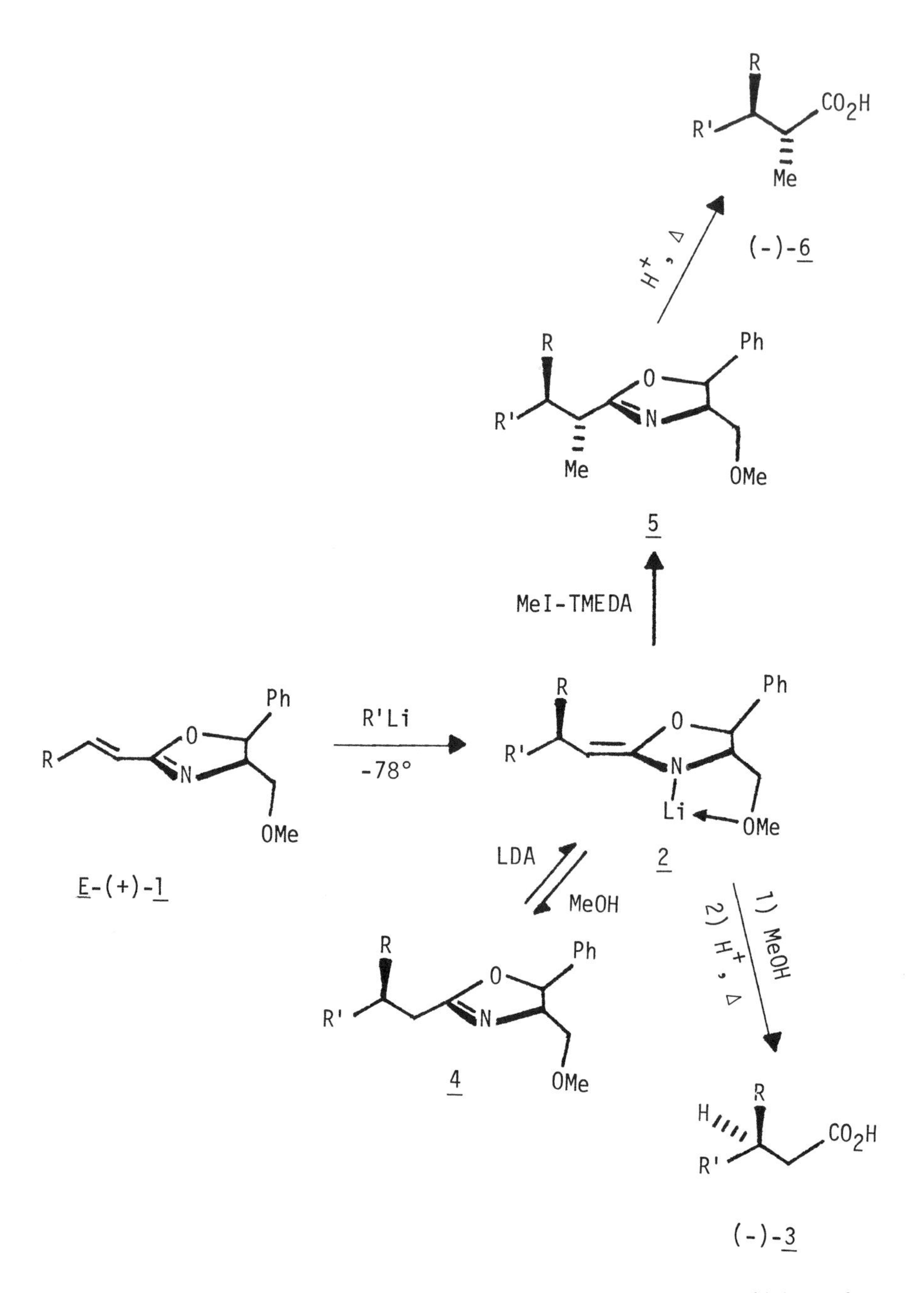

Figure 1. Formation of 2,3-disubstituted alkanoic acids by conjugate addition and alkylation of α,β-unsaturated oxazolines.

of 5. Several examples of this tandem dialkylation are given in Table I. With regard to the absolute configuration of the acids 6, they are assigned on the basis of both known compounds and previous predictions described in earlier work (1). It is interesting to note that quenching adduct 2 to give the 2-alkyloxazoline 4 and then metalation back to 2 followed by alkylation gave the 2,3-disubstituted acids 6 of the same configuration as that obtained from 2 in the sequential dialkylation process. This confirms that the same lithio azaenolate 2 is formed both by conjugate addition (1→2) and metalation (4→2). Since both processes have been performed as separate methods, leading either to chiral 2-substituted carboxylic acids or 3-substituted chiral acids, the absolute configuration of the 2,3-disubstituted acids 6 is consistent with these earlier findings.

Table I. 2,3-Disubstituted Carboxylic Acids 6

R (in 1)	R'Li	Diastereomeric Ratio 5	Yield 5	% Diastereomeric Ratio, 6	
				α-C	β-C
Me	Et	99	65	77 (R)	99 (R)
Me	n-Bu	99	80	82 (R)	99 (R)
t-Bu	n-Bu	99	82	80 (S)	99 (R)
n-Bu	t-Bu	99	75	79 (R)	99 (R)

Aldol Products via Boron Azaenolate

The use of chiral oxazolines as reagents for aldol type products (Fig. 2) rich in erythro or threo β-hydroxy acids has also been accomplished. In earlier work in our laboratory (8) we described the formation of β-hydroxyesters 7 from lithio oxazolines and various aldehydes in 20-25% ee. The absence of an α-alkyl group was considered the major reason for the poor ee's of the product which lacked stringent stereochemical requirements in the transition state. The process was repeated with the 2-ethyloxazoline and gave 8 in much higher selectivity mainly as the threo-isomer and in 75% enantiomeric purity (7). We have now investigated this aldol process using the boron "enolates" of oxazolines 9 and 10 (Fig. 3) (9). It should be noted that boron azaenolate 9 contains the chiral center on the organoborane, whereas 10 contains the chiral center on the oxazoline.

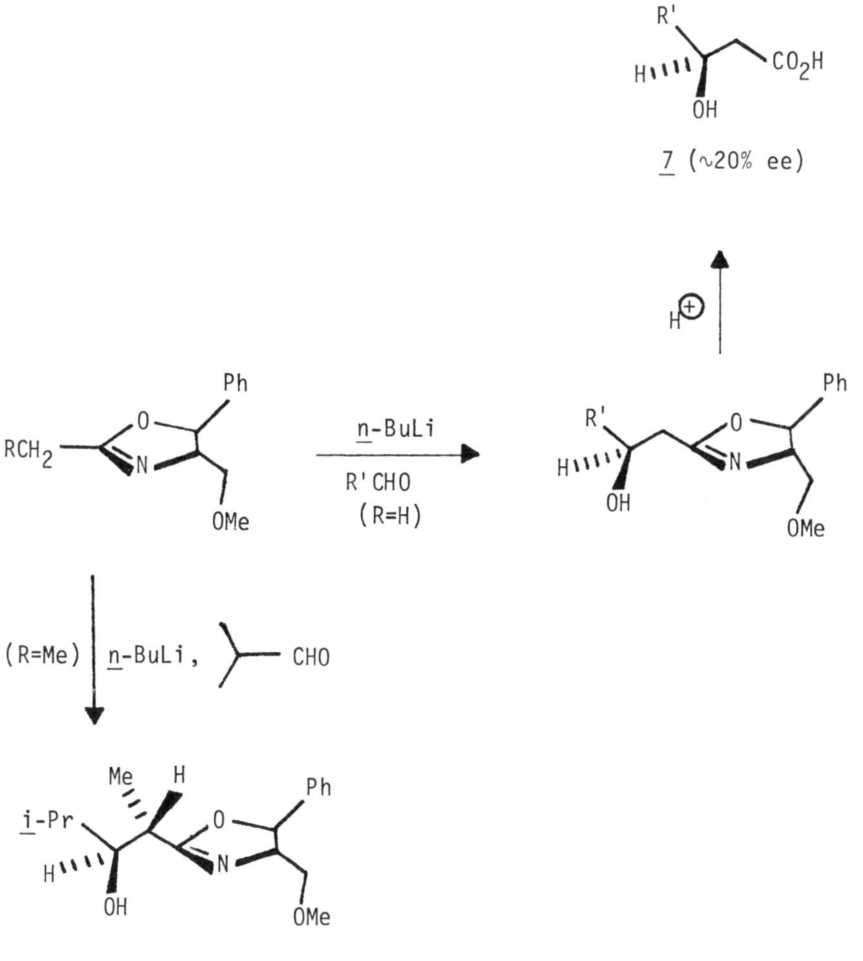

R'
H
CO_2H
OH

7 (~20% ee)

$H^{\oplus}$

Ph
RCH_2
O
N
OMe

n-BuLi
R'CHO
(R=H)

R'
H
OH
O
N
Ph
OMe

(R=Me) n-BuLi, CHO

Me H
i-Pr
H
OH
O
N
Ph
OMe

8 (82% threo, 75% ee)

Figure 2. Chiral oxazolines used as reagents for aldol-type products rich in erythro *or* threo *β-hydroxy acids.*

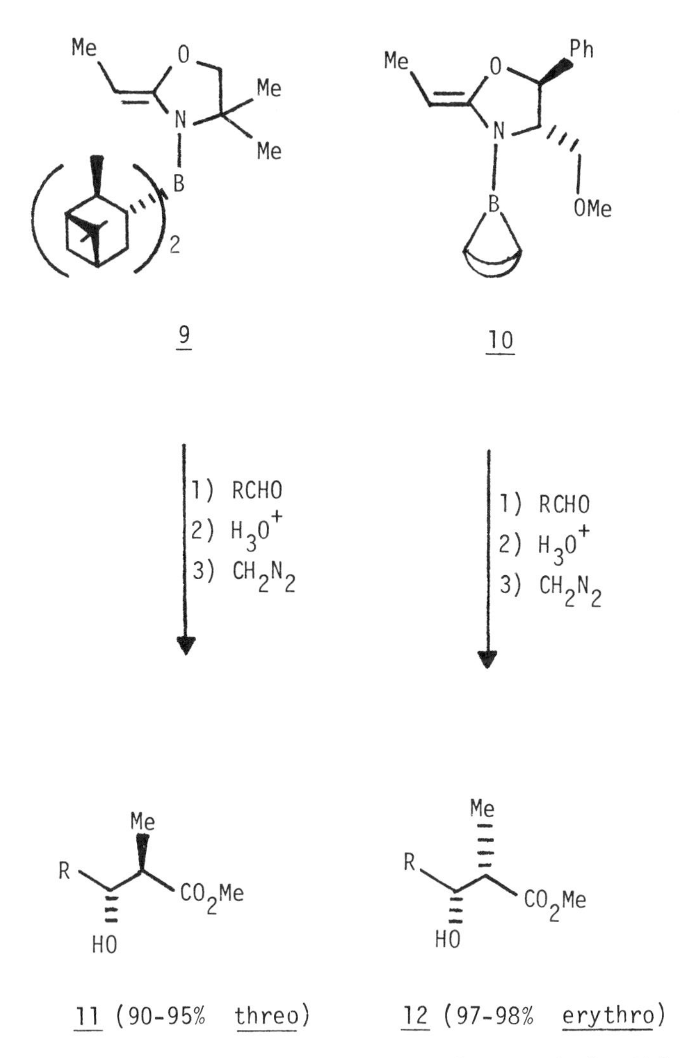

Figure 3. Formation of β-hydroxyesters with an α-alkyl group by the aldol process using boron enolates of oxazolines 9 *and* 10.

Treatment of 9 with various aldehydes gave the threo-β-hydroxyester 11 after hydrolysis and esterification, in 90-95% diastereoselectivity with enantiomeric excess of 77-85% (Table II). On the other hand, the boron enolate 10, when treated similarly with aldehydes now gave the erythro β-hydroxy esters 12 in 97-98% diastereoselectivity though in somewhat poorer ee's (40-60%, Table III). ^{13}C NMR spectroscopy employing 9 and 10 with ^{13}C-enriched methyl groups confirmed that only a single enolate was formed at -78° under the conditions of kinetic control. Equilibration of 9 or 10 took place by warming their ether solutions to -25° showing a steady increase in a second methyl signal until equilibrium was reached at 2:1. Unfortunately, which enolate is formed at -78° is not known at this time. When the condensation

Table II. Chiral threo-hydroxyesters 11 from 9

RCHO	% threo-erythro	% ee	Conf'n. Major Isomer	$[\alpha]_D$ $(CHCl_3)$
EtCHO	92:8	77	2R,3R	-9.9
PrCHO	91:9	77	2R,3R	-2.5
n-PentCHO	90:10	77	2R,3R	-3.1
Me_2CHCHO	91:9	85	2R,3R	-12.5
CyclohexCHO	95:5	84	2R,3R	-8.1
t-BuCHO	94:6	79	2R,3S	-21.2

Table III. Chiral erythro-hydroxyesters 12 from 10

RCHO	% erythro-threo	% ee	Conf'n. Major Isomer	$[\alpha]_D$ $(CHCl_3)$
EtCHO	98:2	40	2S,3R	1.4
Me_2CHCHO	98:2	41	2S,3R	-2.3
t-BuCHO	97:3	60	2S,3S	-6.6

with aldehydes was carried out with equilibrated 9 or 10, the diastereoselectivity in 11 and 12 decreased, as expected, to ∿80:20. Since the stereochemistry of the boron enolates 9 and 10 is not known, it is difficult to advance a reasonable mechanism except to postulate a chair-type six-membered transition state based on the Zimmerman model (10). This correctly predicts the stereochemistry of the product provided the Z boron enolates of 9 and 10 are employed (Fig. 4).

A variety of other oxazolines was investigated (Fig. 5) to probe the nature of structural parameters in determining the erythro threo ratios of β-hydroxy esters. Thus, reaction of boron enolates of 13-16 and 9-BBN also gave high erythro selectivity of 12 (R = i-Pr) when treated with isobutyraldehyde. It is interesting to note that 13 gave high threo selectivity (Table II) when diisopinocampheyl borane enolate 9 was employed while giving high erythro selectivity of 12 (R = i-Pr) using 9-BBN enolate. This implies a major effect on the product due to the nature of the boron substituents. Further work should help clarify this point.

Chiral Phthalides

Aromatic oxazolines have also been utilized (Fig. 6) as vehicles for asymmetric synthesis. Thus the chiral oxazoline 17, used as its lithio or magnesiohalide derivative 17b (from the bromo compound 17a) was treated with several carbonyl compounds to give the adducts 18, whose diastereomer ratios were assessed by ^{1}H-nmr or HPLC (Table IV). The extent of stereoselection was rather poor indicating a sterically undemanding transition state.

Table IV.

R_2CO	% Yield	Diast. Ratio[a]
PhCOMe	71	64:31
PhCHO	60	57:43
o-MeOPhCHO	63	59:41
n-BuCHO	64	51:49

a) Assessed on 18

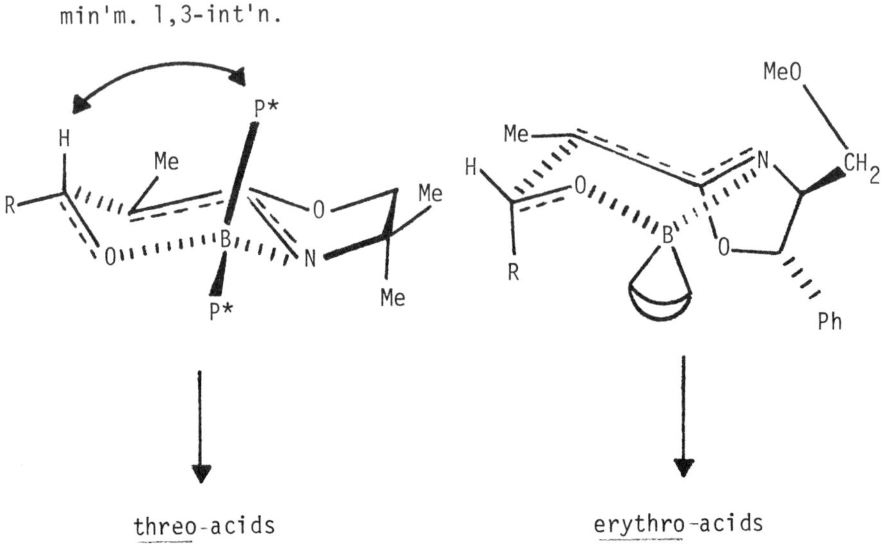

min'm. 1,3-int'n.

threo-acids

erythro-acids

Figure 4. Prediction of stereochemistry of boron enolates 9 *and* 10 *using chair-type six-membered transition state based on the Zimmerman model.*

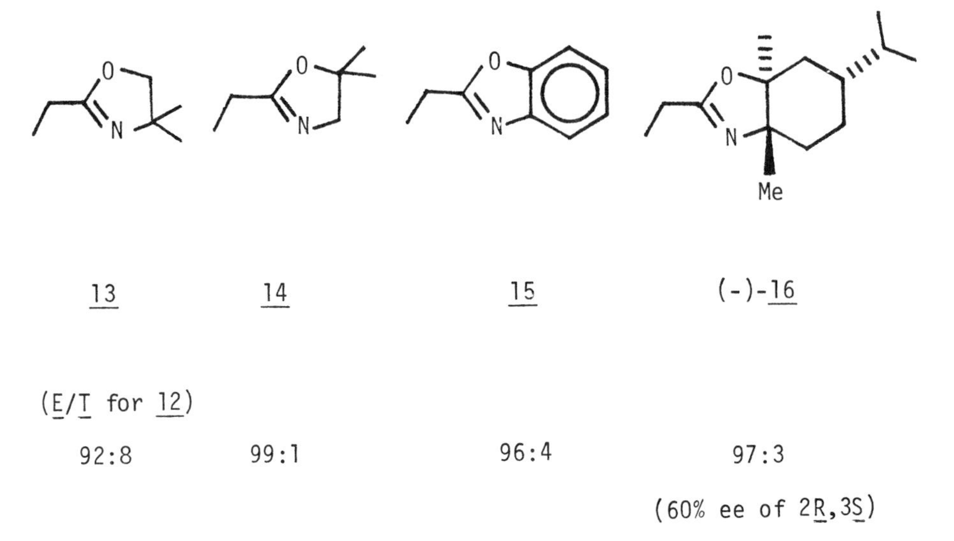

Figure 5. A variety of oxazolines used to determine erythro–threo *ratios of β-hydroxy esters.*

These findings are in strong contrast to the highly successful results (Fig. 7) of Mukaiyama (11) using the proline derivative 20. Hydrolysis of 18 in acidic medium gave the phthalide 19. In order to assess the absolute configuration of the lactones, particularly those with a dialkyl substitution pattern, which are not known in the literature, an x-ray study was performed using the p-bromophenyl acetophenone, 19 (R = Me, R' = p-Bromophenyl, S configuration). Pure enantiomers of 19 were obtained by MPLC-assisted resolutions of 18 followed by hydrolysis with acid.

We next turned to reversal of the nucleophile-electrophile combination by preparing o-acylaryl oxazolines 21 and treating them with organometallics (Fig. 8). Addition of organolithium or Grignard reagent gave the adducts 22 which smoothly rearranged to the iminolactones 23. HPLC analyses of 23 showed the ratio of diastereomers to be rather low again suggesting that addition of organolithium reagents to 21 was perhaps too fast with a low $\Delta\Delta G^{\ddagger}$. However, when methylmagnesium chloride was treated with the o-benzoyloxazoline 24, the reaction proceeded more slowly and, after hydrolysis, the phthalide 25 was recovered almost enantiomerically pure (Fig. 9) (12).

Future efforts will now be directed to Grignard additions to ketooxazolines in the hope that the above reaction possesses some generality. The complexities of this system and factors affecting the transition state will have to be more carefully addressed before a general synthetic approach to chiral phthalides can be achieved.

Chiral Binaphthyls

An asymmetric synthesis of chiral binaphthyls has been accomplished utilizing naphthyloxazolines. The method is based on the facile displacement of an o-methoxyl group in aryloxazolines by various nucleophiles (13). The aromatic substitution process has now also been found to proceed with o-methoxynaphthyloxazolines (Fig. 10). A number of nucleophilic reagents smoothly displaced the methoxyl group to 26 and after hydrolysis led to 1-substituted-2-naphthoic acids 27. Utilization of this efficient coupling reaction with chiral oxazolines was examined in an attempt to reach chiral binaphthyls. Thus, 28 was treated with the Grignard reagent of 1-bromo-2-methoxynaphthalene at room temperature in THF to give the binaphthyl adduct 29 (Fig. 11). Nmr analysis showed that the ratio of diastereomers in 29 was greater than 95:5 indicating a high degree of stereoselection in the coupling reaction. Hydrolysis of 29 followed by hydride reduction of the intermediate ester gave the chiral binaphthyl 30 in ~100% ee (confirmed by LISR-nmr techniques). Two additional naphthyl Grignard reagents were examined (Fig. 12) which led to products whose ratios were not as high as in the methoxy naphthyl system, but, nevertheless, were still formed in 60-70% ee. The crystalline nature of 29

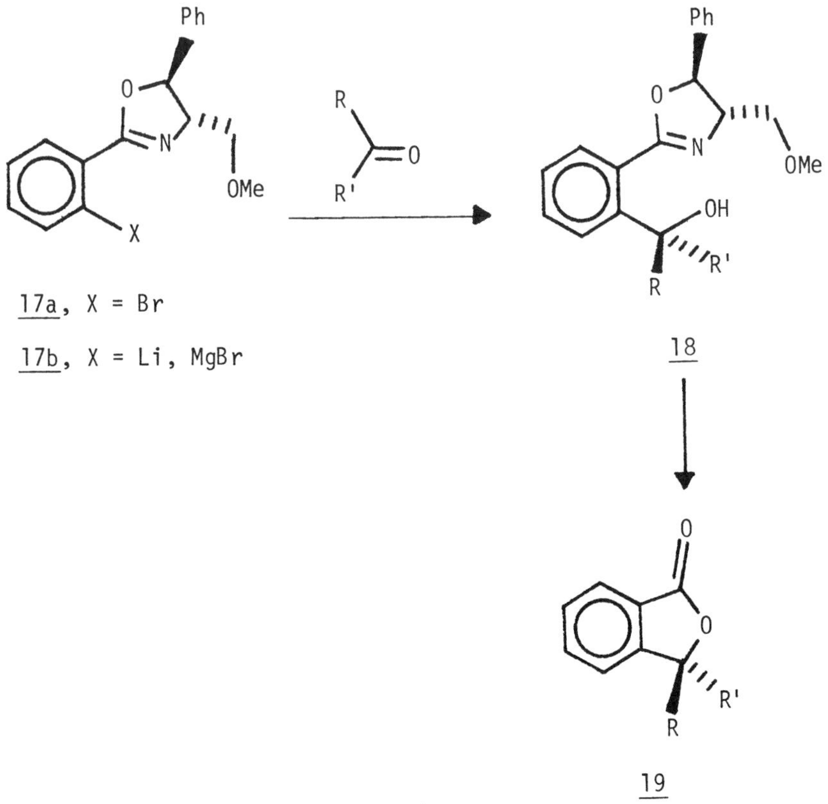
Ph
O
N
OMe
X
R
O
R'
OH
R'
R
17a, X = Br
17b, X = Li, MgBr
18
19

Figure 6. Aromatic oxazolines used as vehicles for asymmetric synthesis.

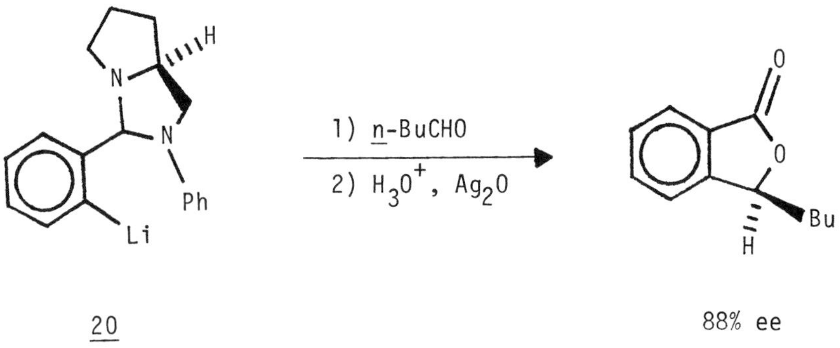
H
N
N
Ph
Li
1) n-BuCHO
2) H_3O^+, Ag_2O
O
O
Bu
H
20
88% ee

Figure 7. Asymmetric synthesis using proline derivative 20. *(11)*

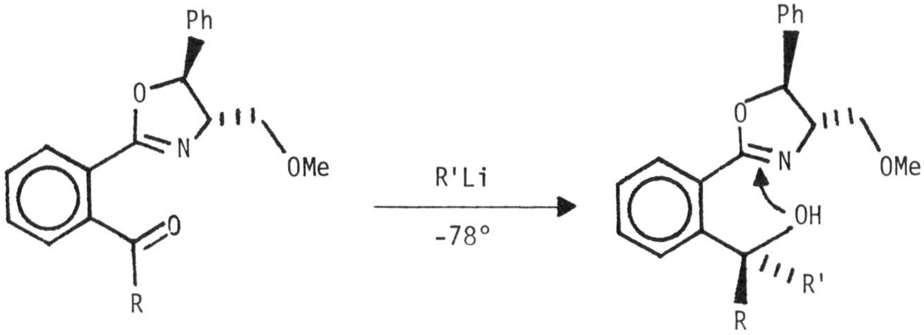

Diastereomeric Ratios of 23

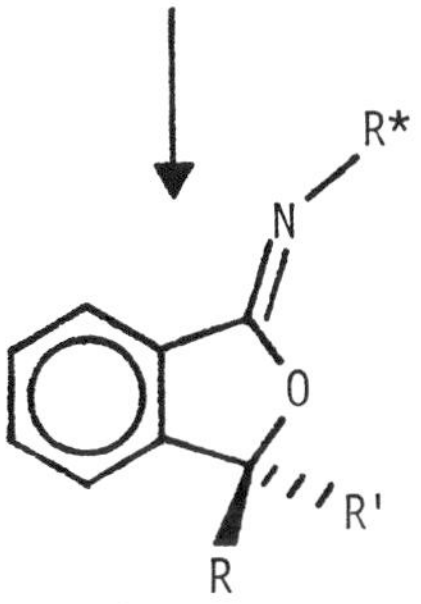

R	R'Li	% Yield	Ratio
t-Bu	Ph	94	2:1
Ph	Me	85	2:1
Ph	n-Bu	86	1.2:1

Figure 8. Synthesis of iminolactones, 23, by reaction of o-*acylaryloxazolines with organolithium or Grignard reagent and rearrangement of product, 22.*

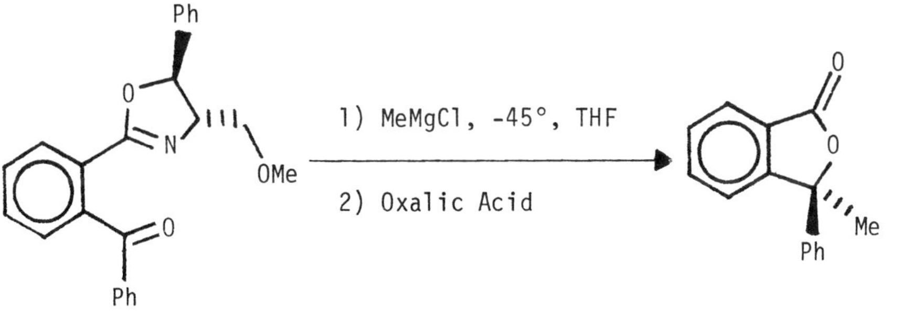

Figure 9. Preparation of phthalide, 25, by treatment of o-*benzoyloxazoline, 24, with Grignard reagent followed by hydrolysis.*

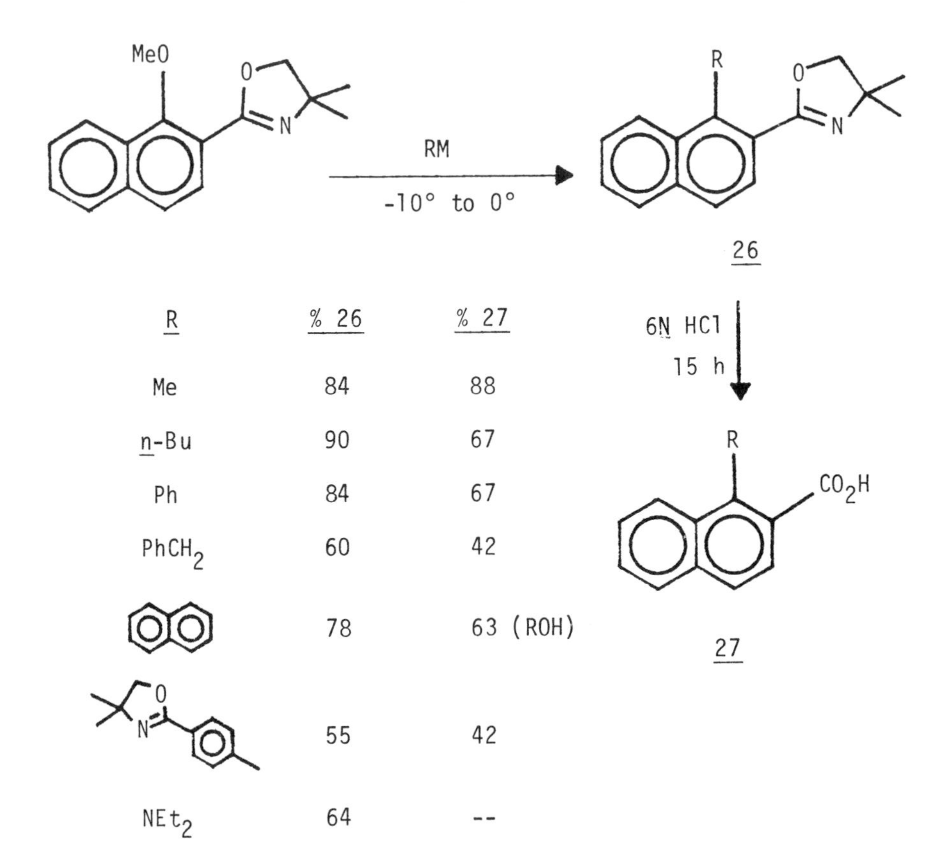

Figure 10. Synthesis of 1-substituted-2-naphthoic acids by aromatic substitution of o-*methoxynaphthyloxazolines followed by hydrolysis.*

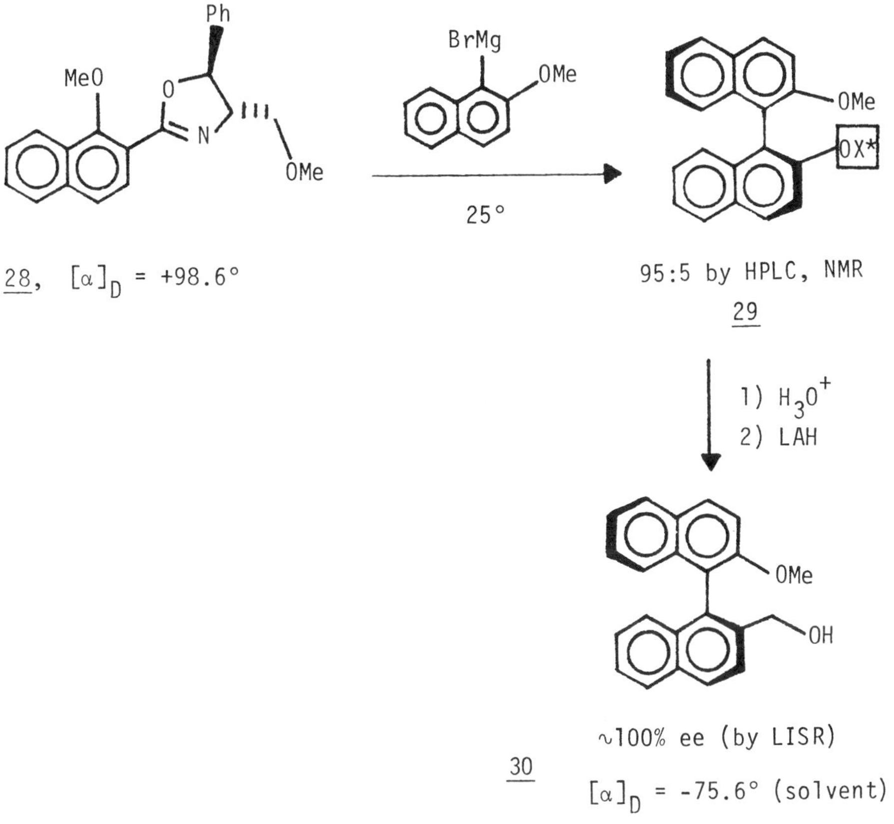

Figure 11. Asymmetric synthesis of chiral binaphthyls, 30, *with high degree of stereoselectivity.*

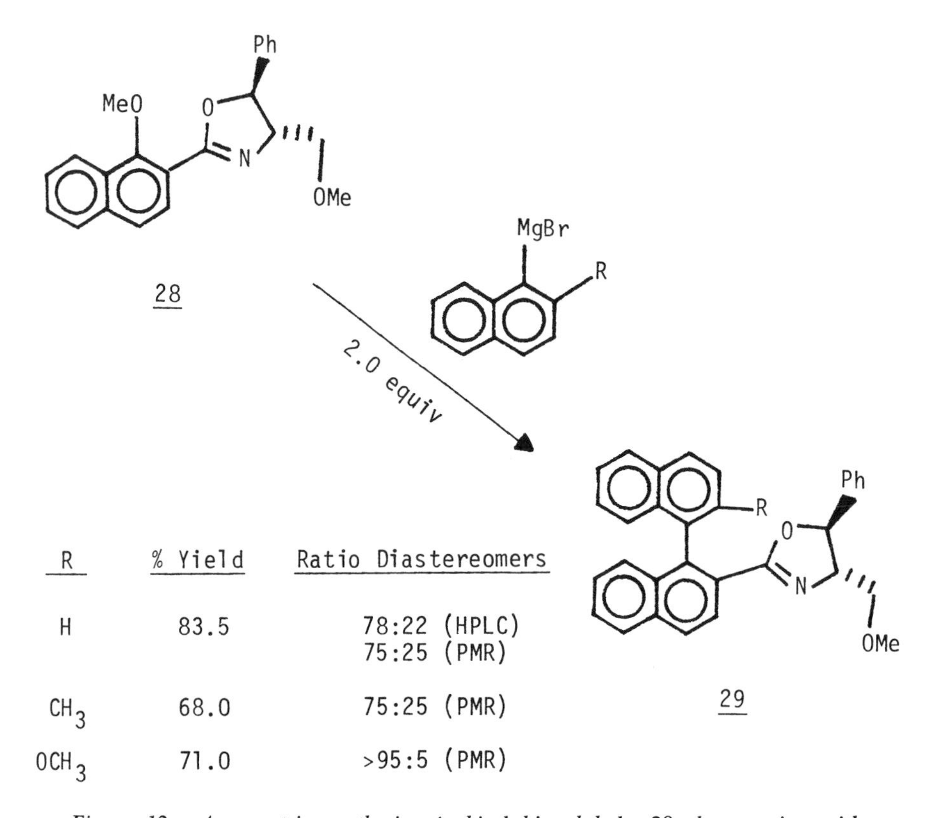

R	% Yield	Ratio Diastereomers
H	83.5	78:22 (HPLC) 75:25 (PMR)
CH_3	68.0	75:25 (PMR)
OCH_3	71.0	>95:5 (PMR)

Figure 12. Asymmetric synthesis of chiral binaphthyls, 29, by reaction with naphthyl Grignard reagents. Ratios of diastereomers are given.

(R = H, Me, MeO) resulted in very easy purification to single diastereomers by simple crystallization. However, since this was not the current aim of the study, care was taken to avoid inadvertent resolution during the workup and purification of 29. The most convenient method to reach chiral binaphthyls was to carry out a tandem hydrolysis-reduction to the hydroxymethyl group (Fig. 13). Thus, the binaphthyloxazolines 29 were only partially hydrolyzed to the aminoesters 31 and then subjected to hydride reduction to the alcohol 30. The absolute configuration of the chiral binaphthyls was determined by correlation methods to known derivatives as well as CD spectral characteristics.

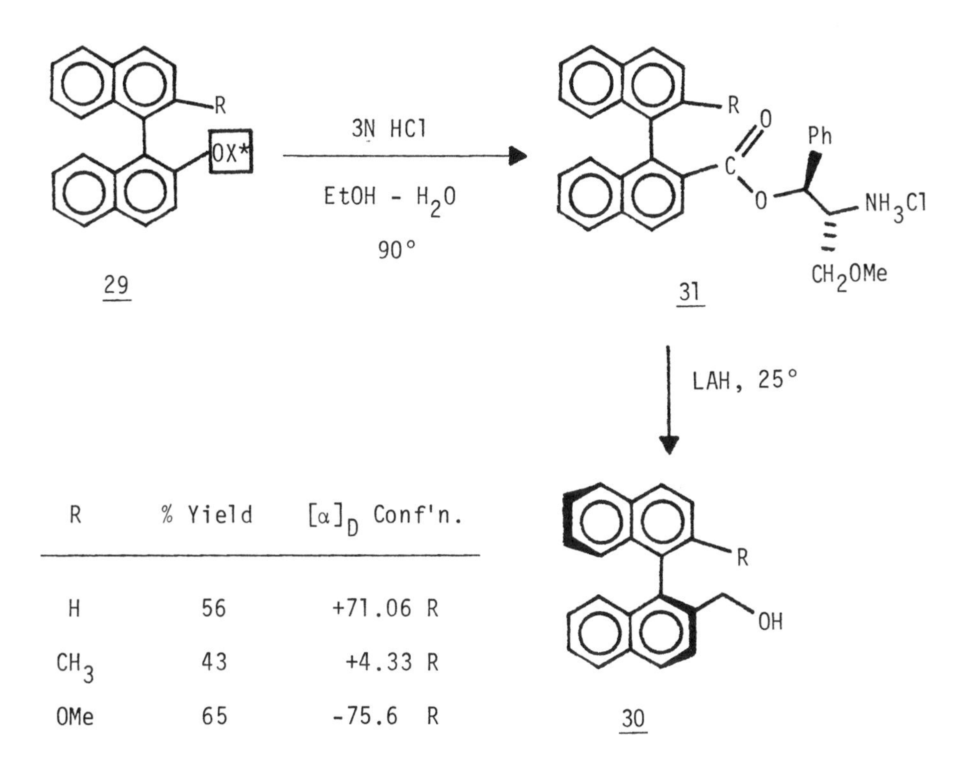

R	% Yield	$[\alpha]_D$	Conf'n.
H	56	+71.06	R
CH_3	43	+4.33	R
OMe	65	-75.6	R

Figure 13. Hydrolysis of binaphthyloxazolines, 29, *to aminoesters,* 31, *and reduction to alcohol producing chiral binaphthyls.*

Literature Cited

1. Meyers, A. I. Accounts of Chem. Res., 1978, 11, 375.
2. Meyers, A. I. Pure & Appl. Chem., 1979, 51, 1255.
3. Meyers, A. I.; Smith, R. K.; Whitten, C. E. J. Org. Chem., 1979, 44, 2250.
4. Meyers, A. I.; Smith, R. K. Tet. Lett., 1979, 2749.
5. Meyers, A. I.; Slade, J.; J. Org. Chem., 1980, 45, 2785.
6. Meyers, A. I.; Yamamoto, Y.; Mihelich. E. D.; Bell, R. A. J. Org. Chem., 1980, 45, 2792.
7. Meyers, A. I.; Reider, P. J. J. Am. Chem. Soc., 1979, 101, 2501.
8. Meyers, A. I.; Knaus, G. Tet. Letters, 1974, 1333.
9. Meyers, A. I.; Yamamoto, Y. J. Am. Chem. Soc., 1981, 103, 4278.
10. Zimmerman, H. E.; Traxler, M. D. J. Am. Chem. Soc., 1957, 79, 1920.
11. Asami, M. and Mukaiyama, T. Chem. Lett., 1980, 17.
12. Meyers, A. I.; Hanagan, M. A., research in progress.
13. Meyers, A. I.; Gabel, R.; Mihelich, E. D. J. Org. Chem., 1978, 43, 1372.

RECEIVED December 14, 1981.

Highly Selective Synthesis with Novel Metallic Reagents

HITOSI NOZAKI, TAMEJIRO HIYAMA, KOICHIRO OSHIMA, and KAZUHIKO TAKAI

Kyoto University, Department of Industrial Chemistry, Kyoto, 606, Japan

Allylchromium reagents as produced from allylic bromides and Cr(II) salts in anhydrous THF or DMF react with carbonyl components to form homoallylic alcohols. The aldehyde adducts, RCH(OH)-CHMeCH=CH_2, are oxidized with various recently described reagents to produce epoxy alcohols with different ways of steric control.--Alkylation of cyclopropane derivatives with R_3Al proceeds from preliminary heterolysis in one case, whereas the reaction introduces alkyl carbanions with S_N2-like inversion in other cases.--Catalysis with Pd(0) makes possible the substitution of an -OPO$(OR)_2$ group on an sp^2 carbon and finds a number of synthetic applications.--Finally, the aliphatic Claisen rearrangement is smoothly performed at room temperature by means of R_2AlX reagents (where X = R, H, or SPh etc.) involving "combined acid-base" attack.

This paper will deal with four topics: the first one is related to allylchromium reagents, while the latter three refer to the behavior of trialkylaluminum or related species in different situations. The authors' main concern here is to describe new reactions useful for selective synthesis.

Allylchromium Reagents in Homoallyl Alcohol Synthesis

Organochromium compounds prepared from halides and Cr(II) species in anhydrous, aprotic, polar solvents provide means of selective synthesis as has been described previously (1,2). In particular, the Grignard type carbonyl addition of allylchromium reagents proceeds much more slowly and selectively than that of organomagnesium compounds.

Scheme 1 indicates threo selectivity in the reaction of

0097-6156/82/0185-0099$05.00/0

Scheme 1

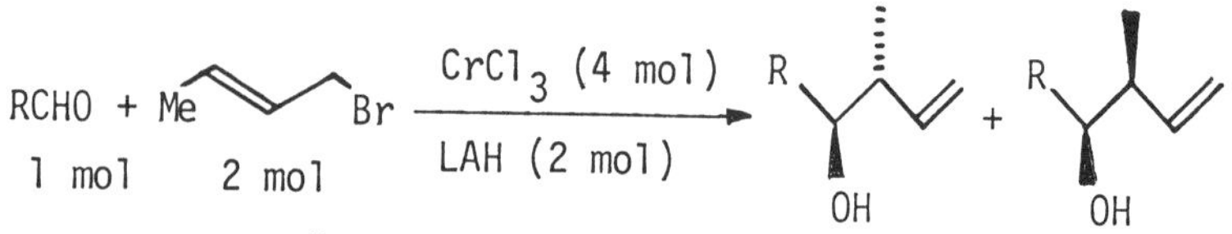

RCHO (1 mol) + Me–CH=CH–CH2Br (2 mol) —[$CrCl_3$ (4 mol), LAH (2 mol)]→ threo + erythro

R	Solvent[a]	Yield (%)	threo (%)	erythro (%)
Ph	THF	96	100	0
Ph[b]	THF	87	100	0
Ph	DMF	92	75	25
n-Pr	THF	59	93	7
i-Pr	THF	55	95	5
i-Pr	DMF	78	66	34
n-Am	THF	70	97	3
n-Am	DMF	77	68	32

[a]The reaction was carried out at room temp for 2 h.
[b]The cis-isomer of crotyl bromide was used.

Scheme 2

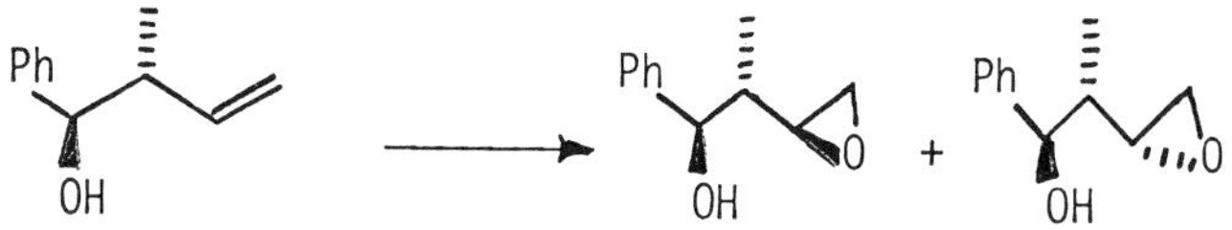

Oxidant	Yield (%)	Isomer Ratio (%)
t-BuOOH/$VO(acac)_2$	52	76 : 24
t-BuOOH/$Al(OBu\text{-}t)_3$	51	18 : 82
mCPBA	69	55 : 45

crotyl bromide (3) as reported also by Heathcock (4). Tetrahydrofuran (THF) as solvent gives higher selectivities but somewhat poorer yields than dimethylformamide (DMF). Epoxidation of the resulting homoallylic alcohols has been investigated (Scheme 2). The Sharpless and related epoxidation techniques (5,6) provide a way to control the stereochemistry of three neighboring carbons. The Cr(II) mediated reaction has been extended further to systems involving aldehydes and 2,2-diiodopropane (with HI loss) as well as vinyl iodides and bromides, all affording allylic alcohols (7).

Alkylation of an sp^3 Carbon with Trialkylaluminum

The reaction of diethyl geranyl phosphate with R_3Al produces quantitatively a mixture of geranyl-R and linalyl-R products in 9:1 ratio, while the corresponding neryl phosphate affords 4-$RCMe_2$-substituted 1-methylcyclohexenes exclusively (8). Evidence for the intermediacy of a carbocation species in the related reaction shown in Scheme 3 is derived from the fact that the optical activity of the starting acetate substrate is completely lost in the ring cleavage reaction product (9,10). A possible explanation is given in Scheme 4. Throughout these and subsequent reactions we use no less than a 2:1 mol ratio of aluminum reagents which are mostly dimeric. We postulate that the leaving acetate group is substantially reduced in nucleophilicity by double complexation with R_3Al, so that the cationic part is almost naked even in the early ion-pair stage. The cyclopropylmethyl cation is isomerized to the more stable benzylic one which is then slowly alkylated by the complex anion part. It should be noted that the anionic complex, but not the Lewis acid itself, participates in this key step. Thus the R_3Al reagent may be called a "combined acid-base."

In sharp contrast, however, Scheme 5 gives an instance of a methyl carbanion being introduced largely in an S_N2-type inversion stereochemistry. Note that the substrate carries a cyclopropane carbon doubly activated by 1,3-dicarbonyl groups. A possible explanation is given in Scheme 6. The reaction can be utilized in the selective synthesis of dl-neonepetalactone and its epimer. The sequence involves (1) enolization (NaH) and phosphorylation ($ClPO(OEt)_2$), (2) methylation (Me_2CuLi), (3) ozonolysis (MeOH, -78°) and reduction, and (4) the final lactonization (PyH.OTs).

Alkylation of an sp^2 Carbon with the R_3Al-Pd(0) System

The methylation [step (2)] in the above sequence proceeds smoothly due to the presence of an ethoxycarbonyl activating group. A new technique (11) is based on the catalysis by a Pd(0) complex and provides a methodology of alkyl substitution of an enol phosphate moiety in the absence of such an activating group. The results are given in Table 1. As the enolization of ketones can be performed regioselectively, the technique furnishes an approach to regioselective olefin formation from ketones.

Scheme 3

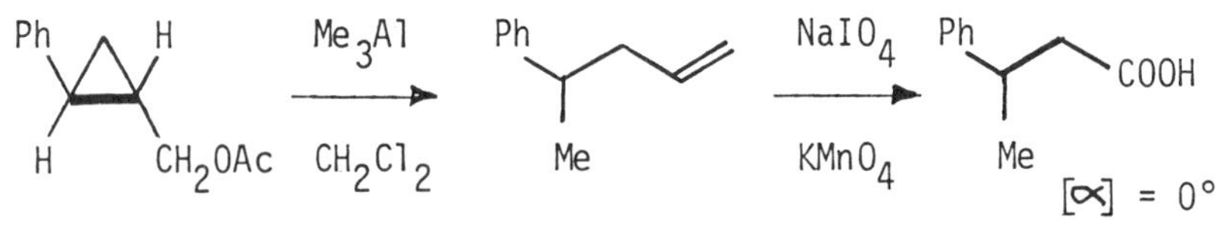

Ph, H, H, CH_2OAc —(Me_3Al, CH_2Cl_2)→ Ph, Me —($NaIO_4$, $KMnO_4$)→ Ph, COOH, Me $[\alpha] = 0°$

Scheme 4

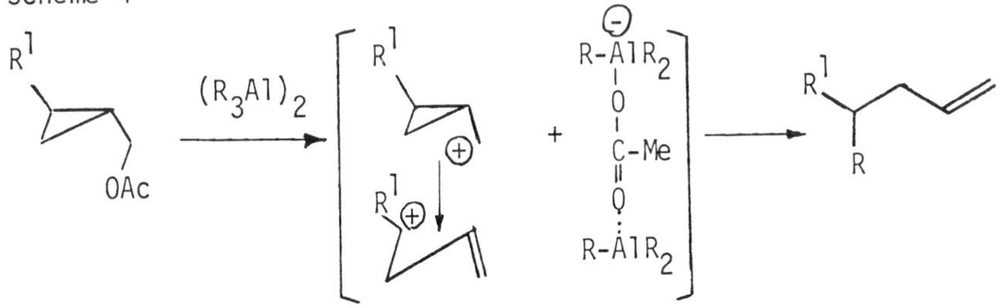

R^1, OAc —($(R_3Al)_2$)→ [R^1 (⊕) + $R-\overset{\ominus}{Al}R_2$ – O – C-Me = O ··· $R-AlR_2$] → R^1, R

Scheme 5

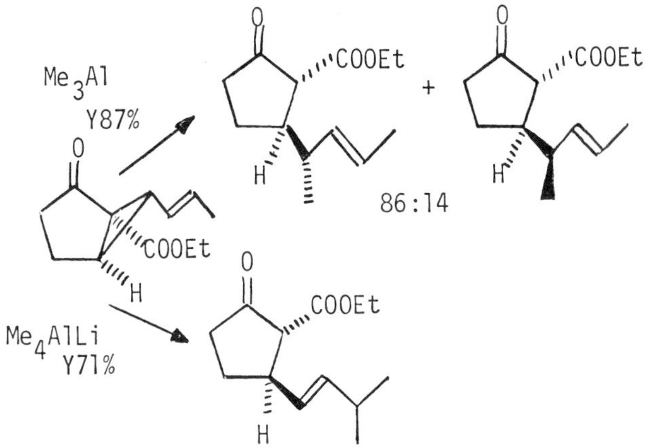

Me_3Al Y87%: COOEt, H + COOEt, H 86:14

O, COOEt, H

Me_4AlLi Y71%: O, COOEt, H

Scheme 6

R_3Al, R_3Al, O, O, H, OEt, $R-AlR_2$'''O=C

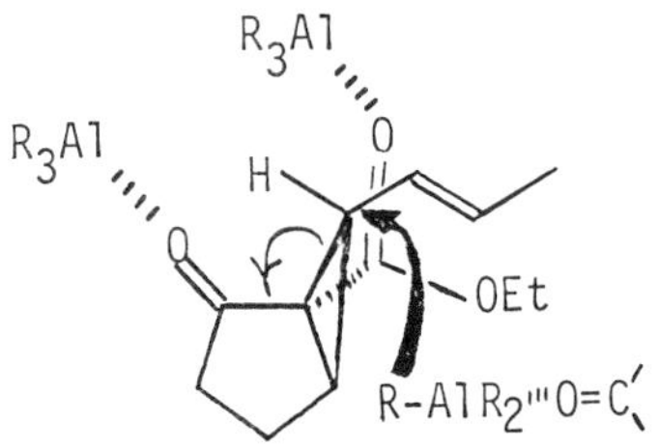

Table 1. Coupling on an sp^2 Carbon

Substrate	Reagent[a]	Time (h)	Product Y (%)
$Me(CH_2)_9C(=CH_2)$-$OPO(OPh)_2$	Me_3Al	2	91
	Et_3Al	3	71
	$Me(CH_2)_4C{\equiv}CAlEt_2$	3	47[b]
	$PhC{\equiv}CAlEt_2$	2	82[b]
	(E)-1-heptenylAlBu^i_2	4	66[c]
$PhC(=CH_2)$-$OPO(OPh)_2$	Me_3Al	2	94[d]
	Et_3Al	2	80
	$PhC{\equiv}CAlEt_2$	3	67[b]
4-t-Bu-1-cyclohexenyl-$OPO(OPh)_2$	Me_3Al	5	72
	$PhC{\equiv}CAlEt_2$	6	70[b]

[a] $Pd(PPh_3)_4$ 0.1/$ClCH_2CH_2Cl$ at 25°.
[b] Enyne product exclusively.
[c] (E)-Diene product only.
[d] G.l.p.c. yield.

In contrast to the analogous sp^2 carbon alkylation procedures (12,13), the present method does not affect coexisting vinyl sulfide groups as shown in Scheme 7 (14). This reaction provides access to ketones $R'CH_2COR$ starting from $R'CH_2COOH$. Yields in parentheses indicate the formation of ethylated (R = Et) plus hydrogenated (R = H) products in the reaction of Et_3Al. In benzene solvent the ratio of these two products is roughly 2:1. In hexane the hydrogenated products are predominant.

Scheme 8 shows the synthesis of 1,3-dialkylated cyclohexenes from 2-cyclohexenones consisting of 1,4-addition of organocuprates, enol phosphorylation, and the final alkylation of the sp^2 carbon. Scheme 9 provides a novel addition to the technique of 1,2-transposition of a carbonyl moiety accompanied by alkylation *in tandem* (15). The desulfurization is best performed by Mukaiyama's $TiCl_4$ method (16).

Treatment of an enone, PhCOCH=CHMe, with RSLi (R = Ph, Et) and subsequent phosphorylation with $ClPO(OPh)_2$ give $PhC[OPO(OPh)_2]$=CH-CHSR-Me. The phosphate group is substituted by methyl by means of the present technique to produce PhCMe=CH-CHSR-Me, the transformation of which into PhCMe=CHCOMe is known (17). In effect the sequence furnishes a new route of 1,3-carbonyl transposition *cum* alkylation.

Aliphatic Claisen Rearrangement at Room Temperature

Sigmatropic rearrangement of allyl vinyl ether substrates usually requires heating at around 200°. Allyl phenyl ether rearranges at room temperature in the presence of Lewis acid reagents, which have, however, turned out to be ineffective with aliphatic ethers. The concept of "combined acid-base attack" previously mentioned (18,19) has motivated several successful experiments as shown in Schemes 10 through 12 (20).

A solution of Me_3Al in hexane (1 *M*, 4.0 mmol) was added to a solution of 1-butyl-2-propenyl vinyl ether (2.0 mmol, Scheme 10) in 1,2-dichloroethane (15 ml) at 25° under an Ar atmosphere and the mixture was stirred for 15 min. Workup and TLC (SiO_2) purification gave the olefinic alcohol (0.28 g, 91% yield), the *E*/*Z* ratio being almost 1:1.

As shown in Scheme 10 (b,c) an alkynyl or alkenyl group is introduced in preference to an alkyl group. Examples of reductive rearrangement are found in Scheme 11. With exception of a single instance producing a 2-phenylethenyl system, the resulting olefinic linkage has shown practically no stereoselectivity. The regular Claisen products, or γ,δ-unsaturated aldehydes, have been produced in the reaction with R_2AlSPh as summarized in Scheme 12. A combination of acid (Et_2AlCl) and base (PPh_3) has turned out to be effective. It is intriguing to note that the rearrangement of 3,4-dihydro-2-vinyl-2H-pyran affording 3-cyclohexenecarbaldehyde (60% yield) takes place in the presence of this couple at room temperature within one hour. The pyrolytic procedure without

Scheme 7

$$R'CH{=}C(OPO(OPh)_2)(SPh) \longrightarrow R'CH{=}C(R)(SPh)$$

R'	R	Time (h)	Y (%)
Ph	Me	1	64
	Et	2	(55)[a]
n-Pr	Me	1	83
	Et	2	(82)[a]
	PhC≡C-	2	83

[a]A mixture of ethylation and hydrogenation product (see text).

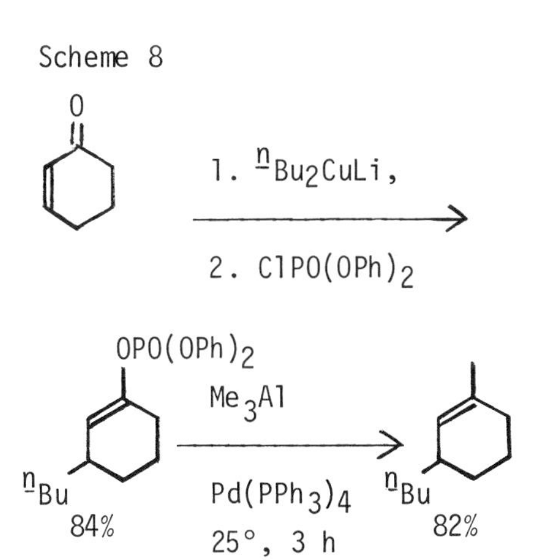

Scheme 9

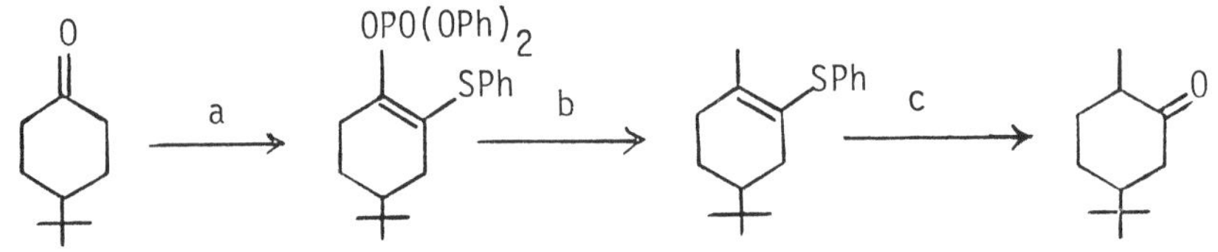

O → (a) → $OPO(OPh)_2$, SPh → (b) → SPh → (c) → O

a) 1. LDA/PhSSPh (Y78%), 2. NaH, $ClPO(OPh)_2$ (Y82%).
b) Me_3Al, $Pd(PPh_3)_4$ (Y80%). c) $TiCl_4$, aq, CH_2Cl_2 (Y78%).

Scheme 10

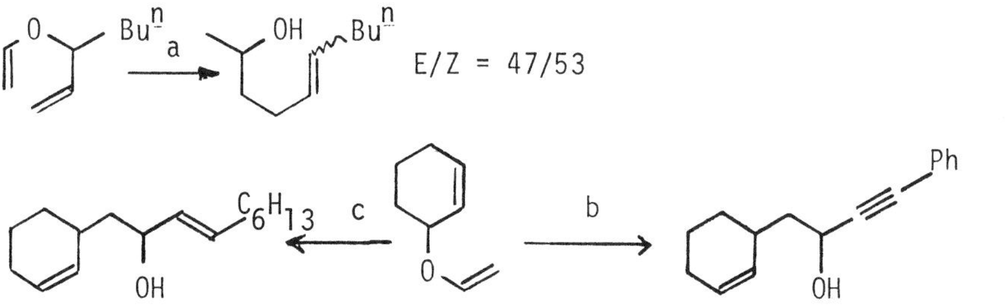

O, Bu^n → (a) → OH, Bu^n E/Z = 47/53

C_6H_{13}, OH ← (c) ← O → (b) → Ph, OH

a) Me_3Al (Y91%). b) $Et_2AlC{\equiv}CPh$ (Y82%). c) $^{i}Bu_2AlCH{=}CHC_6H_{13}$-(E)

Scheme 11

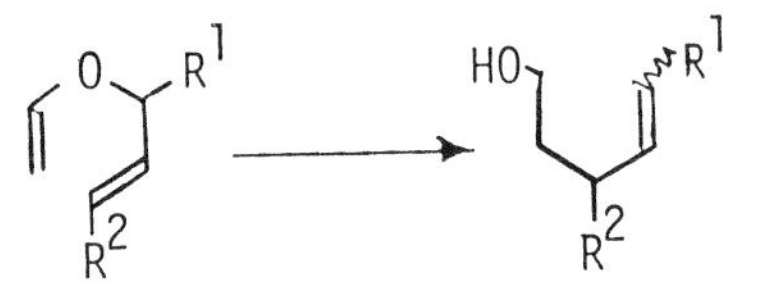

O, R^1, R^2 → HO, R^1, R^2

R^1	R^2	Reagent	Y(%)	E/Z
^{n}Bu	H	A	82	38/62
^{n}Bu	H	B	80	40/60
^{n}Bu	Me	A	89	45/55
Ph	H	A	93	100/0
H	Ph	A	91	---
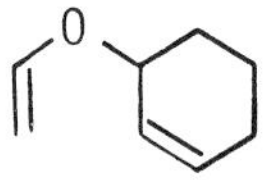 O (cyclohexenyl vinyl ether)		A	90	---

A: $^{i}Bu_3Al$ (2.6).
B: $^{i}Bu_2AlH$ (2.5).

Scheme 12

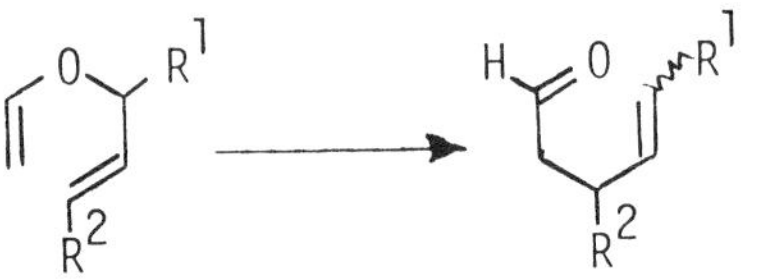

O, R^1, R^2 → H, O, R^1, R^2

R^1	R^2	Reagent	Y(%)	E/Z
^{n}Bu	H	A	84	39/61
^{n}Bu	H	B	81	43/57
^{n}Bu	Me	A	77	52/48
Ph	H	A	67	100/0
H	Ph	A	86	---
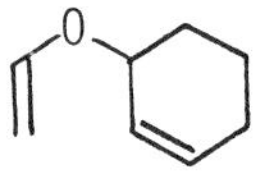 O (cyclohexenyl vinyl ether)		A	78	---

A: Et_2AlSPh (2.5).
B: Et_2AlCl (2) + PPh_3 (2.2).

such a reagent requires heating at 410°. The Et_2AlCl/PPh_3 system can be compared with Mukaiyama's $R_2BOSO_2CF_3/NR'_3$ system (21) or with Tsuji's R_2AlOR'/NR''_3 system (22). The possibility of an $Et_2AlP^+Ph_3$ species being the active reagent in our reaction will be investigated.

Acknowledgments ——— Thanks are given for helpful discussions with Prof. E. L. Eliel on his occasion of visiting Japan in 1978 as well as for valuable contributions by enthusiastic students of this research group, whose names are found in the references. Financial support by the Ministry of Education, Sciences, and Culture, Japanese Government, through Scientific Research Grants (510202, 56430027 etc.) is gratefully acknowledged.

Literature Cited

1. Okude, Y.; Hirano, S.; Hiyama, T.; Nozaki, H. J. Am. Chem. Soc. 1977, 99, 3175.
2. Okude, Y.; Hiyama, T.; Nozaki, H. Tetrahedron Lett. 1977, 3829.
3. Hiyama, T.; Kimura, K.; Nozaki, H. ibid. 1981, 22, 1037.
4. Buse, C.T.; Heathcock, C. H. ibid. 1978, 1635.
5. Takai, K.; Oshima, K.; Nozaki, H. ibid. 1980, 21, 1657.
6. Tomioka, H.; Takai, K.; Oshima, K.; Nozaki, H. ibid. 1980, 21, 4843.
7. Hiyama, T.; Kimura, K.; Takai, K.; Nozaki, H. The 44th Fall Meeting of Chem. Soc. Jpn. at Okayama, 2D12, Oct. 13, 1981.
8. Yamamoto, H.; Nozaki, H. Angew. Chem. Int. Ed. Engl. 1978, 17, 169.
9. Itoh, A.; Ozawa, S.; Oshima, K.; Sasaki, S.; Yamamoto, H.; Hiyama, T.; Nozaki, H. Bull. Chem. Soc. Jpn. 1980, 53, 2367.
10. Hiyama, T.; Morizawa, T.; Yamamoto, H.; Nozaki, H. ibid. in press.
11. Takai, K.; Oshima, K.; Nozaki, H. Tetrahedron Lett. 1980, 21, 2531.
12. Hayashi, T.; Katsuro, Y.; Kumada, M. ibid. 1980, 21, 3915.
13. Okamura, H.; Miura, M.; Takei, H. ibid. 1979, 43.
14. Sato, M.; Takai, K.; Oshima, K.; Nozaki, H. ibid. 1981, 22, 1609.
15. Fristad, W. E.; Bailey, T. R.; Paquette, L. A. J. Org. Chem. 1980, 45, 3028 and ref. cited.
16. Mukaiyama, T.; Kamio, K.; Kobayashi, S.; Takei, H. Bull. Chem. Soc. Jpn. 1972, 45, 3723.
17. Trost, B. M.; Stanton, J. L. J. Am. Chem. Soc. 1975, 97, 4018.
18. Trost, B. M.; Hutchinson, C. R. (ed.); "Organic Synthesis Today and Tomorrow (IUPAC)"; Pergamon Press: Oxford, New York, 1981; p. 241.

19. Oshima, K.; Nozaki, H. Yuki-Gosei-Kagaku (J. Synth. Org. Chem. Jpn.) 1980, 38, 460.
20. Takai, K.; Mori, I.; Oshima, K.; Nozaki, H. Tetrahedron Lett. in press.
21. Inoue, T.; Mukaiyama, T. Bull. Chem. Soc. Jpn. 1980, 53, 174.
22. Tsuji, J.; Yamada, T.; Kaito, M.; Mandai, T. Tetrahedron Lett. 1979, 2257; Bull. Chem. Soc. Jpn. 1980, 53, 1417.

RECEIVED December 14, 1981.

Novel Approaches to the Asymmetric Synthesis of Peptides

IWAO OJIMA

Sagami Chemical Research Center, Nishi-Ohnuma 4-4-1, Sagamihara, Kanagawa 229, Japan

A variety of dehydrodipeptides (N-protected free acids or methyl esters) have been hydrogenated with homogeneous rhodium catalysts bearing a variety of chiral diphosphine ligands. Diastereomer excess is frequently above 95%. The stereoselectivity of the reaction is, in a number of instances, quite different from that in hydrogenation of N-acyldehydroamino acids. The synthesis of acylphenylalanyl-α,β-d_2-alanine methyl ester as a nearly pure diastereomer (and enantiomer) is described.

Dipeptides have also been synthesized by cycloaddition of azidoketene (formed *in situ* from azidoacetyl chloride) to *t*-butyl esters of α-amino acids to give β-lactams which are then chromatographically resolved into diastereomers and cleaved by mild hydrogenolysis over palladium. By an extension of this method, tri-, tetra- and higher oligopeptides can be obtained. A salient feature is the high solubility of the β-lactam intermediates in common organic solvents which facilitates chromatographic purification. By an adaptation of this method, Leucine-Enkephalin (Tyr-Gly-Gly-Phe-Leu) *t*-butyl ester hydrochloride and its analog have been synthesized.

Peptide linkages are generally formed by the coupling of two optically active amino acid components through acyl chloride, acyl azide, mixed anhydride, carbodiimide, or enzymatic methods. These methods have been developed for the synthesis of naturally occurring polypeptides with minimum racemization. Recently, it has been shown that significant modifications of biological activities can be effected through inversion of configuration at one or more chiral centers, or through replacement of one or more "natural" amino acid residues by "unnatural" amino acid components in a bio-

0097-6156/82/0185-0109$07.50/0

logically active polypeptide such as Enkephalin, Vasopressin, Angiotensin II, Gonadoliberin and other hormones (1). In order to obtain such synthetic polypeptides by the conventional methods mentioned above, it is indispensable to prepare chiral amino acids with "unnatural" configuration or "unnatural" substituents. As an approach to the synthesis of chiral oligo- and polypeptides with desired structures, we have been trying to develop facile approaches to obtaining chiral building blocks. We will describe here such approaches involving i) catalytic asymmetric hydrogenation, and ii) the use of β-lactams as synthetic intermediates.

Synthesis of Chiral Dipeptides by Means of Asymmetric Hydrogenation of Dehydrodipeptides

As precursors of modified peptides, naturally occurring dehydropeptides may be interesting candidates since catalytic asymmetric hydrogenation can, in principle, convert the dehydroamino acid residue into an amino acid residue with either R or S configuration. Indeed, the homogeneous asymmetric hydrogenation of dehydro-α-amino acids catalyzed by rhodium complexes with chiral diphosphine ligands has turned out to be quite effective for the synthesis of chiral α-amino acids (2). An interesting point in this reaction is whether the chiral center of the dehydrodipeptide exerts a strong influence on the asymmetric induction, i.e., whether the optical purity of the newly formed chiral center is or is not affected by the already existing chiral center, and whether, in fact, we can synthesize dipeptides having the desired configurations.

N-acyldehydrodipeptides were readily prepared either by the condensation of N-acyldehydro-α-amino acids with α-amino acid esters or by the reaction of the azlactones of dehydro-α-amino acid with α-amino acid esters (eq. 1). Asymmetric hydrogenation of the N-acyldehydrodipeptides thus obtained (eq. 2) was carried out by using rhodium complexes with a variety of chiral diphosphines such as p-Br-Phenyl-CAPP (3), Ph-CAPP (3), (-)BPPM (4), (+)BPPM (4), (-)DIOP (5), (+)DIOP (5), diPAMP (6), Chiraphos (7), Prophos (8), BPPFA (9) and CBZ-Phe-PPM (Fig. 1) (10). The chiral catalysts were prepared *in situ* from chiral diphosphine ligand with $[Rh(NBD)_2]^+ClO_4^-$ (NBD = norbornadiene). Typical results are summarized in Tables I-V.

As Table I shows, the efficiency of each chiral diphosphine ligand exhibited in the asymmetric hydrogenation of dehydrodipeptides is considerably different from that reported for the reaction of N-acyldehydroamino acids, especially in the case of Chiraphos and BPPFA, which are known to lead to much better enantioselectivity than DIOP in the dehydroamino acid case (2, 7, 9). When Ac-ΔPhe-(S)Phe-OH was employed as substrate, Chiraphos induced S configuration (Entry 17) and BPPFA led to R configuration (Entry 19) with low stereoselectivities; in both cases, the directions of asymmetric induction are opposite to those observed for α-acet-

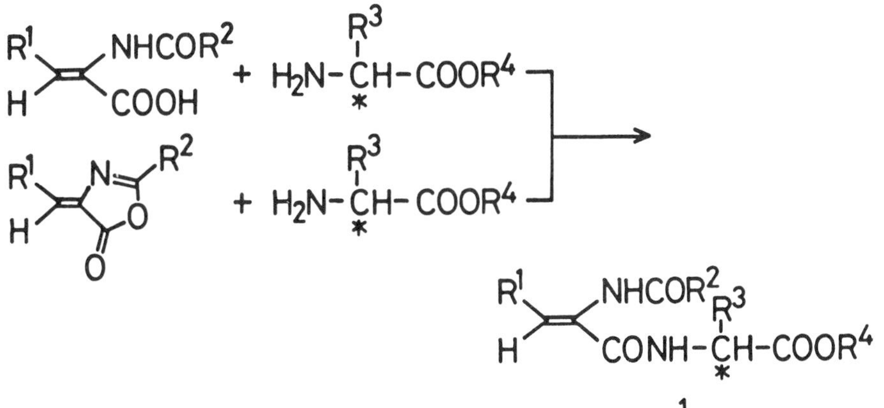

$$R^1(H)C{=}C(NHCOR^2)COOH + H_2N\text{-}\overset{R^3}{\underset{*}{C}}H\text{-}COOR^4$$

$$\text{(4-}R^1CH{=}\text{oxazolone, 2-}R^2) + H_2N\text{-}\overset{R^3}{\underset{*}{C}}H\text{-}COOR^4$$

$$\longrightarrow R^1(H)C{=}C(NHCOR^2)CONH\text{-}\overset{R^3}{\underset{*}{C}}H\text{-}COOR^4 \quad \mathbf{1}$$

Equation 1.

$$R^1(H)C{=}C(NHCOR^2)CONH\text{-}\overset{R^3}{\underset{*}{C}}H\text{-}COOR^4 \ (\mathbf{1}) \xrightarrow[L^*\text{-}Rh]{H_2} R^2CONH\text{-}\overset{CH_2R^1}{\underset{*}{C}}H\text{-}CONH\text{-}\overset{R^3}{\underset{*}{C}}H\text{-}COOR^4 \quad \mathbf{2}$$

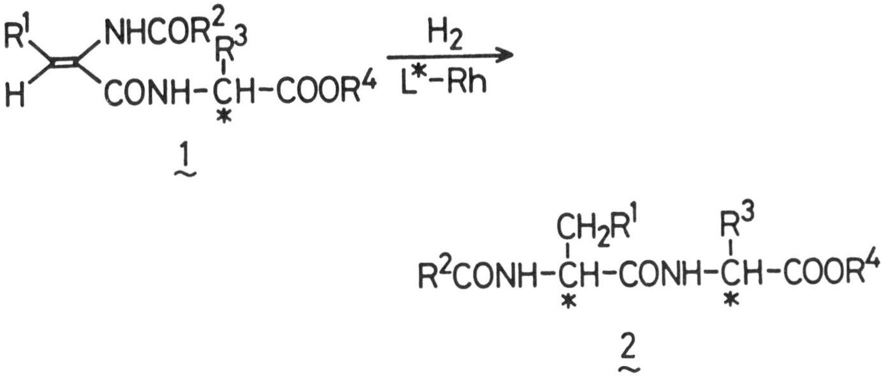

Equation 2.

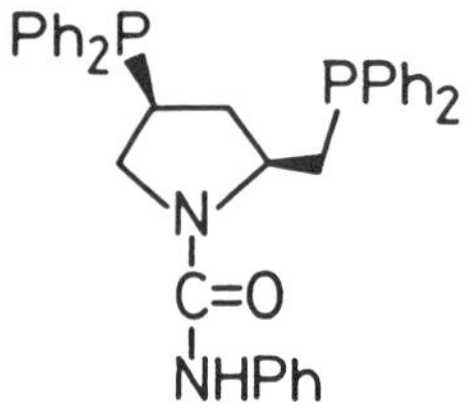

Ph-CAPP 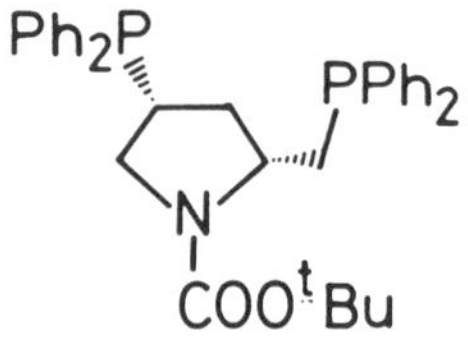 (+)BPPM 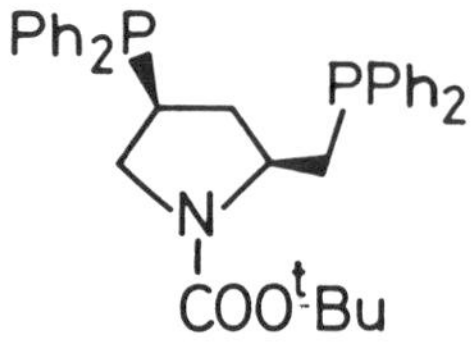 (–)BPPM

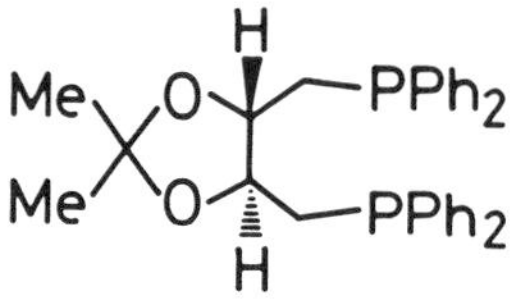

(+)DIOP 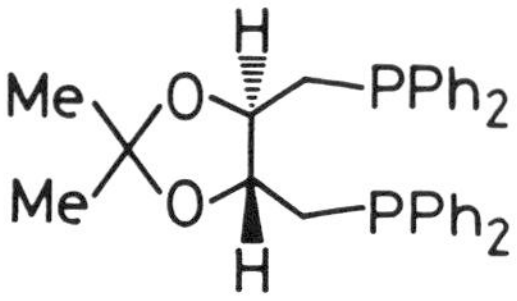 (–)DIOP

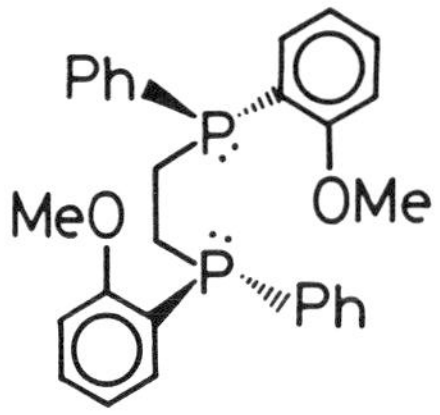

diPAMP 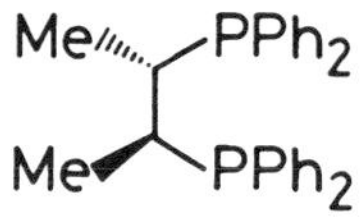 Chiraphos

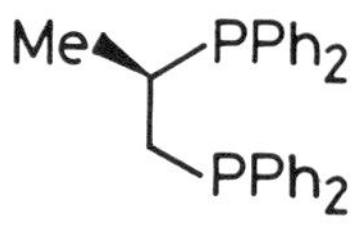

Prophos 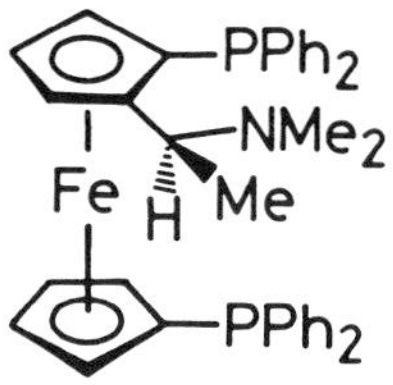 BPPFA

Figure 1. Typical chiral diphosphine ligands.

Table I. Efficiency of Chiral Diphosphine Ligands in the Asymmetric Hydrogenation of Typical Dehydrodipeptides[a]

Entry	Substrate	Ligand	Conditions (H_2 press., Temp., Time)	Conversion (%)[b]	Dipeptide (R,S)/(S,S)[b]
1	Bz-ΔPhe-(S)Phe-OMe	p-Br-phenyl-CAPP	1 atm, 40°C, 3h	100	99.2/0.8
2		(-)BPPM	1 atm, 40°C, 1h	100	98.7/1.3
3		(+)BPPM	1 atm, 40°C, 1h	100	0.9/99.1
4		(-)DIOP	5 atm, 25°C, 18h	100	84.1/15.9
5		(+)DIOP	5 atm, 25°C, 18h	100	15.0/85.0
6		diPAMP	10 atm, 50°C, 15h	100	2.2/97.8
7		Chiraphos	5 atm, 40°C, 10h	82	85.1/14.9
8		Prophos	5 atm, 40°C, 10h	99	4.1/95.9
9		BPPFA	5 atm, 40°C, 10h	51	18.7/81.3
10		dppb	1 atm, 40°C, 5h	85	37.8/62.2
11	Ac-ΔPhe-(S)Phe-OH	Ph-CAPP	5 atm, 40°C, 20h	100	98.0/2.0
12		(-)BPPM	10 atm, 50°C, 20h	100	96.2/3.8
13		(+)BPPM	10 atm, 50°C, 20h	97	0.6/99.4
14		(-)DIOP	5 atm, 40°C, 20h	100	81.8/18.2
15		(+)DIOP	5 atm, 40°C, 20h	89	5.9/94.1
16		diPAMP	5 atm, 50°C, 20h	86	1.4/98.6
17		Chiraphos	10 atm, 50°C, 20h	96	39.1/60.9
18		Prophos	10 atm, 50°C, 20h	95	18.8/81.2
19		BPPFA	50 atm, 50°C, 20h	23	61.2/38.8
20		dppb	10 atm, 50°C, 20h	99	34.1/65.9

a All reaction were run with 5.0×10^{-4} mol of the substrate and 5.0×10^{-6} mol of the catalyst.
b Determined by HPLC.

amidocinnamic acid. Prophos induced high stereoselectivity with Bz-ΔPhe-(S)Phe-OMe (Entry 8) whereas it was no longer a very good chiral ligand for Ac-ΔPhe-(S)Phe-OH (Entry 18). Pyrrolidinodiphosphines and diPAMP achieved extremely high stereoselectivities. There seems to be a trend that the chiral ligands which form seven membered ring chelates with rhodium give rise to much better results than those forming rigid five membered ring chelates or quasi five membered ring chelates except diPAMP. The results may imply that the seven membered ring chelate has flexibility for "induced-fit" action like an enzyme, which is quite an important factor for a chiral complex catalyst when the substrate is polyfunctional (11).

As for the influence of the chiral center in the substrate on asymmetric induction, considerable double asymmetric induction was observed on using a dehydrodipeptide bearing a free acid terminus such as Ac-ΔPhe-(S)Phe-OH (see Entry 12, 13, and 14, 15). To realize the extent of double asymmetric induction in a quantitative manner, one has to look not at the difference of the relative amounts of diastereomers in percent but at the ratio of two diastereomers, which is related to $\Delta\Delta G^{\ddagger}$: For BPPM, (R,S)/(S,S) = 25.3 (Entry 12, (-)BPPM), (S,S)/(R,S) = 165.7 (Entry 13, (+)BPPM); for DIOP, (R,S)/(S,S) = 4.49 (Entry 14, (-)DIOP), (S,S)/(R,S) = 15.9 (Entry 15, (+)DIOP). Thus, the extent of double asymmetric induction turns out to be more pronounced for BPPM than for DIOP although the apparent difference in optical purity is much smaller for BPPM compared with that for DIOP. The results concerning the double asymmetric induction indicate that the formation of the (S,S)-isomer is preferred in these systems. An experiment using an achiral diphosphine ligand, bis(diphenylphosphino)butane (dppb), gave a consistent result (Entry 20), i.e. 31.8% asymmetric induction favoring the formation of the (S,S)-isomer of Ac-Phe-Phe-OH was observed. On the other hand, when dehydrodipeptide methyl esters were employed, only a slight effect of the existing chiral center was observed as far as DIOPs were concerned, e.g., Bz-Phe-Phe-OMe: (+)DIOP, (R,S)/(S,S) = 16.4/83.6, (-)DIOP, (R,S)/(S,S) = 84.1/15.9; Ac-Phe-Phe-OMe: (+)DIOP, (R,S)/(S,S) = 19.4/80.6, (-)DIOP, (R,S)/(S,S) = 81.6/18.4; Bz-Phe-Val-OMe: (+)DIOP, (R,S)/(S,S) = 20.6/79.4, (-)DIOP, (R,S)/(S,S) = 83.0/17.0.

The results could be interpreted by assuming exclusive coordination of the N-acyldehydroamino acid moiety with the rhodium complex in which the rest of the molecule, i.e. the α-amino ester moiety, is located in the outer sphere of the chiral coordination site: this may be the reason why virtually no double asymmetric induction was observed. However, a simple asymmetric hydrogenation using dppb as achiral ligand (Entry 10) disclosed preferential formation of Bz-(S)Phe-(S)Phe-OMe with 24.4% asymmetric induction, which is consistent with the result using Ac-ΔPhe-(S)Phe-OH as substrate (Entry 20). Accordingly, it seems that the results of using DIOPs are rather exceptional. In this context, we further looked at the effect of the chiral center on the catalytic asymmetric induction by using Ac-ΔPhe-(R)Phe-OMe and Ac-ΔPhe-(S)Phe-OMe as substrates, and CBZ-(S)Phe-PPM (**3a**), CBZ-(S)Val-PPM (**3f**) and CBZ-(S)Pro-PPM (**3d**) as chiral ligands for the cationic rhodium complex. The results are listed in Table II. As Table II shows, there is only a slight difference between the two substrates in percent asymmetric induction from a synthetic point of view, since the reactions achieve quite high stereoselectivities, yet there is a significant difference in $\Delta\Delta G^{\ddagger}$ since it is observed that the (R,R)/(S,R) ratio is three to four times larger than the (R,S)/(S,S) ratio in every case examined. Moreover, similar results were obtained in the asymmetric hydrogenation of Ac-ΔPhe-(R)Phe-OMe by using (-)BPPM and (+)BPPM as shown in Table III. Thus, the reaction using (-)-BPPM led to 99.6% production of Ac-(R)Phe-(R)Phe-OMe while that using (+)BPPM produced 98.5% of the (S,R)-isomer: (R,R)/(S,R) = 249 for (-)BPPM: (S,R)/(R,R) = 65.7 for (+)BPPM. Consequently, it may be said that there is a significant extent of double asymmetric induction for the reaction of dehydrodipeptide methyl esters, too, and the case of DIOP is rather exceptional.

Table III summarizes typical results for the asymmetric hydrogenation of a variety of N-acyldehydrodipeptides with pyrrolidinodiphosphines and diPAMP. As Table III shows, (R,S), (S,S), (S,R) or (R,R)-dipeptides in high optical purities can be readily synthesized by using these chiral ligands, and, in one recrystallization, easily lead to optically pure dipeptides.

Since catalytic asymmetric hydrogenation can generate either S or R configuration at the position of the dehydroamino acid residue, this method could be potentially useful for the specific labeling of certain amino acid residues in a polypeptide. The regiospecific and stereoselective labeling of an amino acid residue is difficult to achieve with the conventional stepwise peptide synthesis. We carried out the dideuteration of Ac-ΔPhe-(S)Ala-OMe with the use of the cationic rhodium complexes with (-)BPPM and (+)BPPM (Scheme 1), which gave Ac-(R,R)Phe(d_2)-(S)Ala-OMe [(R,R,S)/(S,S,S) = 98.7/1.3] and Ac-(S,S)Phe(d_2)-(S)Ala-OMe [(R,R,S)/(S,S,S) =0.5/99.5], respectively, without any scrambling of deuterium.

As it has been shown that the introduction of deuterium at the chiral center of certain amino acids, e.g., 3-fluoro-2-deuterio-(R)-alanine, changes biological activity remarkably (12), this stereoselective dideuteration may provide a convenient device for this kind of modification of biological activity.

Since pyrrolidinodiphosphines, e.g., Ph-CAPP, p-Br-Phenyl-CAPP and BPPM, gave excellent stereoselectivities, we prepared a series of new chiral pyrrolidinodiphosphines, in which the nitrogen atom of PPM (4, 11) is linked up with a variety of α-aminoacyl groups. The rhodium complexes with these ligands may serve as good biomimetic models of reductase when they are anchored on polymers especially polyamides. α-Aminoacyl-PPMs (**3**) were prepared by the condensation of PPM with an N-CBZ-α-amino acid or an N-CBZ-dipeptide in the presence of dicyclohexylcarbodiimide (DCC) and 1-hydroxybenztriazole (HOBT)(eq. 3).

Table II. Effect of Chiral Center in Dehydrodipeptides on Stereoselectivity[a]

Entry	Ligand	Substrate	Ac-Phe-Phe-OMe (R,S)/(S,S)[b]	or (R,R)/(S,R)[b]
1	CBZ-(S)Phe-PPM(3a)	Ac-ΔPhe-(S)Phe-OMe	98.0/2.0	
2	CBZ-(S)Phe-PPM(3a)	Ac-ΔPhe-(R)Phe-OMe		99.5/0.5
3	CBZ-(S)Pro-PPM(3d)	Ac-ΔPhe-(S)Phe-OMe	98.1/1.9	
4	CBZ-(S)Pro-PPM(3d)	Ac-ΔPhe-(R)Phe-OMe		99.5/0.5
5	CBZ-(S)Val-PPM(3f)	Ac-ΔPhe-(S)Phe-OMe	98.9/1.1	
6	CBZ-(S)Val-PPM(3f)	Ac-ΔPhe-(R)Phe-OMe		96.2/3.8

a All reactions were run with 5.0×10^{-4} mol the substrate and 5.0×10^{-6} mol of the catalyst at 40°C and 1 atm of hydrogen for 2h. Conversion was 100% for every case examined.

b Determined by HPLC.

Table III. Typical Results on the Asymmetric Hydrogenation of Dehydrodipeptides

Substrate	Ligand	Conditions (H_2 press., Temp., Time)	Conversion (%)	Dipeptide (R,S)/(S,S)[a] or (R,R)/(S,R)[b]
Bz-ΔPhe-(S)Phe-OMe	p-Br-phenyl-CAPP	1 atm, 40°C, 3h	100	99.2/0.8
	(+)BPPM	1 atm, 40°C, 1h	100	0.9/99.1
Bz-ΔPhe-(S)Val-OMe	p-Br-phenyl-CAPP	10 atm, 40°C, 20h	100	98.0/2.0
	diPAMP	10 atm, 40°C, 24h	100	1.0/99.0
Ac-ΔPhe-(S)Phe-OMe	Ph-CAPP	1 atm, 40°C, 46h	100	99.0/1.0
	(+)BPPM	1 atm, 40°C, 1h	100	0.6/99.4
Ac-ΔPhe-(S)Phe-OH	Ph-CAPP	5 atm, 40°C, 20h	100	98.0/2.0
	(+)BPPM	10 atm, 40°C, 20h	100	0.6/99.4
Ac-ΔPhe-(S)Val-OH	Ph-CAPP	10 atm, 50°C, 20h	100	96.3/3.7
	diPAMP	10 atm, 50°C, 20h	100	3.0/97.0
Bz-ΔLeu-(S)Phe-OMe	(-)BPPM	1 atm, 40°C, 24h	100	95.3/4.7
	(+)BPPM	1 atm, 40°C, 16h	100	3.6/96.4
Ac-ΔPhe-(R)Phe-OMe	(-)BPPM	1 atm, 40°C, 2h	100	99.6/0.4[b]
	(+)BPPM	1 atm, 40°C, 2h	100	1.5/98.5[b]
Ac-ΔPhe-(R)Phy-OMe	(-)BPPM	5 atm, 40°C, 24h	100	95.7/4.3[b]
	diPAMP	10 atm, 40°C, 24h	100	2.6/97.4[b]
Bz-ΔPhe-(R)Phe-OMe	CBZ-(S)Phe-PPM	1 atm, 40°C, 1h	100	99.5/0.5[b]
Ac-Δ(Ac)Tyr-(R)Ala-OMe[c]	Ph-CAPP	5 atm, 40°C, 24h	100	99.8/0.2[b]
Ac-Δ(AcO)(MeO)Phe-(R)Ala-OMe[d]	Ph-CAPP	5 atm, 40°C, 24h	100	99.4/0.6[b]
Ac-Δ(F)Phe-(S)Leu-OMe[e]	(+)BPPM	5 atm, 40°C, 64h	100	0.9/99.1

a Determined by HPLC. (R,S)/(S,S) unless otherwise noted. *b* (R,R)/(S,R). *c* (Ac)Tyr = 4-acetoxytyrosyl. *d* (AcO)(MeO)Phe = 3-methoxy-4-acetoxyphenylalanyl. *e* (F)Phe = 4-fluorophenylalanyl.

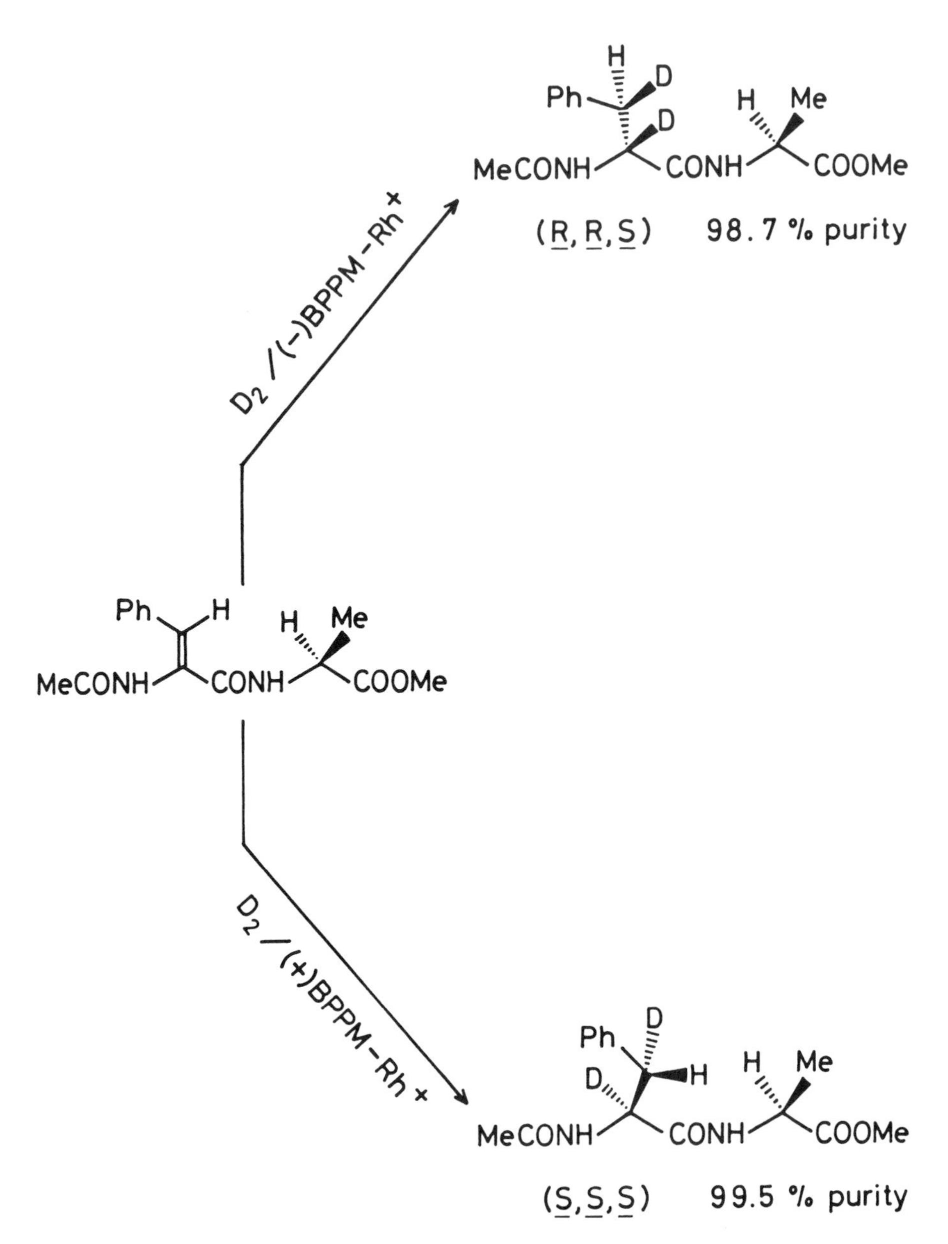
Ph
H
D
D
H
Me
MeCONH
CONH
COOMe
(R, R, S)
98.7 % purity
D2 / (–)BPPM-Rh+
Ph
H
H
Me
MeCONH
CONH
COOMe
D2 / (+)BPPM-Rh+
Ph
D
D
H
H
Me
MeCONH
CONH
COOMe
(S, S, S)
99.5 % purity

Scheme 1.

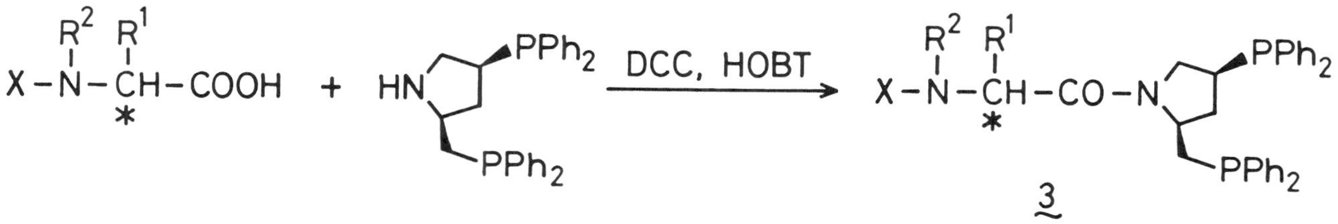

$$X{-}N(R^2){-}\overset{*}{C}H(R^1){-}COOH \;+\; \text{HN(pyrrolidine)}(PPh_2)(CH_2PPh_2) \xrightarrow{\text{DCC, HOBT}} X{-}N(R^2){-}\overset{*}{C}H(R^1){-}CO{-}N(\text{pyrrolidine})(PPh_2)(CH_2PPh_2)$$

3

3a $R^1 = PhCH_2$, $R^2 = H$, $X = CBZ$: CBZ-(S)Phe-PPM

3b $R^1 = Me$, $R^2 = H$, $X = CBZ$: CBZ-(S)Ala-PPM

3c $R^1 = Me$, $R^2 = H$, $X = CBZ$: CBZ-(R)Ala-PPM

3d R^1, $R^2 = {+}CH_2{+}_3$, $X = CBZ$: CBZ-(S)Pro-PPM

3e R^1, $R^2 = {+}CH_2{+}_3$, $X = CBZ$: CBZ-(R)Pro-PPM

3f $R^1 = {}^{i}Pr$, $R^2 = H$, $X = CBZ$: CBZ-(S)Val-PPM

3g $R^1 = R^2 = H$, $X = CBZ$: CBZ-Gly-PPM

3h $R^1 = {}^{i}Pr$, $R^2 = $ CBZ-(S)Phe, $X = H$: CBZ-(S)Phe-(S)Val-PPM

Equation 3.

Typical results on using Bz-ΔPhe-(S)Phe-OMe as substrate are listed in Table IV. Table V summarizes the results on using CBZ-(S)Phe-PPM (3a) as chiral ligand for the reaction of several dehydrodipeptides. As Table IV shows, (i) the stereoselectivities attained by these α-aminoacyl-PPMs are as high as by other pyrrolidinodiphosphines, and (ii) there is no significant difference in stereoselectivity by changing the amino acid residue except in the case of CBZ-(R)Pro-PPM (3e), which shows lower stereoselectivity and a lower hydrogenation rate than CBZ-(S)Pro-PPM (3d) and other α-aminoacyl-PPMs. It should be noted that the reaction rate is considerably higher than that realized by other pyrrolidinodiphosphines such as Ph-CAPP and BPPM. These results may indicate a promising activity of the corresponding polymer anchored catalyst.

Peptide Synthesis by the Use of β-Lactams as Building Blocks

Although synthesis of β-lactams has been extensively studied in connection with the naturally occurring antibiotics, little attention has been paid to the β-lactam ring as a synthetic intermediate. The cleavage of β-lactam rings usually takes place at the N-C(O) bond (type a), e.g., hydrolysis gives α-amino acids. However, conceptually, other types of cleavage are possible, i.e., cleavage at $N-C^4$, C^3-C^4, C^2-C^3 or metathesis (Scheme 2).

Among these possibilities, we found that exclusive cleavage of the $N-C^4$ bond (type b) takes place in a palladium catalyzed hydrogenolysis when an aryl substituent was attached to C^4 (13). For instance, 1-(1-methoxycarbonyl-1-phenyl)methyl-3-benzyloxy-4-phenylazetidin-2-one (Scheme 3) has three benzylic positions capable of being cleaved. It is well known that the cleavage of the benzyl-oxygen bond is by far faster than that of the benzyl-nitrogen bond; in particular, the benzyl-nitrogen bond in N-benzylamides can hardly be cleaved under ordinary conditions. It is therefore reasonable to anticipate that cleavage of the benzyl-oxygen bond will be the only reaction observed. To our surprise, however, the cleavage of the β-lactam ring was much faster than that of the benzyl-oxygen bond; the other benzyl-nitrogen bond remains intact as expected (14). The result clearly indicates that the ring strain of the β-lactam greatly accelerates the $N-C^4$ cleavage.

Metallic palladium was found to be an effective catalyst for the selective hydrogenolysis of the $N-C^4$ bond, whereas Rh-C, Pt-C, Ru-C, $PdCl_2$, $PdCl_2(PhCN)_2$, and $Pd(PPh_3)_4$ were inactive. Raney-Ni showed some activity, but its activity and selectivity were by far lower than those found for palladium.

As 3-substituted-4-arylazetidin-2-ones can easily be synthesized by cycloaddition of a Schiff base to a ketene which is generated *in situ* from an acyl chloride and triethylamine (15), the type b cleavage can serve as a new synthetic route to functionalized amides, α-amino acids and α-hydroxy acids (Scheme 4).

We carried out the asymmetric synthesis of an α-aminocarboxamide using (3,4-dimethoxyphenyl)methylidene-(R)-1-phenyl-

Table IV. Asymmetric Hydrogenation of Bz-ΔPhe-(S)Phe-OMe by Using α-Aminoacyl-PPMs (3) as Chiral Ligand[a]

Entry	Ligand	Conditions (H_2 press., Temp., Time)	Conversion[b] (%)	Bz-Phe-Phe-OMe (R,S)/(S,S)[b]
1	CBZ-(S)Phe-PPM (3a)	1 atm, 40°C, 1h	100	98.0/2.0
2	CBZ-(S)Ala-PPM (3b)	1 atm, 40°C, 1h	100	97.7/2.3
3	CBZ-(R)Ala-PPM (3c)	1 atm, 40°C, 1h	100	97.6/2.4
4	CBZ-(S)Pro-PPM (3d)	1 atm, 40°C, 1h	100	98.1/1.9
5	CBZ-(R)Pro-PPM (3e)	1 atm, 40°C, 18h	100	83.0/17.0
6	CBZ-(S)Val-PPM (3f)	1 atm, 40°C, 1h	100	96.2/3.8
7	CBZ-Gly-PPM (3g)	1 atm, 40°C, 4h	100	97.8/2.2
8	CBZ-(S)Phe-(S)Val-PPM (3h)	1 atm, 40°C, 2h	100	96.4/3.6

[a] All reactions were run with 5.0×10^{-4} mol of the substrate and 5.0×10^{-6} mol of the catalyst in 15 mL of ethanol.

[b] Determined by HPLC.

Table V. Asymmetric Hydrogenation of Dehydrodipeptides by Using CBZ-(S)Phe-PPM (3a) as Chiral Ligand[a]

Entry	Substrate	Conditions (H_2 press., Temp., Time)	Conversion[b] (%)	Dipeptide (R,S)/(S,S)[b]
1	Ac-ΔPhe-(S)Phe-OMe	1 atm, 40°C, 2h	100	97.9/2.1
2	Bz-ΔPhe-(S)Val-OMe	1 atm, 40°C, 3h	100	98.2/1.8
3	Ac-ΔPhe-(S)Ala-OMe	1 atm, 40°C, 1h	99	98.3/1.7
4	Ac-ΔPhe-(S)Phe-OH	1 atm, 40°C, 3h	99	97.6/2.4
5	Ac-ΔPhe-(S)Val-OH	1 atm, 40°C, 24h	97	96.0/4.0

a All reactions were run with 5.0×10^{-4} mol of the substrate and 5.0×10^{-6} mol of the catalyst in 15 mL of ethanol.

b Determined by HPLC.

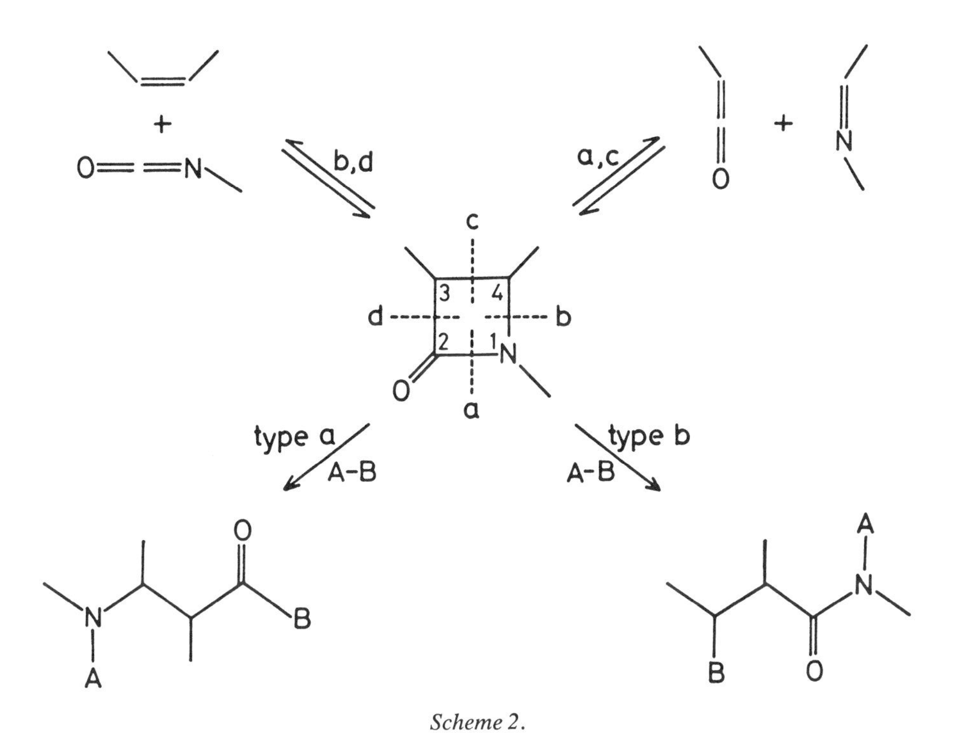

Scheme 2.

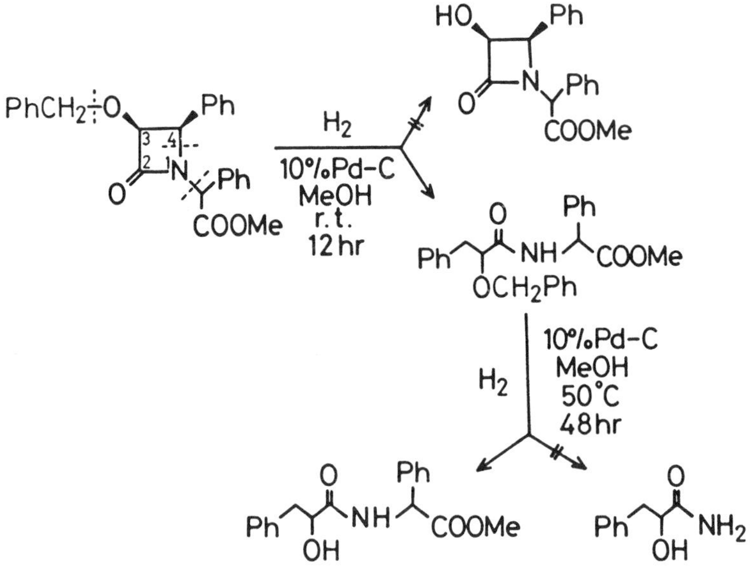

PhCH2–O Ph 3 4 2 1 N Ph O COOMe

H2 10%Pd-C MeOH r.t. 12 hr

HO Ph O N Ph COOMe

O Ph Ph NH COOMe OCH2Ph

H2 10%Pd-C MeOH 50°C 48hr

O Ph Ph NH COOMe OH

O Ph NH2 OH

Scheme 3.

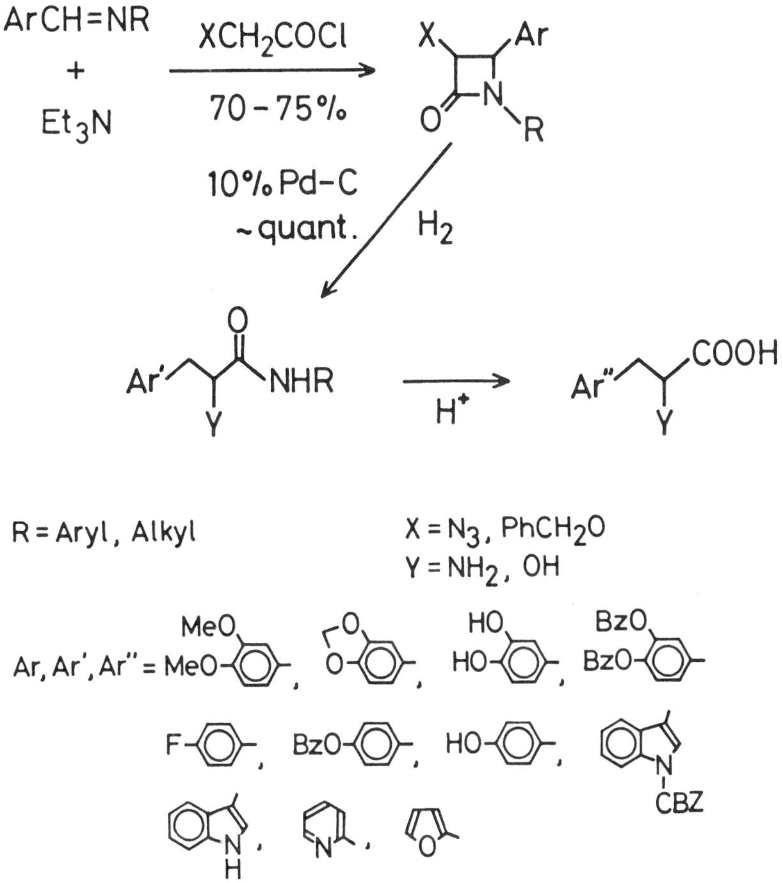

$ArCH{=}NR + Et_3N \xrightarrow[70-75\%]{XCH_2COCl}$ (β-lactam: X, Ar, N–R, O)

10% Pd–C, ~quant., H_2

$Ar'CH_2CH(Y)CONHR \xrightarrow{H^+} Ar''CH_2CH(Y)COOH$

R = Aryl, Alkyl

$X = N_3$, $PhCH_2O$
$Y = NH_2$, OH

Ar, Ar′, Ar″ = 3,4-(MeO)₂C₆H₃–, 3,4-methylenedioxyphenyl–, 3,4-(HO)₂C₆H₃–, 3,4-(BzO)₂C₆H₃–, F–C₆H₄–, BzO–C₆H₄–, HO–C₆H₄–, N-CBZ-indol-3-yl, indol-3-yl, 2-pyridyl, 2-furyl

Scheme 4.

ethylamine as the Schiff base (13). The extent of chirality transfer was 40%, the (3*R*,4*R*,1'*R*)-isomer being predominantly produced. Hydrogenolysis of the resulting β-lactam afforded *N*-[(*R*)-1-phenylethyl]-2-amino-3-(3,4-dimethoxyphenyl)propionamide in 94% yield: (2*S*,1'*R*)/(2*R*,1'*R*) = 70/30. The two diastereomers were readily separated by column chromatography on silica gel. The (2*S*,1'*R*)-isomer was easily hydrolyzed to give L-DOPA (Scheme 5).

In a similar manner, we synthesized dipeptides (Scheme 6) and their hydroxy analogs (depsipeptide unit) (Scheme 7) starting from the Schiff bases of α-amino acid esters (14). Optically pure dipeptides were synthesized via separation of the two diastereomers of the azido-β-lactam by HPLC as typified in Scheme 8. It turned out that no racemization at all took place during the ring opening (14).

Since a β-lactam derived from the Schiff base of an α-amino acid ester was thus proved to be the synthetic equivalent of dipeptide, we prepared tripeptide and tetrapeptide "synthons" involving one or two β-lactams, which were further submitted to hydrogenolysis to give the corresponding tri- and tetrapeptides (Scheme 9)(16).

The tripeptide synthon (**B**) was prepared by the coupling of an N-protected α-amino acid with an amino-β-lactam such as **8**. The tetrapeptide synthon (**C**) was obtained by the coupling of a β-lactam carboxylic acid (**9a**) with an amino-β-lactam (**8a**) as typified in Scheme 10. The tripeptide synthon (**D**) was prepared by azidoketene cycloaddition to a β-lactam Schiff base such as **26** (Scheme 14). The tandem style bis-β-lactam (**C**) was further transformed to a tetra-β-lactam, which gave an octapeptide upon hydrogenolysis (Scheme 11).

A striking feature of these di-, tri-, or tetrapeptide synthons is that they are highly soluble in regular organic solvents such as ether, ethyl acetate, chloroform etc., and even the octapeptide synthon is readily soluble in chloroform. Thus, these compounds can be chromatographed on an ordinary silica gel column in conventional fashion unlike other known peptide precursors. This characteristic should provide an important advantage in peptide synthesis.

The above-described "β-lactam method" has been applied to the synthesis of Leucine-Enkephalin (Tyr-Gly-Gly-Phe-Leu) (17). We prepared the Tyr-Gly synthon (**15**) and the Gly-Phe-Leu synthon (**17**) following the procedure mentioned above, and coupled them by using DCC, HOBT to give bis-β-lactam **18** (98%) (Scheme 12) in which two β-lactams were connected with a Gly-Gly chain. This β-lactam displayed a high solubility in common organic solvents, and was purified by column chromatography on silica gel. The hydrogenolysis of the bis-β-lactam on Pd-C in the presence of hydrochloric acid gave Leucine-Enkephalin *t*-butyl ester hydrochloride (84%). In a similar manner, a Leucine-Enkephalin analog, (L)Tyr-

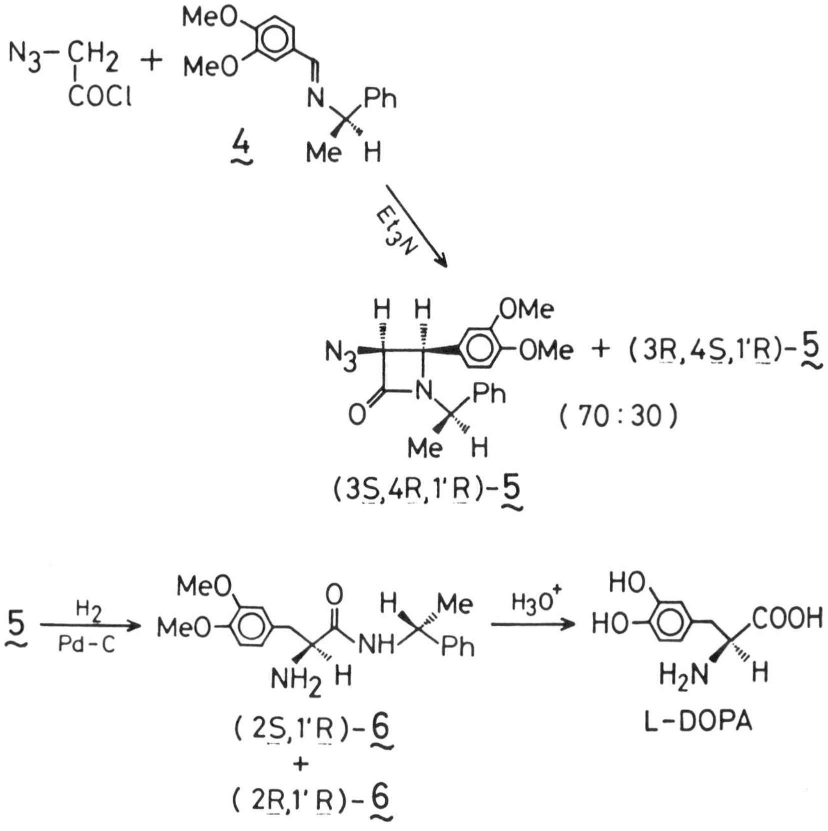

$N_3-CH_2(COCl)$ + MeO / MeO (aryl imine) N–CH(Me)(H)Ph

4

Et_3N

H H OMe

N_3 OMe + (3R,4S,1'R)-5

O N Ph

Me H

(70 : 30)

(3S,4R,1'R)-5

5 $\xrightarrow[Pd-C]{H_2}$ MeO / MeO — O — H Me — NH — Ph — NH_2 H $\xrightarrow{H_3O^+}$ HO / HO — COOH — H_2N H

(2S,1'R)-6

+

(2R,1'R)-6

L-DOPA

Scheme 5.

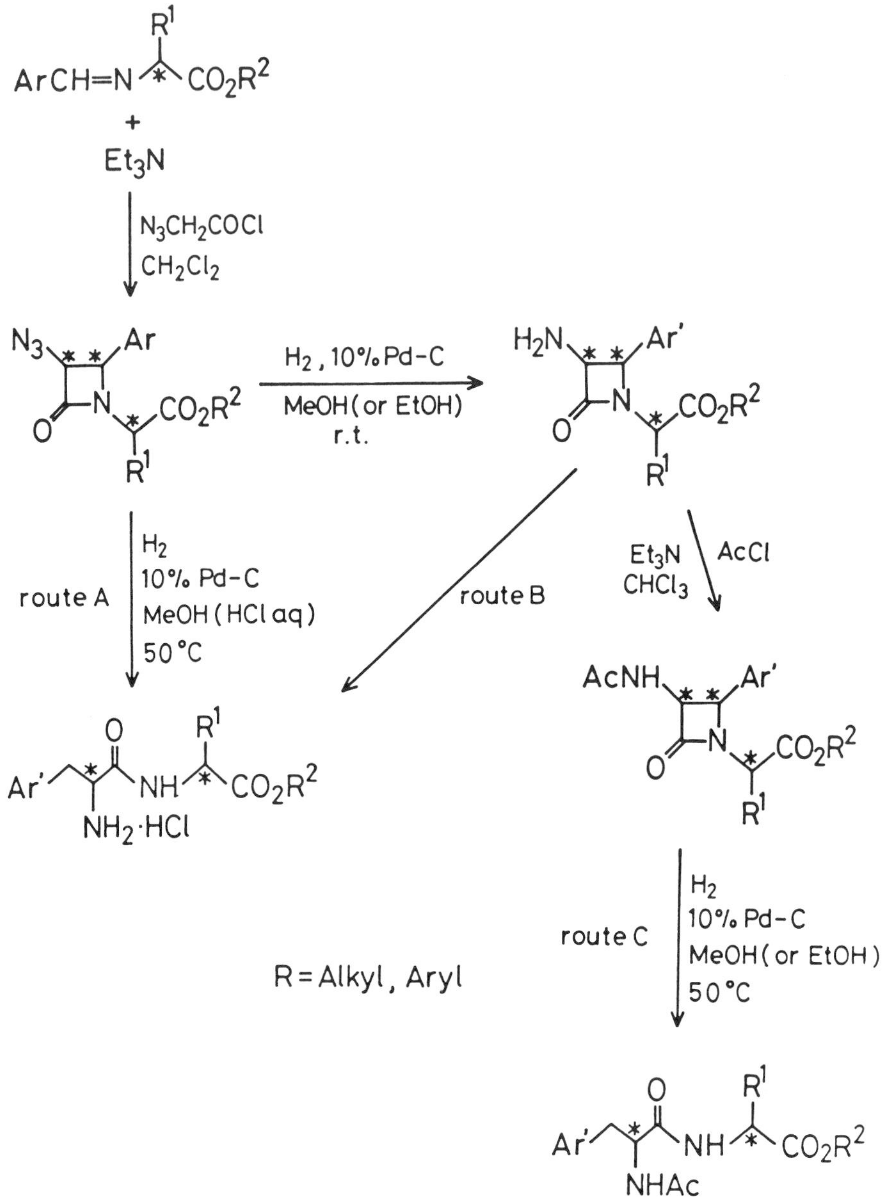

$ArCH{=}N\overset{*}{C}H(R^1)CO_2R^2$

+

Et_3N

N_3CH_2COCl

CH_2Cl_2

N_3 Ar CO_2R^2 R^1

H_2, 10% Pd-C

MeOH (or EtOH)

r.t.

H_2N Ar′ CO_2R^2 R^1

route A

H_2

10% Pd-C

MeOH (HCl aq)

50 °C

route B

Et_3N

$CHCl_3$

AcCl

AcNH Ar′ CO_2R^2 R^1

Ar′ O R^1 NH CO_2R^2 $NH_2{\cdot}HCl$

route C

H_2

10% Pd-C

MeOH (or EtOH)

50 °C

R = Alkyl, Aryl

Ar′ O R^1 NH CO_2R^2 NHAc

Scheme 6.

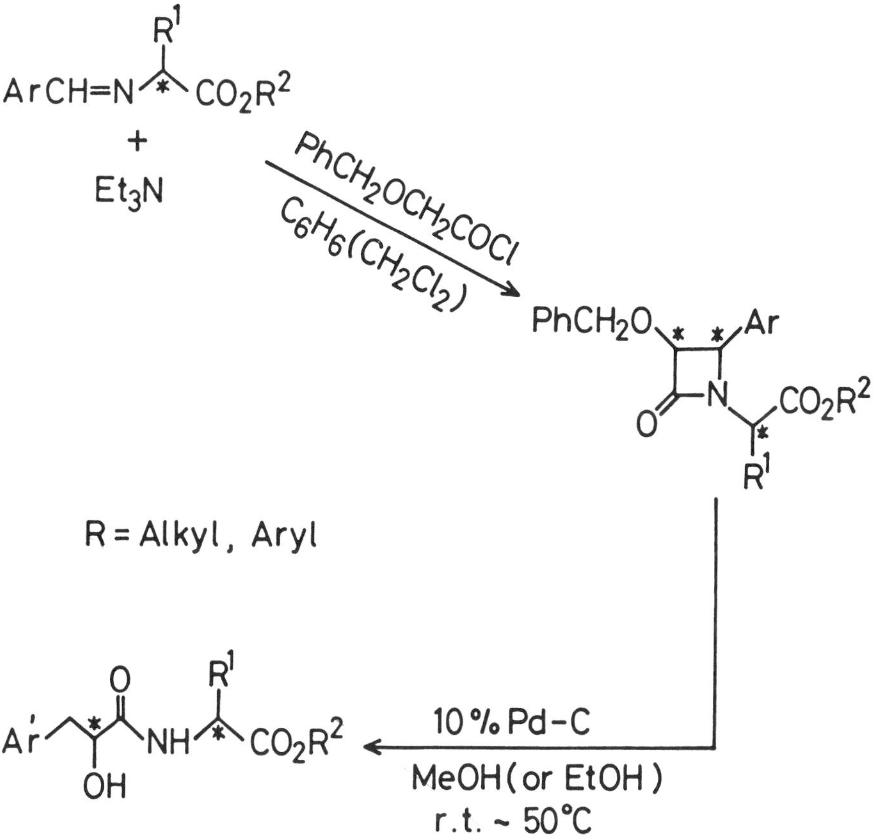
ArCH=N
R1
CO2R2
+
Et3N
PhCH2OCH2COCl
C6H6(CH2Cl2)
PhCH2O
Ar
N
CO2R2
R1
R = Alkyl, Aryl
Ar'
O
OH
NH
R1
CO2R2
10 % Pd-C
MeOH (or EtOH)
r.t. ~ 50°C

Scheme 7.

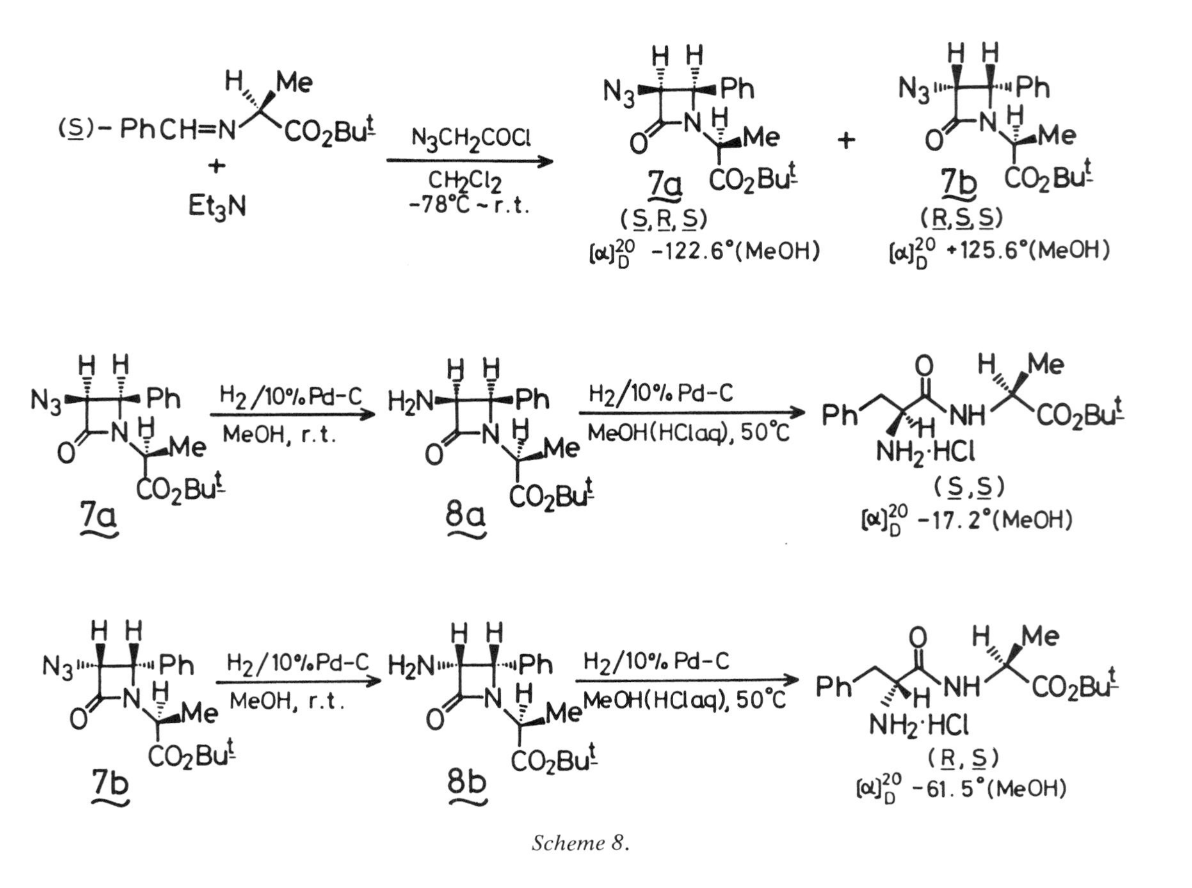
(S) - PhCH=N
+
Et3N
N3CH2COCl
CH2Cl2
-78°C ~ r.t.
7a
(S,R,S)
[α]D20 -122.6°(MeOH)
+
7b
(R,S,S)
[α]D20 +125.6°(MeOH)
H2/10% Pd-C
MeOH, r.t.
8a
H2/10% Pd-C
MeOH(HClaq), 50°C
NH2·HCl
(S,S)
[α]D20 -17.2°(MeOH)
7b
H2/10% Pd-C
MeOH, r.t.
8b
H2/10% Pd-C
MeOH(HClaq), 50°C
NH2·HCl
(R,S)
[α]D20 -61.5°(MeOH)

Scheme 8.

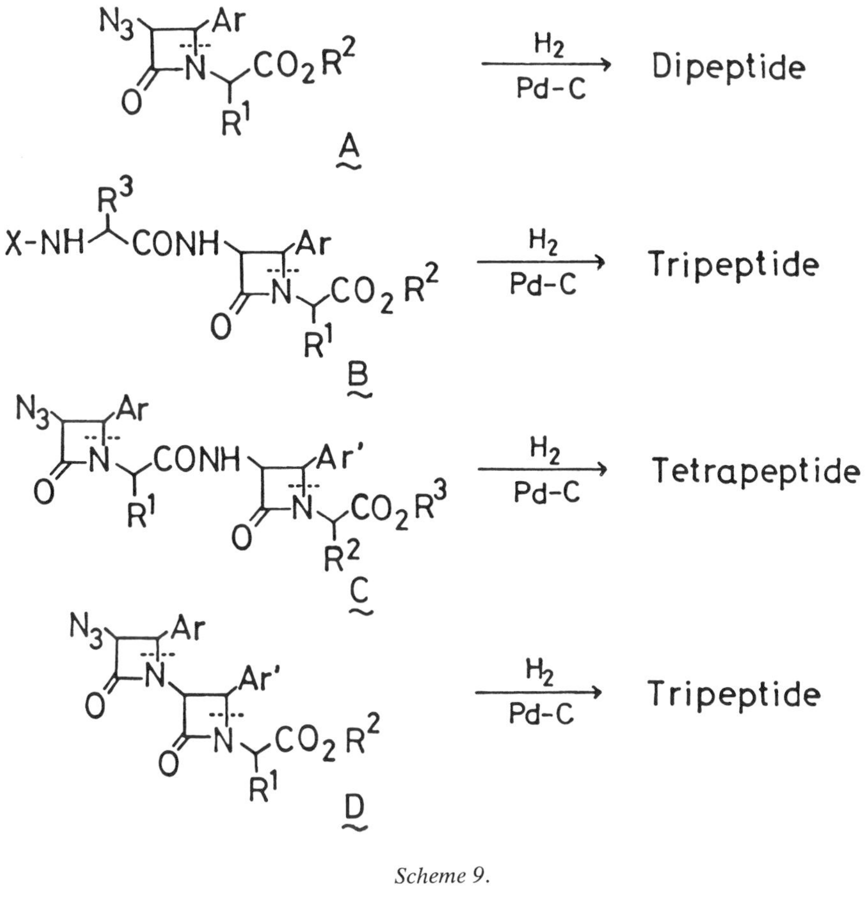

A $\xrightarrow[\text{Pd-C}]{H_2}$ Dipeptide

B $\xrightarrow[\text{Pd-C}]{H_2}$ Tripeptide

C $\xrightarrow[\text{Pd-C}]{H_2}$ Tetrapeptide

D $\xrightarrow[\text{Pd-C}]{H_2}$ Tripeptide

Scheme 9.

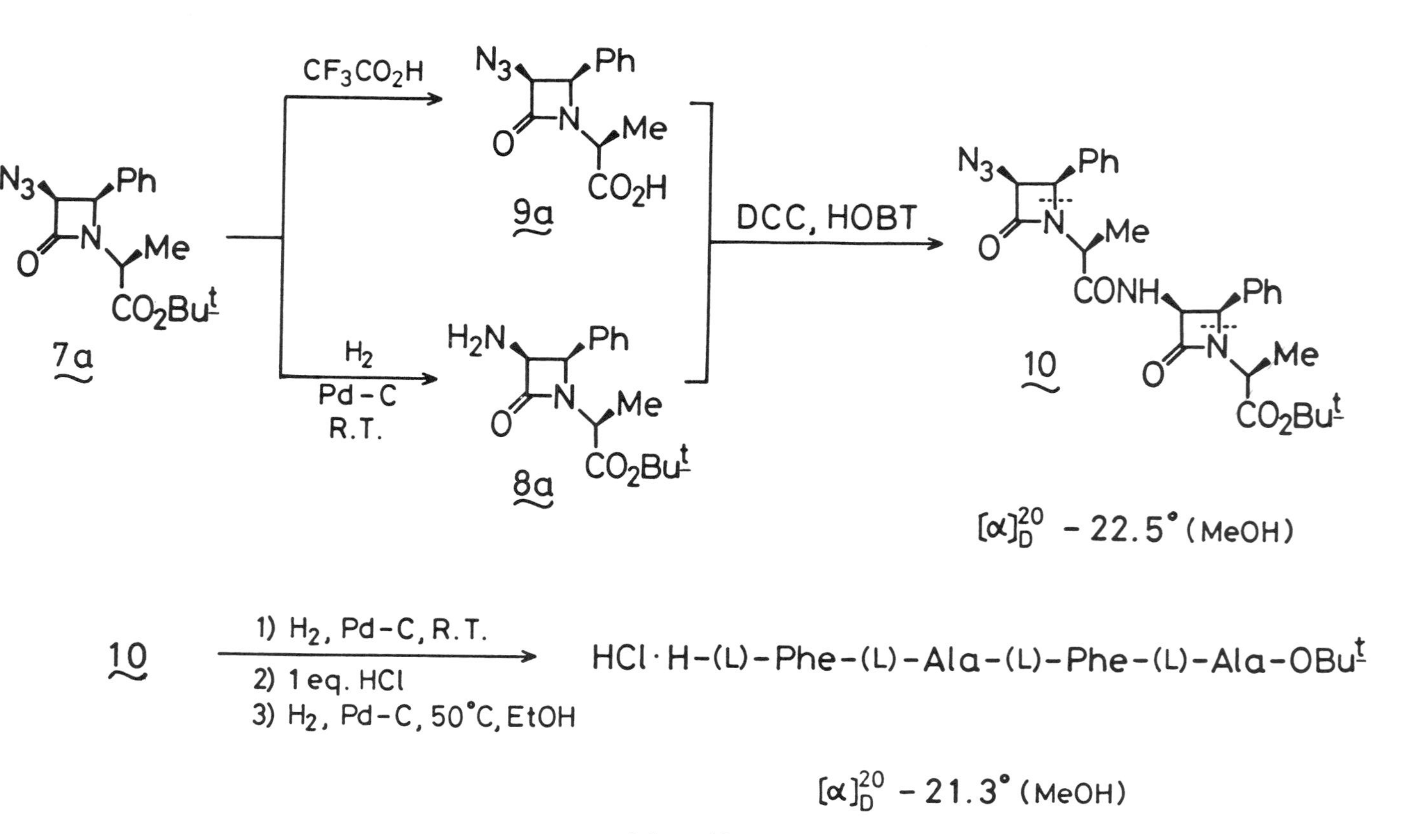

N3
Ph
O
N
Me
CO2Bu^t
7a
CF3CO2H
H2
Pd-C
R.T.
9a
CO2H
H2N
8a
DCC, HOBT
CONH
10
[α]D20 - 22.5° (MeOH)
1) H2, Pd-C, R.T.
2) 1 eq. HCl
3) H2, Pd-C, 50°C, EtOH
HCl · H-(L)-Phe-(L)-Ala-(L)-Phe-(L)-Ala-OBu^t
[α]D20 - 21.3° (MeOH)

Scheme 10.

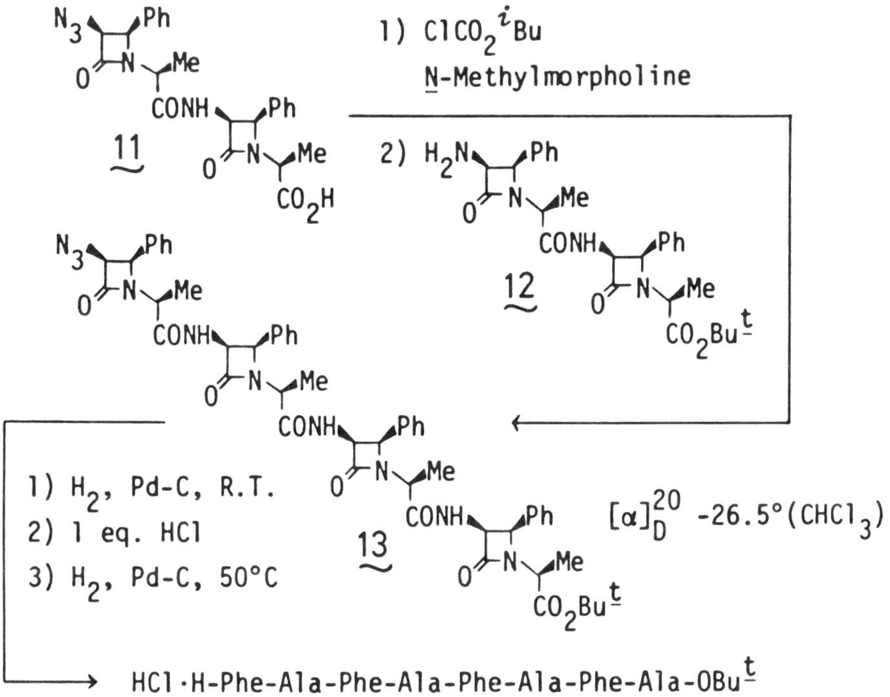

1) $ClCO_2{}^{i}Bu$
N-Methylmorpholine

2) H_2N ... 12

11

13

$[\alpha]_D^{20}$ -26.5°($CHCl_3$)

1) H_2, Pd-C, R.T.
2) 1 eq. HCl
3) H_2, Pd-C, 50°C

HCl·H-Phe-Ala-Phe-Ala-Phe-Ala-Phe-Ala-OBu^t

Scheme 11.

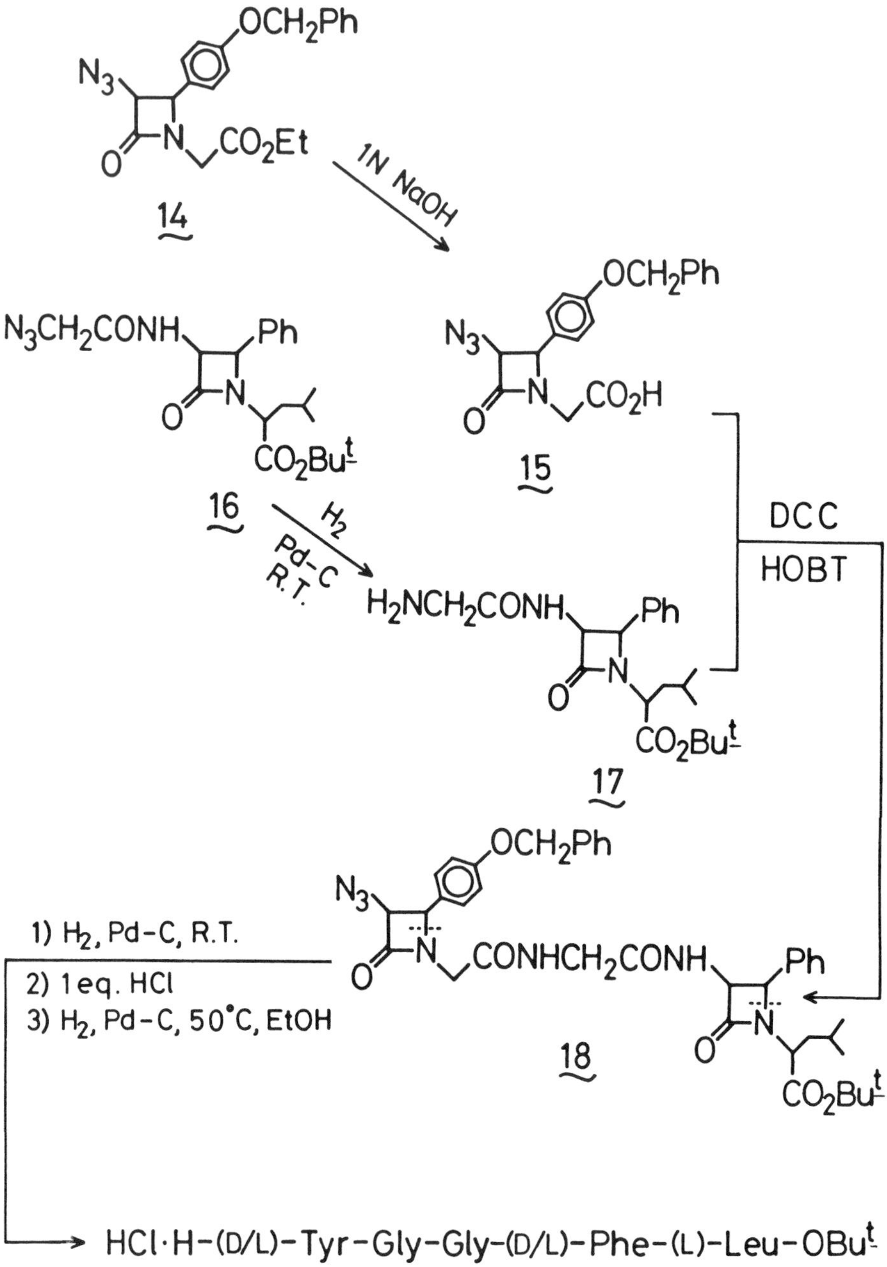

OCH_2Ph
N_3
CO_2Et
14

1N NaOH

OCH_2Ph
N_3
CO_2H
15

N_3CH_2CONH Ph
CO_2Bu^t
16

H_2 Pd-C R.T.

H_2NCH_2CONH Ph
CO_2Bu^t
17

DCC
HOBT

OCH_2Ph
N_3
$CONHCH_2CONH$ Ph
CO_2Bu^t
18

1) H_2, Pd-C, R.T.
2) 1eq. HCl
3) H_2, Pd-C, 50°C, EtOH

HCl·H-(D/L)-Tyr-Gly-Gly-(D/L)-Phe-(L)-Leu-OBu^t

Scheme 12.

(D)Ala-Gly-(L)Phe-(L)Leu-ol which has been shown to have stronger opioid activity than the parent Enkephalin, was synthesized in good yield (Scheme 13).

For the synthesis of bis-β-lactam (**D**) (Scheme 14), the synthetic equivalent of a tripeptide, a quite effective asymmetric reaction in the formation of β-lactam was found: The azidoketene cycloaddition to a diastereomerically pure 3-benzylideneamino-β-lactam (**26a** or **26b**) proceeds in a completely stereoselective manner to give only one of the possible two isomers of (**D**) (18). The stereochemistry of the cycloaddition was readily established by submitting the resulting bis-β-lactam (**27a** or **27b**) to hydrogenolysis. The tripeptides thus obtained were fully identified with authentic samples by HPLC analysis. It becomes clear, thus, that the stereochemistry of the newly added β-lactam is opposite to that of the parent β-lactam.

In conclusion, we have been developing two new methods for peptide synthesis and modification, viz., catalytic asymmetric hydrogenation of dehydrodipeptides and the use of β-lactams as building blocks. Of course, each method has advantages and limitations. We believe, however, that the combination of these two methods provides novel and effective approaches to chiral peptides having a variety of biological activities.

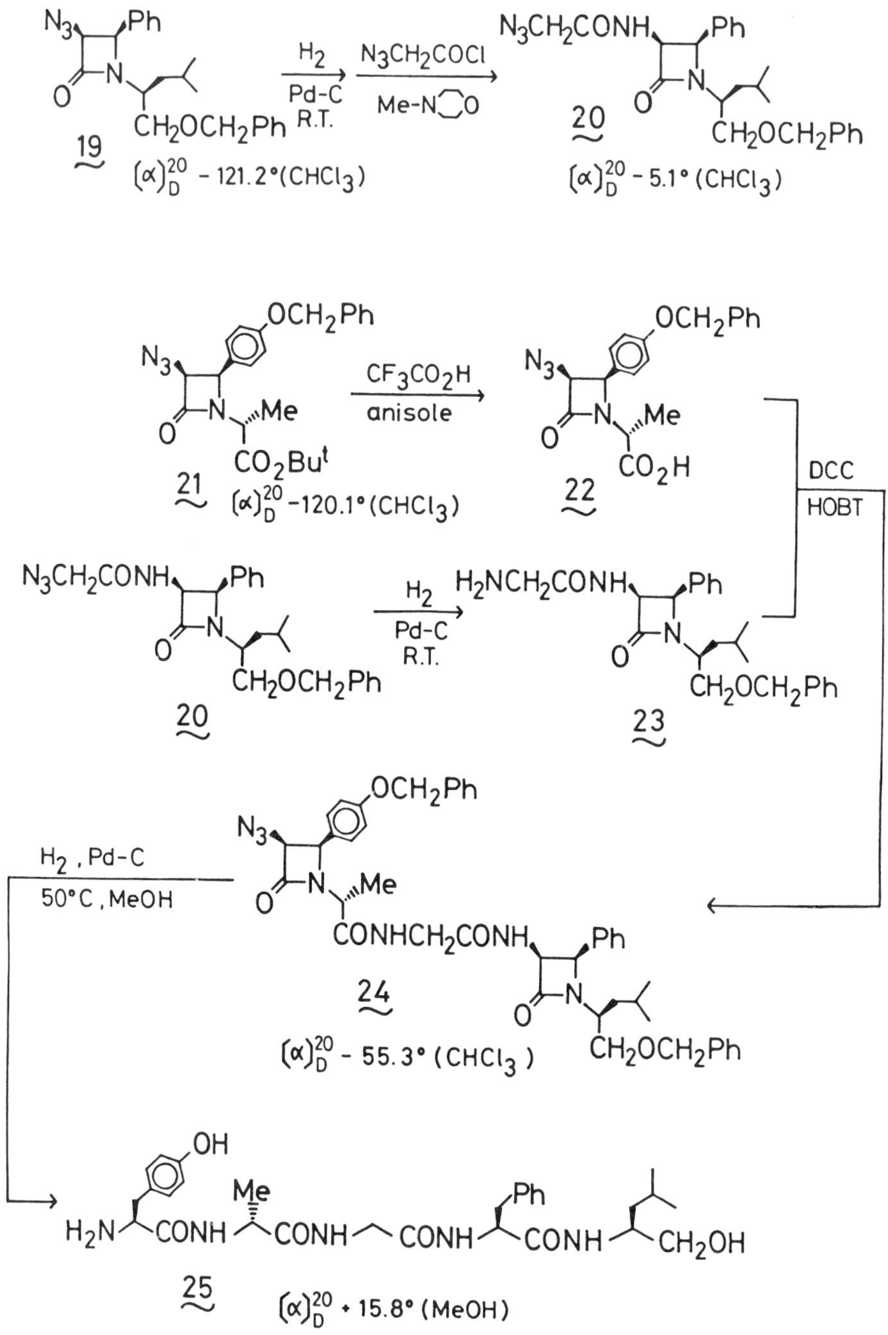

19: $[\alpha]_D^{20}$ − 121.2° ($CHCl_3$)

H_2, Pd-C, R.T.; N_3CH_2COCl, Me-N‹morpholine›O

20: $[\alpha]_D^{20}$ − 5.1° ($CHCl_3$)

21: $[\alpha]_D^{20}$ −120.1° ($CHCl_3$)

CF_3CO_2H, anisole

22

20 → H_2, Pd-C, R.T. → 23

22 + 23: DCC, HOBT → 24

24: $[\alpha]_D^{20}$ − 55.3° ($CHCl_3$)

H_2, Pd-C, 50°C, MeOH

25: H_2N-Tyr-CONH-CH(Me)-CONH-CH_2-CONH-CH(CH_2Ph)-CONH-CH($CH_2CH(CH_3)_2$)-CH_2OH

$[\alpha]_D^{20}$ + 15.8° (MeOH)

Scheme 13.

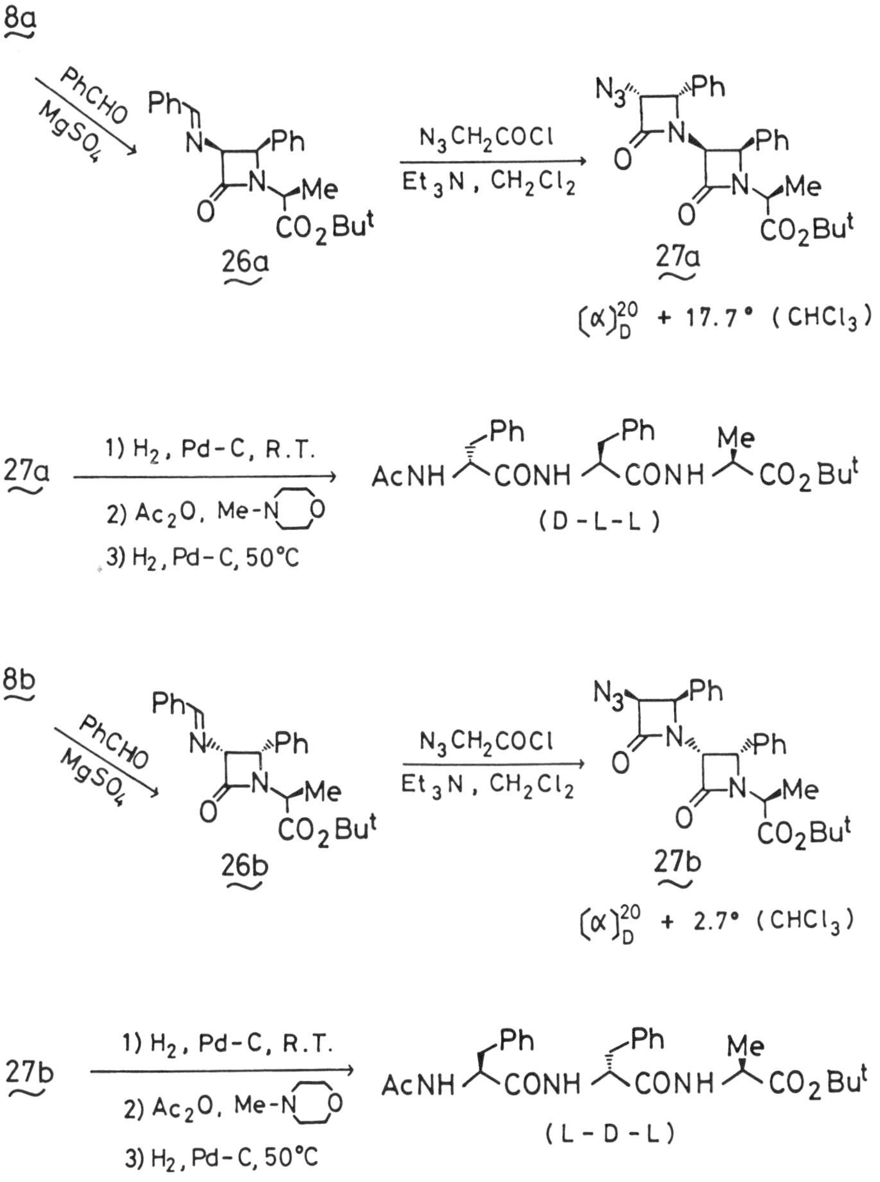

8a → (PhCHO, $MgSO_4$) → 26a → (N_3CH_2COCl, Et_3N, CH_2Cl_2) → 27a

$[\alpha]_D^{20}$ + 17.7° ($CHCl_3$)

27a → 1) H_2, Pd-C, R.T. 2) Ac_2O, Me-N(morpholine) 3) H_2, Pd-C, 50°C → AcNH–CH(CH₂Ph)–CONH–CH(CH₂Ph)–CONH–CH(Me)–CO_2Bu^t (D-L-L)

8b → (PhCHO, $MgSO_4$) → 26b → (N_3CH_2COCl, Et_3N, CH_2Cl_2) → 27b

$[\alpha]_D^{20}$ + 2.7° ($CHCl_3$)

27b → 1) H_2, Pd-C, R.T. 2) Ac_2O, Me-N(morpholine) 3) H_2, Pd-C, 50°C → AcNH–CH(CH₂Ph)–CONH–CH(CH₂Ph)–CONH–CH(Me)–CO_2Bu^t (L-D-L)

Scheme 14.

Acknowledgments. The author is grateful to his coworkers, Dr. Tetsuo Kogure, Noriko Yoda, Momoko Yatabe, Tadashi Suzuki, Toshiyuki Tanaka, Dr. Naoto Hatanaka, Rumiko Abe, and Shigemi Suga for their efforts and critical contributions.

Literature Cited

1. For example, Gross E.; Meienhofer J. in "The Peptides", Vol. 1, Academic Press, Inc., New York, 1979, Chapter 1.
2. (a) Čaplar V.; Comisso G.; Šunjić V. Synthesis, 1981, 85. (b) Valentine, Jr. D.; Scott J. W. ibid., 1978, 329.
3. Ojima I.; Yoda N. Tetrahedron Lett., 1980, 21, 8265.
4. Achiwa K. J. Am. Chem. Soc., 1976, 98, 8265.
5. Kagan H. B.; Dang T.-P. J. Am. Chem. Soc., 1972, 94, 6429.
6. Vineyard B. D.; Knowles W. S.; Sabacky M. J.; Bachman G. L.; Weinkauff D. J. J. Am. Chem. Soc., 1977, 99, 5946.
7. Fryzuk M. D.; Bosnich B. J. Am. Chem. Soc., 1977, 99, 6262.
8. Fryzuk M. D.; Bosnich B. J. Am. Chem. Soc., 1978, 100, 5491.
9. Hayashi T.; Mise T.; Mitachi S.; Yamamoto K.; Kumada M. Tetrahedron Lett., 1976, 1133.
10. Preliminary communication: Ojima I.; Suzuki T. Tetrahedron Lett., 1980, 21, 1239. See also (a) Meyer D.; Poulin J.-P.; Kagan H. B.; Levine-Pinto H.; Morgat J.-L.; Fromageot P. J. Org. Chem., 1980, 45, 4680 and (b) Onuma K.; Ito T.; Nakamura A. Chem. Lett., 1980, 481.
11. Ojima I.; Kogure T.; Yoda N. J. Org. Chem., 1980, 45, 4728.
12. Kollonitsch J.; Barash L. J. Am. Chem. Soc., 1976, 98, 5591.
13. Ojima I.; Suga S.; Abe R. Chem. Lett., 1980, 853.
14. Ojima I.; Suga S.; Abe R. Tetrahedron Lett., 1980, 21, 3907.
15. For example, (a) Bose A. K.; Manhas M. S.; Chib J. S.; Chawla H. P. S.; Dayal B. J. Org. Chem., 1974, 39, 2877. (b) Mukerjee A. K.; Singh A. K. Tetrahedron, 1978, 34, 1731.
16. Hatanaka N.; Ojima I. Chem. Lett., 1981, 231.
17. Hatanaka N.; Abe R.; Ojima I. Chem. Lett., 1981, 1297.
18. Hatanaka N.; Ojima I. J. C. S. Chem. Comm., 1981, 344.

RECEIVED December 14, 1981.

9

Asymmetric Carbon–Carbon Bond Formation Using Enantiomerically Pure Vinylic Sulfoxides

GARY H. POSNER, JOHN P. MALLAMO, KYO MIURA, and MARTIN HULCE

Johns Hopkins University, Department of Chemistry, Baltimore, MD 21218

A new, general method is developed for preparation of various 3-substituted carbonyl compounds of high enantiomeric purity. Application of this method is made to asymmetric synthesis of either enantiomer of 3-methylalkanoic acids, of enantiomerically pure 3-methylcyclopentanone, 3-methylcyclohexanone, 3-naphthylcyclopentanone 16 and 3-vinylcyclopentanone 18. 9,11-Seco steroid 16 and steroid intermediate 18 are precursors of enantiomerically pure steroids equilenin and estrone of natural absolute configuration. The basis for this asymmetric synthetic method rests on the transfer of chirality from the sulfoxide sulfur atom to the β-carbon carbon atom during organometallic β-addition to enantiomerically pure α-carbonyl α,β-ethylenic sulfoxides, and the amount of asymmetric induction is highest (*i.e.*, > 98%) with cyclopentenone sulfoxide (*S*)-(+)-10.

Stimulated by the optical activity of most naturally-occurring compounds and by the complete asymmetric induction in most chemical reactions occurring in biological systems, organic chemists have long sought ways to prepare optically active compounds directly without using resolution techniques and ways to mimic the absolute stereocontrol in enzymic reactions. In recent years, progress in this area of asymmetric synthesis has been extraordinary (1). Two industrially important processes exemplifying this type of recent advance include asymmetric catalytic hydrogenation using chiral rhodium complexes (2) and asymmetric steroid synthesis using natural amino acids as chiral directors (3). Many literature reports within the past 5 years document the phenomenal success of the organic chemist in achieving often very high asymmetric inductions during formation of carbon-carbon bonds via nucleophilic addition to electrophilic olefins (1c,d,4).

0097-6156/82/0185-0139$05.00/0

During the past three years, we have had excellent success in achieving some asymmetric syntheses (5). We have focused attention specifically on faithful transfer of chirality from the sulfur atom of some α-carbonyl α,β-ethylenic sulfoxides to the β-carbon atom during organometallic β-addition reactions. This type of high asymmetric induction in forming carbon-carbon bonds has led to successful preparation of several classes of optically active synthetic intermediates such as 3-methylalkanoic acids and 3-methylcycloalkanones. In addition, this asymmetric methodology has been applied successfully to preparation of more complex, enantiomerically pure molecules such as steroids and steroid intermediates.

The first literature report of asymmetric β-addition to an enantiomerically pure α,β-ethylenic sulfoxide appeared in 1971 and involved β-addition of piperidine to propenyl sulfoxide 1 (eq. 1) (6). The absolute stereochemistry of this reaction was rationalized by Stirling in terms of transition state 1a in which the nucleophile approached the β-carbon atom on that side of the double bond remote from the bulky tolyl group in the conformation shown in model 1a (6).

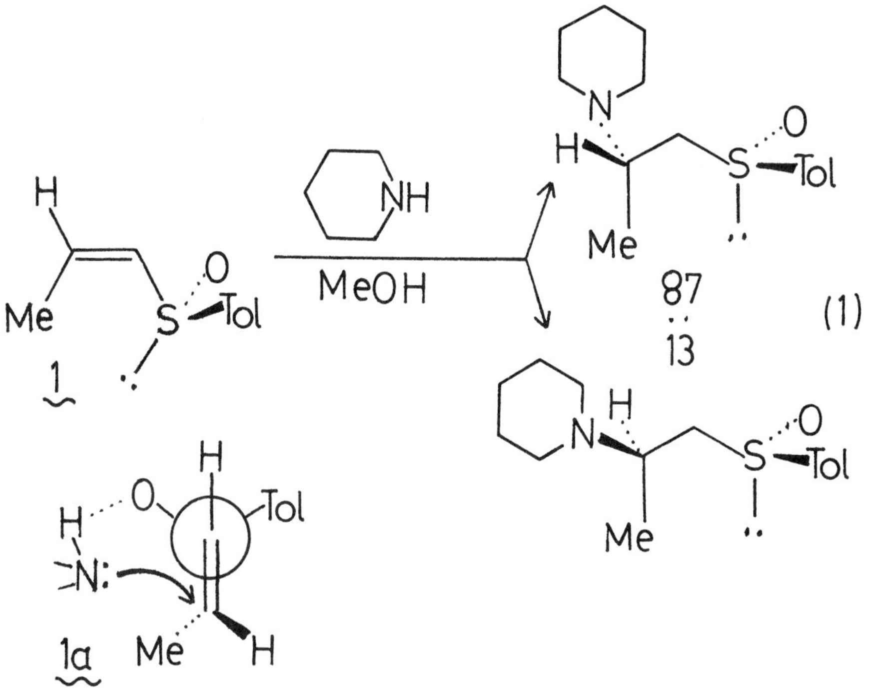

In 1973 Tsuchihashi reported asymmetric induction during carbon-carbon bond formation between nucleophilic malonate and electrophilic enantiomerically pure styryl sulfoxide **2**, producing intermediate diastereomeric carbanions **3a** and **3b** (eq 2) (7). Selective formation of diastereomer **3a** in this irreversible, kinetically controlled addition was rationalized in terms of the preference for an α-sulfinyl carbanion to have the carbon lone-pair orbital trans to the sulfinyl oxygen orbital in a polar solvent.

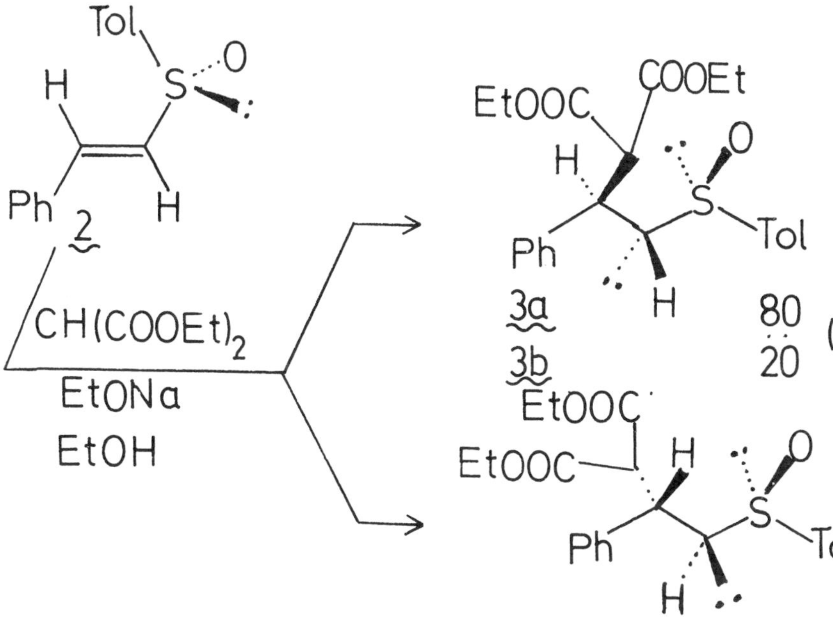

Pursuing these two reports as well as our own interest in organometallic additions to unsaturated sulfur compounds, (8) we examined the behavior of some 1-alkenyl aryl sulfoxides toward relatively non-basic organocopper reagents with the aim of attaching a hydrocarbon group β to the sulfur atom in a stereocontrolled fashion. To our surprise, rather than addition to the carbon-carbon double bond, metalation occurred regiospecifically at the 1-position generating a vinylmetallic species; likewise, methyllithium and several lithium amides produced such vinylmetallic species which reacted successfully with a variety of electrophiles to give various 1-substituted 1-alkenyl sulfoxides (e.g., eq 3) (9).

$$\text{Ph-S(=O)-CH=CH-}\underline{n}\text{-Hex} \xrightarrow[\text{2. MeI}]{\text{1. } Me_2CuLi} \text{Ph-S(=O)-C(Me)=CH-}\underline{n}\text{-Hex} \quad (90\%) \qquad (3)$$

Using enantiomerically pure 1-alkenyl aryl sulfoxides (E)-(+)-4 and (Z)-(-)-4, we found that 1-deprotonation and then reprotonation of the (E)-(+)-4 isomer produced no double bond isomerization and no racemization, whereas similar treatment of the (Z)-(-)-4 isomer produced double bond isomerization and some racemization (eqs. 4,5) (5a).

$$\underline{n}\text{-Non-CH=CH-S(=O)Ph} \;\; (E)\text{-}(+)\text{-}\mathbf{4}\; [\alpha]_D +95.7^\circ \xrightarrow[\text{2. } H^+]{\text{1. MeLi}} (E)\text{-}(+)\text{-}\mathbf{4}\; [\alpha]_D +95.2^\circ \qquad (4)$$

$$\underline{n}\text{-Non-CH=CH-S(=O)Ph} \;\; (Z)\text{-}(-)\text{-}\mathbf{4}\; [\alpha]_D -200^\circ \xrightarrow[\text{2. } H^+]{\text{1. MeLi}} (E)\text{-}(+)\text{-}\mathbf{4}\; [\alpha]_D +67.3^\circ \qquad (5)$$

Carboxylation of such a 1-lithio 1-alkenyl sulfoxide led to a diastereomerically and enantiomerically pure α-carboxyl α,β-ethylenic sulfoxide such as 5a after protonation of the intermediate lithium carboxylate and to the corresponding methyl ester 5b after methylation with methyl iodide-hexamethylphosphoramide (HMPA) (eq. 6) (5a).

(S)-(+)-5a, R= H, >95%

Ph S=O, n-Hex, H; 1) i-Pr_2NLi, 2) CO_2; H^+; MeI, HMPA; Ph S=O, n-Hex, COOR (6)

(S)-(+)-5b, R = Me, 80%

We reasoned that metal ion complexation with α-carboxyl α,β-ethylenic sulfoxides such as 5a and 5b should produce a chelate such as 6, locking the sulfoxide group into the conformation shown. Approach of a nucleophilic methylmetallic species toward the β-carbon atom should now occur from the unshielded side of the carbon-carbon double bond and should lead therefore to (R)-3-methylalkanoates with high asymmetric induction. Working model 6 further suggested that aromatic groups bulkier than phenyl and metal ions that form strong complexes might possibly lead to complete asymmetric induction.

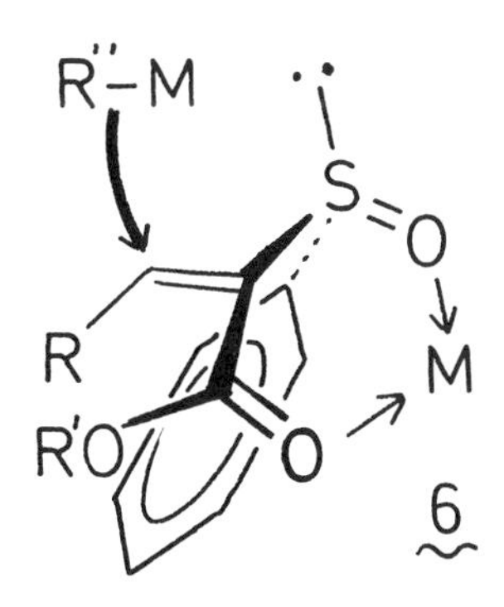

α-Carboxyl α,β-ethylenic sulfoxide 5a reacted with dimethylcoppermagnesium iodide in a conjugate manner; sodium amalgam reductive cleavage of the intermediate α-sulfinyl carboxylic acid produced (R)-(+)-3-methylnonanoic acid in 61% enantiomeric excess (eq. 7). Likewise, α-methoxycarbonyl α,β-ethylenic sulfoxide 5b reacted with dimethylcopperlithium followed by reductive sulfur-carbon bond cleavage and saponification to produce (R)-(+)-3-methylnonanoic acid in 65% enantiomeric purity (eq. 8, 53% overall yield). Reversing the order of introducing the larger and the smaller alkyl groups at the prochiral β-carbon atom afforded mainly that enantiomer having opposite absolute stereochemistry. Thus (E)-1-propenyl sulfoxide (+)-7 reacted with di-n-butylcopperlithium and then underwent reductive carbon-sulfur bond cleavage and saponification to form (S)-(-)-3-methylheptanoic acid (8) in 59% enantiomeric purity (eq. 9) (5a). Higher asymmetric inductions, however, have been achieved recently by Meyers, by Mukaiyama and by Koga in synthesis of 3-methylalkanoic acids (4).

In sharp contrast to these highly successful methods for enantioselective synthesis of some acyclic systems, virtually no general method has been reported for enantio-controlled preparation of cyclic compounds. Because so many optically pure carbocycles are found in nature and are important synthetic intermediates, the need for effective and highly asymmetric syntheses of such compounds is obvious. More specifically, although many enantiomerically pure naturally-occurring 3-alkylcarbocycles with small 3-alkyl groups are known, asymmetric synthesis of these compounds via attachment of the small alkyl group is usually an extremely difficult process. Despite attempts at asymmetric induction during organometallic conjugate addition to 2-cycloalkenones using optically active solvents (10) or optically active ligands, (11) only poor enantioselectivity has been achieved.

We reasoned that some cyclic enone sulfoxides should form an even more rigid chelate than that formed from the corresponding acyclic alkenyl sulfoxides when complexed with metal ions; model **9** exemplifies the case for a cyclopentenone sulfoxide and suggests a high degree of stereocontrol during the nucleophilic addition reaction.

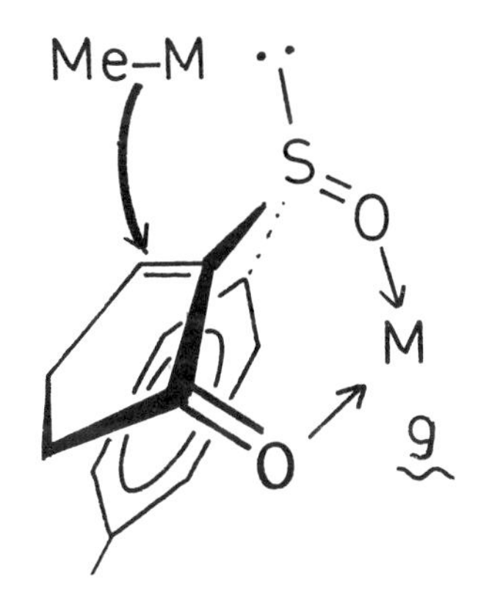

Cyclopentenone sulfoxide (S)-(+)-**10** was prepared via eq. 10 in good yield on a few mg as well as on a 10-gm scale (5a). This enone sulfoxide, which is crystalline and stable at least for several months, reacted with methylmagnesium iodide [in the absence of copper (I)] in a conjugate manner; aluminum amalgam carbon-sulfur bond reductive cleavage produced (R)-(+)-3-methylcyclopentanone (**11**) in 71% chemical yield and in 80% enantiomeric purity (eq. 11). The absolute stereochemistry of this asymmetric induction is consistent with working model **9** and approach of the methyl nucleophile from the pro-(R) direction. Likewise, dimethylcopperlithium reacted with cyclopentenone sulfoxide (S)-(+)-**10** to give, after reductive sulfur-carbon bond cleavage, (R)-(+)-

3-methylcyclopentanone (11) in 91% chemical yield and 80% enantiomeric purity.

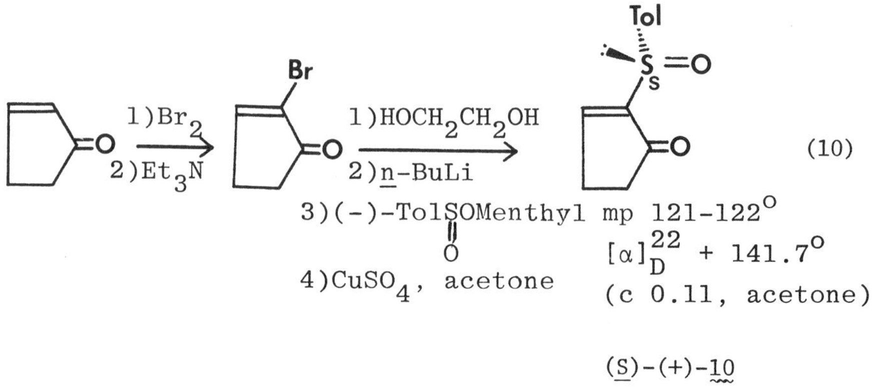

1) Br_2 2) Et_3N → 1) $HOCH_2CH_2OH$ 2) n-BuLi 3) (-)-TolSOMenthyl (S=O) 4) $CuSO_4$, acetone → (10)

mp 121-122°
$[\alpha]_D^{22}$ + 141.7°
(c 0.11, acetone)

(S)-(+)-10

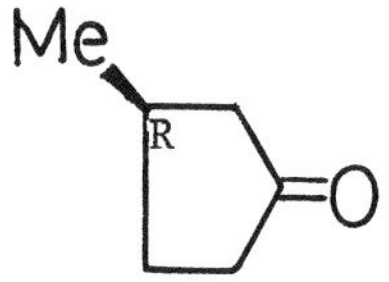

11

			% Yield	% e.e.
(S)-(+)-10	MeMgI →	Al/Hg →	77	80
(S)-(+)-10	Me_2CuLi →	Al/Hg →	91	80

(11)

To preform a strong enone sulfoxide-metal ion complex and thus possibly to increase the amount of asymmetric induction, several metal dibromides were added to cyclopentenone sulfoxide (S)-(+)-10. As shown in eq. 12, only zinc dibromide was highly effective in raising the extent of asymmetric induction during methyl Grignard conjugate addition.

$$(S)\text{-}\mathbf{10} \xrightarrow{MBr_2} \xrightarrow{MeMgX} \xrightarrow{Al\text{-}Hg} \text{(R)-3-methylcyclopentanone} \quad (12)$$

M	X	% Yield	% e.e.
Ni	Br	53	70
Co	Br	73	70
Pd	Br	58	71
Mg	I	81	73
Zn	I	89	87

The best stereochemical results, however, were obtained with the new and bulky methylmetallic reagent, methyl triisopropoxy-titanium, (12) and with methylmagnesium chloride (eq. 13). Presumably, the more electrophilic chloromagnesium species formed a stronger complex with the bidendate enone sulfoxide than did the bromo or the iodomagnesium species (13) and thus forced the β-addition to proceed entirely through the chelated and therefore locked conformation shown in model 9.

$$(S)\text{-}\mathbf{10} \xrightarrow{Me\text{-}M} \xrightarrow{Al\text{-}Hg} \text{(R)-3-methylcyclopentanone} \quad (13)$$

Me-M	% Yield	% e.e.
$MeTi(OPr\text{-}i)_3$	90	90
MeMgCl	91	95-100

We have also prepared (R)-(+)-3-methylcyclohexanone (13) via methylmetallic conjugate addition to enantiomerically pure cyclohexenone sulfoxide (S)-(+)-12 (eq. 14). Equations 13 and 14 represent highly successful asymmetric syntheses of 3-methylcyclopentanone and 3-methylcyclohexanone and illustrate a general new method for preparation of 3-alkylcarbocycles of high or virtually complete enantiomeric purity (14).

(S)-(+)-12 $\xrightarrow{(CH_3)_2CuLi}$ $\xrightarrow{Al\text{-}Hg}$ 13 (14)
90% yield, 72% e.e.

$\xrightarrow{1.\ ZnBr_2,\ 2.\ CH_3MgBr}$ $\xrightarrow{Al\text{-}Hg}$ 100% Yield, 83% e.e.

Besides conjugate addition of the small methyl group, cyclopentenone sulfoxide (S)-(+)-10 also underwent conjugate addition of a large naphthyl group. As shown in scheme I, we have applied this reaction which proceeds with complete asymmetric induction to efficient construction of 3-naphthylcyclopentanone 14 having the natural absolute steroid configuration at carbon 14 (steroid numbering). Reductive cleavage of the sulfinyl group using dimethylcopperlithium allowed regiospecific formation of enolate ion 15 which underwent carbon alkylation to give only 9,11-seco steroid 16 having the desired 13S-14S absolute stereochemistry! Synthetic seco steroid 16 was identical by HPLC, NMR, IR, mass spectrometry, melting point (116.5-118°C), mixed melting point and optical rotation $[[\alpha]^{22°}_{365} = +168°$ (c 0.36, $CHCl_3$)] to a sample of 16 prepared by degradation of natural estradiol (5a). Because we have previously converted racemic 16 into the racemic steroid equilenin 17, (15) preparation of enantiomerically pure 16 amounts to a formal total synthesis of enantiomerically pure equilenin 17.

3-Vinylcyclopentanone 18 and the corresponding enol silyl ether 19 have been used recently in some elegant, creative, and efficient constructions of estrones via intramolecular Diels-Alder cycloaddition reactions of intermediate O-quinodimethanes (Scheme II) (16-18). Only one report, however, has appeared

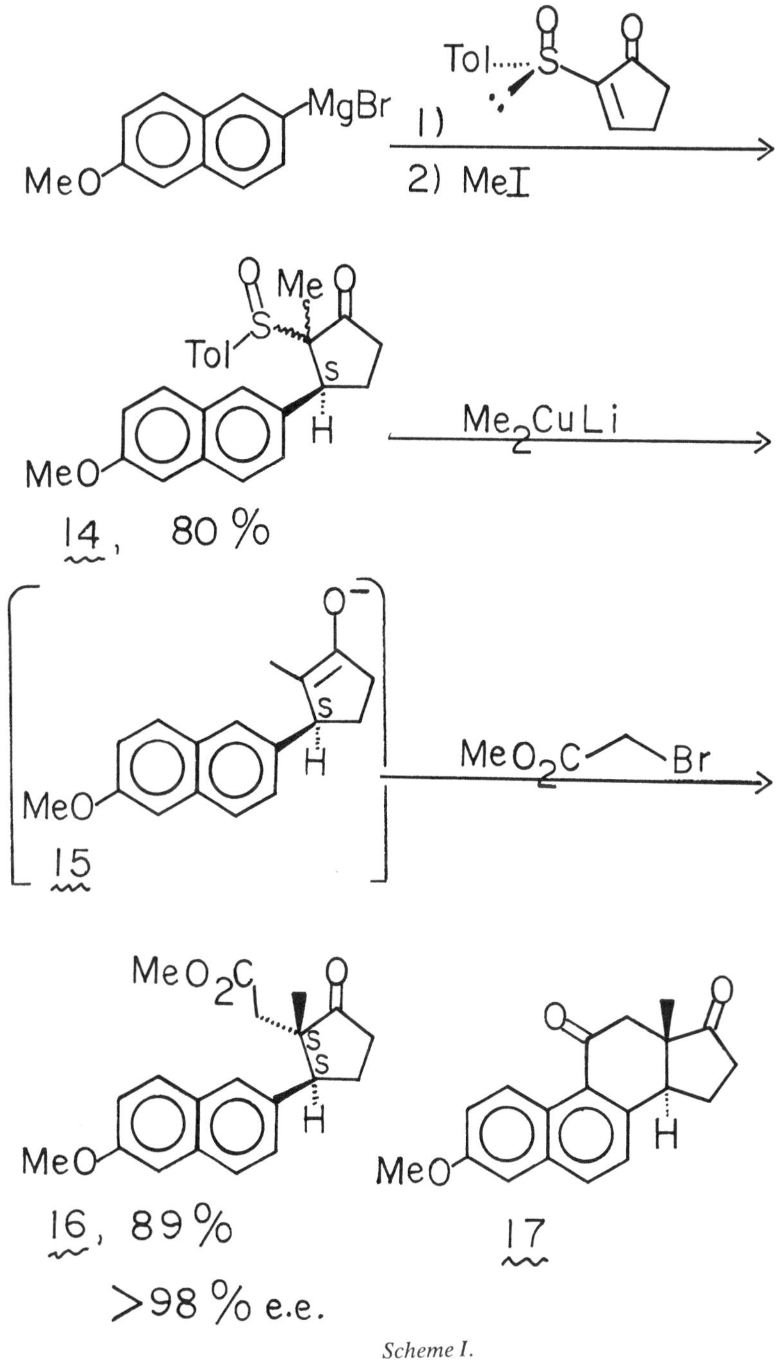

MeO, MgBr; 1) Tol, S, O; 2) MeI

14, 80%

Me_2CuLi

15

MeO_2C, Br

16, 89%

>98% e.e.

17

Scheme I.

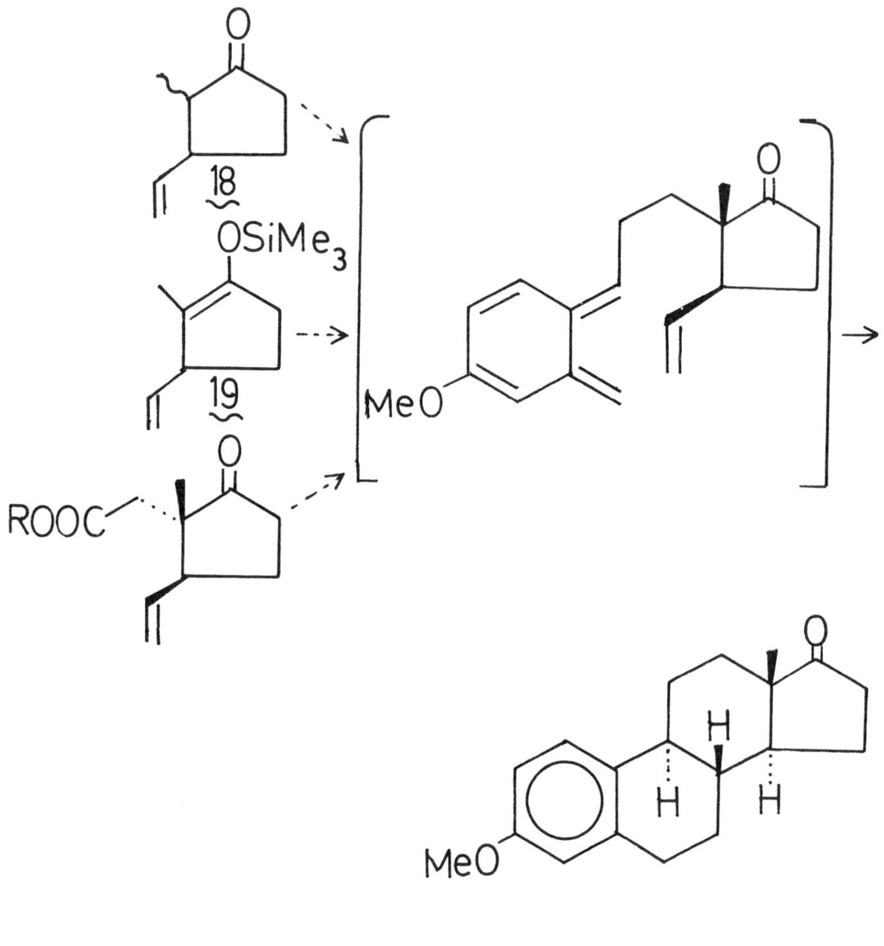

18

OSiMe3

19

ROOC

MeO

MeO

Scheme II.

involving asymmetric synthesis of optically active steroid intermediate 18 used in preparation of optically active estrones via generalized scheme II (16i).

We found that enantiomerically pure cyclopentenone sulfoxide (S)-(+)-10 reacted with vinylmagnesium bromide in the presence of a catalytic amount of cuprous bromide and then with methyl iodide to give 2,2,3-trisubstituted cyclopentanone 20 (Scheme III). Trisubstituted cyclopentanone 20, however, could be formed in better yield (∿75%) via the corresponding sodio enolate. Aluminum amalgam reductive cleavage produced 3-vinylcyclopentanone (S)-18 in 80% enantiomeric purity. The amount of asymmetric induction was improved dramatically, however, by first complexing cyclopentenone sulfoxide (S)-10 with zinc dibromide and then adding vinylmagnesium bromide. In this way, following scheme III, 3-vinylcyclopentanone (S)-18 was formed in >98% enantiomeric purity and in 55-60% overall yield! Reductive cleavage of α-sulfinylcyclopentanone 20 using dimethylcopperlithium followed by addition of trimethylsilyl chloride gave enantiomerically pure enol silyl ether (S)-(+)-19 in 54% overall yield (5b). This complete asymmetric induction in synthesis of steroid intermediates (S)-18 and and (S)-19 amounts to a formal total synthesis of enantiomerically pure estrone!

It is clear from the results summarized here that some very successful, general, and highly useful asymmetric syntheses of carbon-carbon bonds can be performed using enantiomerically pure 1-carbonyl 1-alkenyl sulfoxides and various organometallic reagents. These results add significantly to the rapidly growing number of new, rationally designed, and highly stereocontrolled C-C bond-forming synthetic methods and should be especially useful in asymmetric synthesis of enantiomerically pure 3-substituted carbocycles.

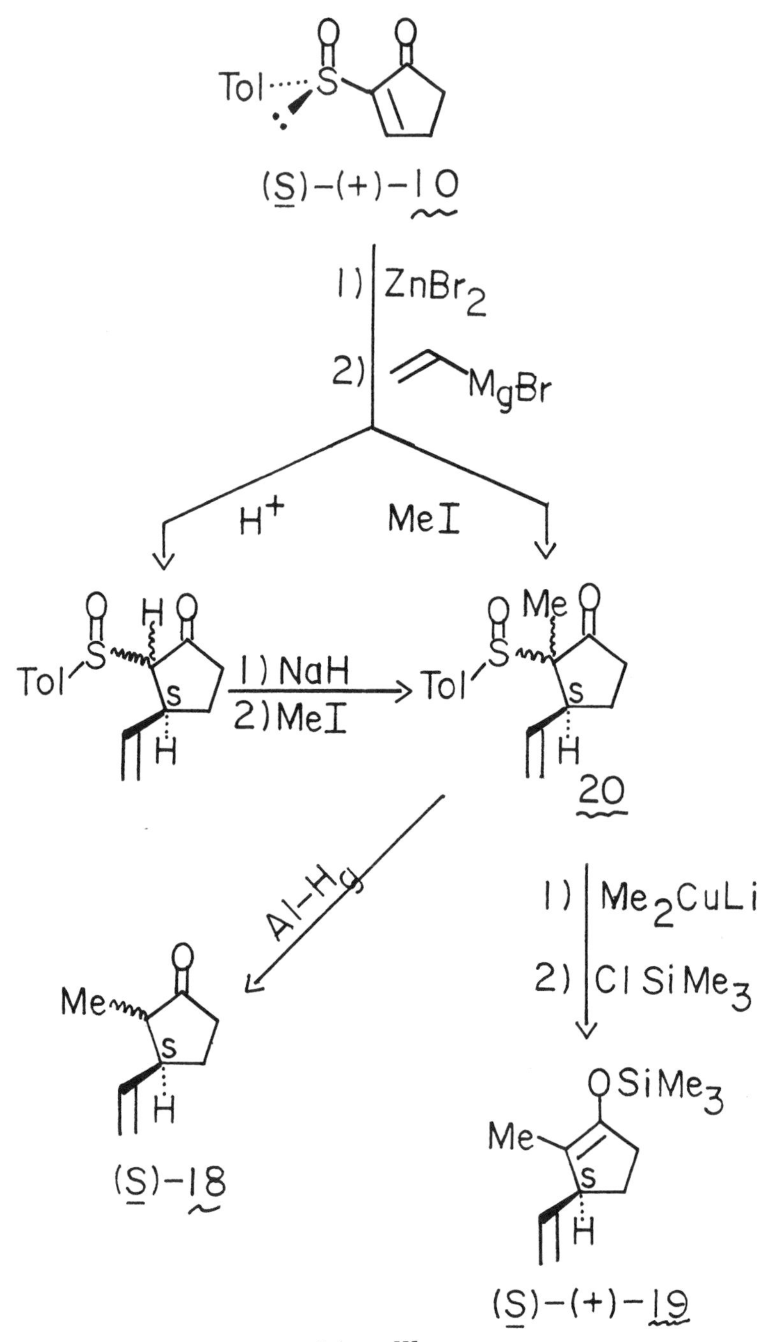
Tol
(S)-(+)-10
1) ZnBr2
2) MgBr
H+
MeI
Tol
1) NaH
2) MeI
Tol
Me
H
20
Al-Hg
1) Me2CuLi
2) Cl SiMe3
Me
H
(S)-18
OSiMe3
Me
H
(S)-(+)-19

Scheme III.

Acknowledgement

We gratefully acknowledge financial support from the National Science Foundation (CHE 79-15161), from the Donors of the Petroleum Research Fund, administered by the American Chemical Society, from G. D. Searle and Co., and from Merck, Sharp, and Dohme. We warmly acknowledge experimental help from P-W. Tang and A. Y. Black.

Literature Cited

1. (a) Morrison, J.D.; Mosher, H.S. "Asymmetric Organic Reactions"; Prentice-Hall; Engelwood Cliffs, N.J.; 1971.
   (b) Scott, J.W.; Valentine, D., Jr. Science 1974, 184, 943.
   (c) Valentine, D., Jr.; Scott, J.W. Synthesis 1978, 329.
   (d) Meyers, A.I. Accts. Chem. Res. 1978, 11, 375.
2. (a) Knowles, W.S.; Sabacky, M.J.; Vineyard, B.D.; Weinkauf, D. J. Amer. Chem. Soc. 1975, 97, 2569.
   (b) Kagan, H.B; Dang, T.P. ibid., 1972, 94, 6429.
   (c) Gelbard, G.; Kagan, H.B.; Stern, R. Tetrahedron, 1976, 32, 233.
   (d) Fryzuk, M.D.; Bosnich, B. J. Amer. Chem. Soc. 1977, 99, 6262.
3. Cohen, N. Accts. Chem. Res. 1976, 9, 412.
4. (a) Hashimoto, S.I.; Yamada, S.I.; Koga, K. J. Amer. Chem. Soc. 1978, 98, 7450.
   (b) Mukaiyama, T.; Takeda, T.; Osaki, Chem. Lett. 1977, 1165.
   (c) Meyers, A.I.; Smith, R.K.; Whitten, C.E. J. Org. Chem. 1979, 44, 2250.
   (d) Hashimoto, S.; Komeshima, N.; Yamada, S.; Koga, K. Chem. Pharm. Bull. 1979, 27, 2437.
   (e) Isobe, M.; Kitamura, M.; Goto, T. Tetrahedron Lett. 1981, 22, 239.
5. (a) Posner, G.H.; Mallamo, J.P.; Miura, K. J. Amer. Chem. Soc. 1981, 103, 2886.
   (b) Posner, G. H.; Hulce, M.; Mallamo, J. P.; Drexler, S. A.; Clardy, J; J. Org. Chem. 1982, 47, 000.
6. Abbott, D.J.; Colonna, S.; Stirling, C.J.M. Chem. Comm. 1971, 471.
7. Tsuchihashi, G.; Mitamura, S.; Inoue, S.; Ogura, K. Tetrahedron Lett. 1973, 323.
8. Posner, G.H.; Brunelle, D.J. J. Org. Chem. 1973, 38, 2747.
9. Posner, G.H.; Tang, P-W.; Mallamo, J.P. Tetrahedron Lett. 1978, 3995.
10. Langer, W.; Seebach, D. Helv. Chim. Acta 1979, 62, 1710.
11. (a) Gustafson, B.; Hallnemo, G.; Ullenius, C. Acta Chem. Scand. 1980, B34, 433.
   (b) Zweig, J.S.; Luche, J.L.; Barreiro, E.; Crabbe, P. Tetrahedron Lett. 1975, 2355.

12. (a) Weidmann, B.; Wildler, L.; Olivero, A.G.; Maycock, C.D.; Seebach, D. Helv. Chim. Acta 1981, 64, 357.
(b) Reetz, M.T.; Steinbach, R.; Westermann, J.; Peter, R. Angew. Chem. Inst. Ed. Engl. 1980, 19, 1011.
13. Ashby, E.C.; Laemmle, J.; Newmann, H.M. Accts. Chem. Res. 1974, 7, 272, and references therein.
14. For a recent synthetic approach to optically active 3-alkylated cyclopentanones and cyclohexanones, see Taber, D.F.; Saleh, S.A.; Korsmeyer, T.W. J. Org. Chem. 1980, 45, 4699.
15. Posner, G.H.; Chapdelaine, M.J.; Lentz, C.M. J. Org. Chem. 1979, 44, 3661.
16. (a) Oppolzer, W.; Petrzilka, M.; Battig, K. Helv. Chim. Acta 1977, 60, 2965.
(b) Kametani, T.; Nemoto, H.; Fukumoto, K. J. Amer. Chem. Soc. 1977, 99, 3461.
(c) Funk, R.L; Vollhardt, K.P.C. J. Amer. Chem. Soc. 1977, 99, 5483 and 1979, 101, 215.
(d) Oppolzer, W.; Battig, K.; Petrzilka, M. Helv. Chim. Acta 1978, 61, 1945.
(e) Nicolaou, K.C.; Barnette, W.E.; Ma, P. J. Org. Chem. 1980, 45, 1463.
(f) Djuric, S.; Sarkan, T.; Magnus, P. J. Amer. Chem. Soc. 1980, 102, 6885.
(g) Ito, Y.; Nakatsuka, M.; Saegusa, T. J. Amer. Chem. Soc. 1981, 103, 476.
(h) Quinkert, G.; Weber, W-D.; Schwartz, U.; Durner, G. Angew. Chem. Int. Ed. Engl. 1980, 19, 1027.
(i) Quinkert, G.; Schwartz, U.; Stark, H.; Weber, W-D.; Baier, H.; Adam, F.; Durner, G. Angew. Chem. Int. Ed. Engl. 1980, 19, 1029.
17. For a review, see Oppolzer, W. Synthesis 1978, 793.
18. A resolved cyclopentanone acetic acid has been used in synthesis of two optically pure estrone derivatives: Oppolzer, W.; Roberts, D.A. Helv. Chim. Acta 1980, 63, 1703.

RECEIVED December 14, 1981.

10

Asymmetric Reactions: A Challenge to the Industrial Chemist

GABRIEL SAUCY and NOAL COHEN

Hoffmann-La Roche Incorporated, Chemical Research Department, Nutley, NJ 07110

A variety of established industrial processes for the manufacture of and new synthetic approaches to certain optically active compounds such as pharmaceuticals, vitamins, and fine chemicals are surveyed. Among the techniques for obtaining optically pure intermediates covered in this review are classical or modified optical resolutions, the utilization of starting materials from the chiral pool, as well as stoichiometric and catalytic asymmetric transformations.

The development of practical and economical processes for large scale industrial preparation of certain optically active compounds such as pharmaceuticals, fine chemicals, and vitamins has been and continues to be a major challenge. Historically, a number of efficient industrial processes have evolved which are based on classical resolution (e.g. D-biotin (1,2) and D-pantothenic acid (3)) or the use of optically active starting materials (e.g., vitamin C (4)). More recently, attractive processes utilizing asymmetric reactions have been designed (5). From an industrial point of view, the use of chiral catalysts to generate asymmetry is particularly advantageous. Unfortunately, our lack of understanding of the quantitative aspects which govern the degree of asymmetry created is a serious problem. The development of chiral catalysts useful to industry is presently very much dependent on the empirical approach. New insight and knowledge are needed to design rational approaches in asymmetric synthesis. This review is intended to show how industry has solved or, at least, confronted the problem of producing certain optically active target compounds in a practical and economical manner, using selected examples from the area of pharmaceuticals, vitamins, and fine chemicals.

0097-6156/82/0185-0155$05.00/0

L-Dopa

Prior to the pioneering development of the asymmetric hydrogenation process for producing L-Dopa by Knowles and co-workers at Monsanto (5), we had investigated an alternative approach involving hydrogenation of the chiral substrate 1 using an achiral catalyst (6). This produced the mixture of epimeric amides 2 and 3 which could be converted, in 89% overall yield, to the desired isomer 3 via simultaneous base catalyzed equilibration - crystallization. Unfortunately, hydrolysis of 3 to L-Dopa gave unacceptably low yields, of the order of only 50%.

19-Norsteroids

Very substantial asymmetric induction at C-13 was found to take place upon condensation of the optically active hydroxy vinyl ketone 4 ($R=C_2H_5$) with 2-methylcyclopentane-1,3-dione (5) giving predominantly the dienol ether 6 (7). The exploitation of this fortuitous result enabled us to design several efficient routes to optically active 19-norsteroids and estrone (8). The key chiral annulating agents 4 were secured by various schemes relying on classical resolutions or microbiological reduction of δ-keto acids giving optically active δ-lactones.

Of particular interest is the regio- and enantiospecific reduction of diketo acid 7 with Margarinomyces bubaki affording the keto lactone 8. The latter serves as the chiral starting point in an asymmetric total synthesis (+)-estr-4-ene-3,17-dione via key intermediate 9 (9).

A most impressive example of catalytic asymmetric synthesis forms the basis for still another and very efficient approach to 19-norsteroids (10,11). The exact mechanism responsible for the extremely high asymmetric induction noted in the crucial conversion of prochiral 10 to ketol 11 and (S)-enedione 12 still needs to be clarified (12,13). Nonetheless, these chiral aldol products serve very effectively as steroid CD-ring synthons (8,14-21).

Zeaxanthin

Zeaxanthin, which occurs in corn and many other plants, is an important carotenoid. Its synthesis in optically active form can be achieved on the basis of the three approaches depicted.

In the first case (22), optical activity is introduced by asymmetric reduction of the enedione 13 with yeast, giving dione 14. Stereo- and regioselective reduction then produced the key ketol 15. In the second approach (23), the cyclohexa-

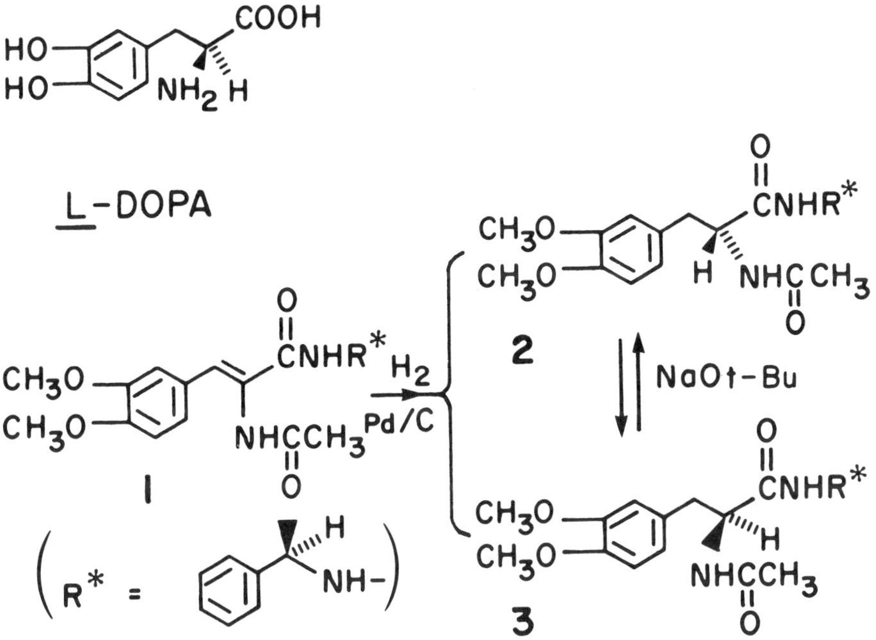

HO
HO
COOH
NH_2 H

L-DOPA

O
CNHR*
CH_3O
CH_3O
H NHCCH_3
O
2

O
CNHR*
H_2
CH_3O
CH_3O
$NHCCH_3$
Pd/C
O
1

NaOt-Bu

(R* = H NH-)

O
CNHR*
CH_3O
CH_3O
H
$NHCCH_3$
O
3

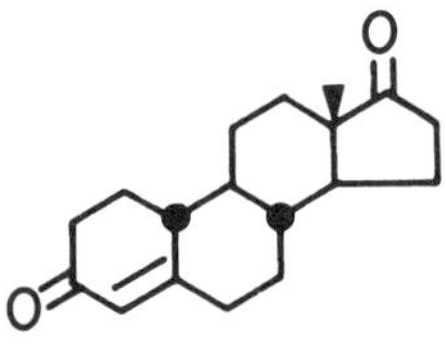

19-NORANDROSTENEDIONE

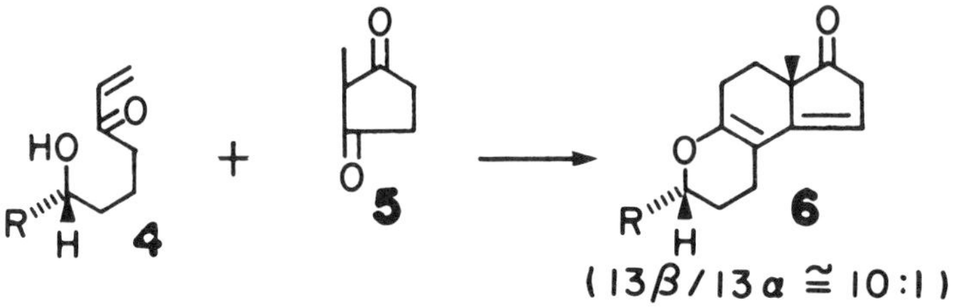

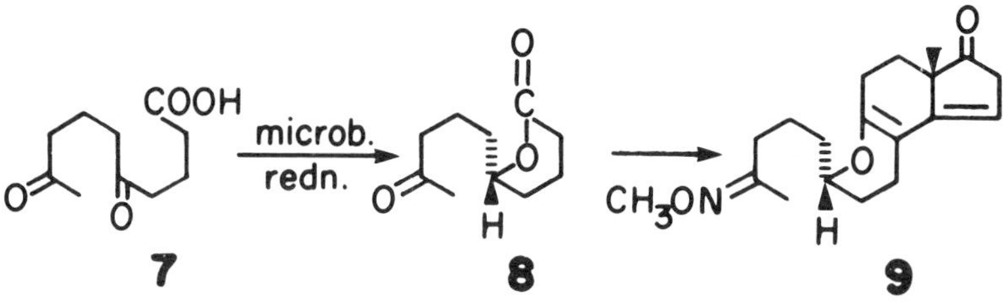

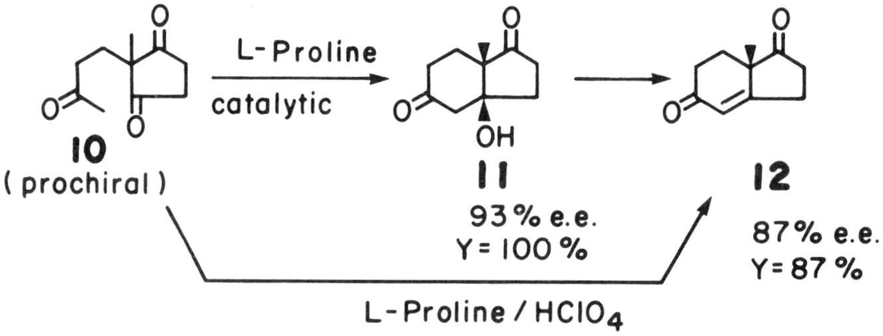

HO R H **4** + **5** → R H **6**

(13β/13α ≅ 10:1)

COOH **7** —microb. redn.→ **8** → CH_3ON **9**

10 (prochiral) —L-Proline catalytic→ **11** 93% e.e. Y= 100% → **12** 87% e.e. Y= 87%

L-Proline / $HClO_4$

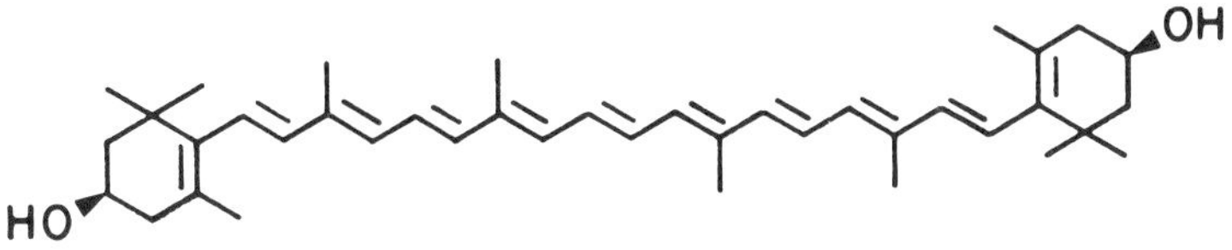
OH
HO

ZEAXANTHIN

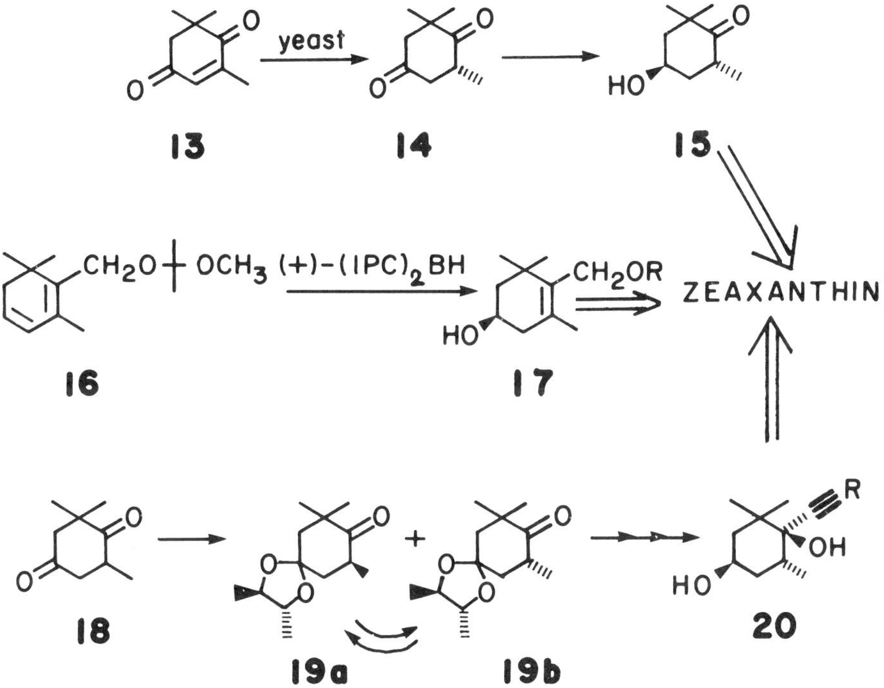
O
O
yeast
O
O
HO
O
13
14
15
CH2O
OCH3
(+)-(IPC)2BH
CH2OR
HO
ZEAXANTHIN
16
17
O
O
O
O
O
+
O
O
O
R
OH
HO
18
19a
19b
20

diene 16, available from safranal, is subjected to an asymmetric hydroboration with (+)-diisopinocampheylborane giving intermediate 17. In this context, it should be noted that asymmetric hydroboration of dienes has also been applied at Hoffmann-La Roche in a synthesis of prostaglandin intermediates having industrial potential (24).

The third approach to zeaxanthin (25) exploits the special features of keto acetal 19b which is theoretically available in quantitative yield starting from the diketone 18 and (2R,3R)-2,3-butanediol. Fortunately, 19b is crystalline and less soluble than its epimer 19a. Equilibration with sodium hydroxide thus favors the desired epimer. Diastereomer 19b is then transformed into the diol 20 in two stereospecific steps.

Pantothenic Acid

The present industrial processes used to produce the crucial intermediate (R)-(-)-pantolactone (22) are based on resolution of racemic material (3). A different and very promising approach has been reported by a Japanese group (26). Independently, Roche workers also investigated this approach which involves asymmetric reduction of ketolactone 21 using rhodium catalysts derived from chiral phosphines (27). In this manner, 22 can be obtained in very high chemical and optical yields.

D-Biotin

The original synthesis of D-biotin, which involves a classical resolution with efficient recycling of the unwanted enantiomer (1), has recently been advantageously modified (28). The key feature of the new Sumitomo route involves preparation of the chiral imide 24 from symmetrical diacid 23. Hydride reduction of 24 occurs with high asymmetric induction, generating hydroxy amide 25 having excellent optical purity, in 65% yield. Treatment with HCl converts 25 to the corresponding γ-lactone and ultimately D-biotin by the established route. The chiral aminopropanediol (R*) is recovered and recycled. Other novel approaches to D-biotin have been studied at Hoffmann-La Roche in recent years (29,30).

Vitamin E ((2R,4'R,8'R)-α-Tocopherol)

The development of a practical total synthesis of natural (2R,4'R,8'R)-α-tocopherol (26) is a major challenge. While much progress has been made in this area, a technically feasible synthetic approach to this form of vitamin E remains an elusive goal and isolation from soybean oil continues to be the major source of 26 (31).

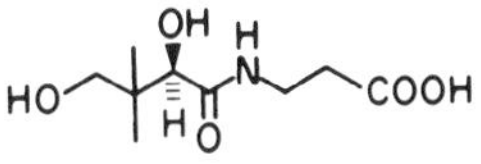

PANTOTHENIC ACID

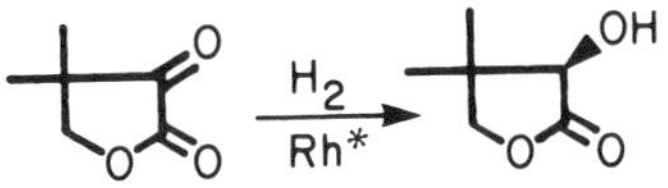

H_2 / Rh*

21 Ketolactone

22 (R)-(–)-Pantolactone

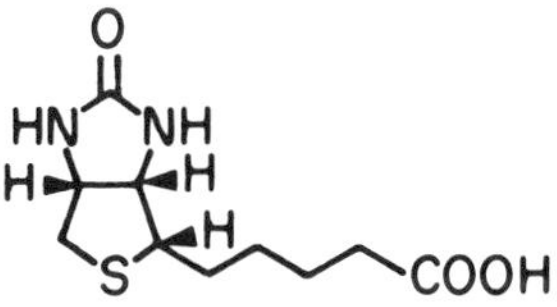

D-BIOTIN

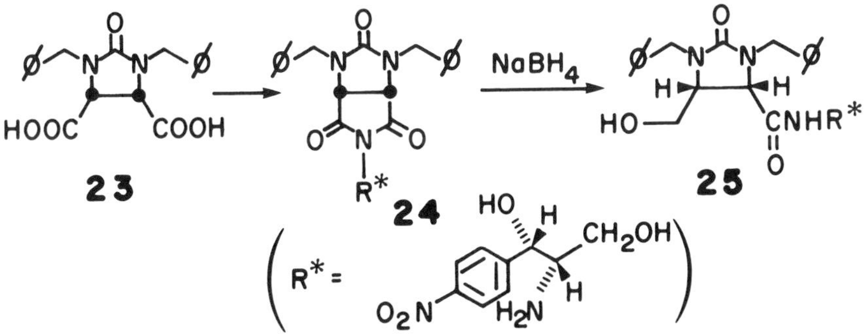

NaBH4

23

24

25

(R* = ...)

Much of our synthetic work aimed at 26 has been summarized in a recent review (32) and will not be covered in detail here. The two homologous strategies employed are depicted by the bond dissections "a" and "b". In the former, a C_{14}-chroman unit is coupled with a C_{15}-side chain intermediate in the penultimate step while in the latter, C_{15}-chroman and C_{14}-side chain synthons are united.

Regarding the side chain, recent developments in our laboratories involve applications of asymmetric hydride reductions (e.g., 27 → 29 and 31) to provide chiral Claisen rearrangement substrates 32, 33, 35, and 36 which, in turn, afford optically active ester 34 or its enantiomer 37 with essentially complete chirality transfer (33). In another approach, catalytic asymmetric hydrogenation of geranic acid (38) yields the C_{10}-intermediate 40 in 70% e.e. (34). Many other, often quite ingenious routes to chiral side chain precursors have been reported recently by various groups (35-39).

Progress has also been made with regard to the accessibility of key chroman intermediates. Thus methods were developed which allow utilization of the unwanted enantiomers of chroman-2-carboxylic and chroman-2-acetic acids (41c, 42c) obtained along with the desired antipodes (41b, 42b) by classical resolution of the racemic forms (41a, 42a) (enantioconvergence 40, 41). For example, a four stage inversion sequence provides a route for transforming 41c into the (S)-enantiomer 41b required for synthesis of 26 (42). Similarly, the homologous, unwanted (R)-chroman-2-acetic acid 42c can be utilized by means of a racemization-recycling process (43). While these approaches still rely on classical resolutions, the modifications incorporated substantially improve the overall efficiency in terms of obtaining optically pure intermediates.

A significant offshoot of our synthetic studies aimed at 26 has been the exploitation of the resulting methodology for preparing all seven stereoisomers of 26 (44). Employing a variation of a gas chromatographic method recently developed for separating the diastereomers of α-tocopherol (45), we were able to demonstrate that all of our synthetic stereoisomers were of high (93-99%) diastereomeric purity (44). The availability of these compounds in pure form will now allow a precise determination of the relationship between stereochemistry and vitamin E biopotency in the α-tocopherol molecule. During the course of this work, it was established for the first time that naturally occurring d-α-tocopherol from soybean oil is a single enantiomer (2R,4'R,8'R), that synthetic d,l-α-tocopherol is an equimolar mixture of four racemates, and that natural (E)-(7R,11R)-phytol is enantiomerically homogeneous (44).

CH3 a
HO H3C H H3C H CH3
2
H3C O 4' 8' CH3
CH3 CH3 b

26

(2R,4'R,8'R)-α-Tocopherol

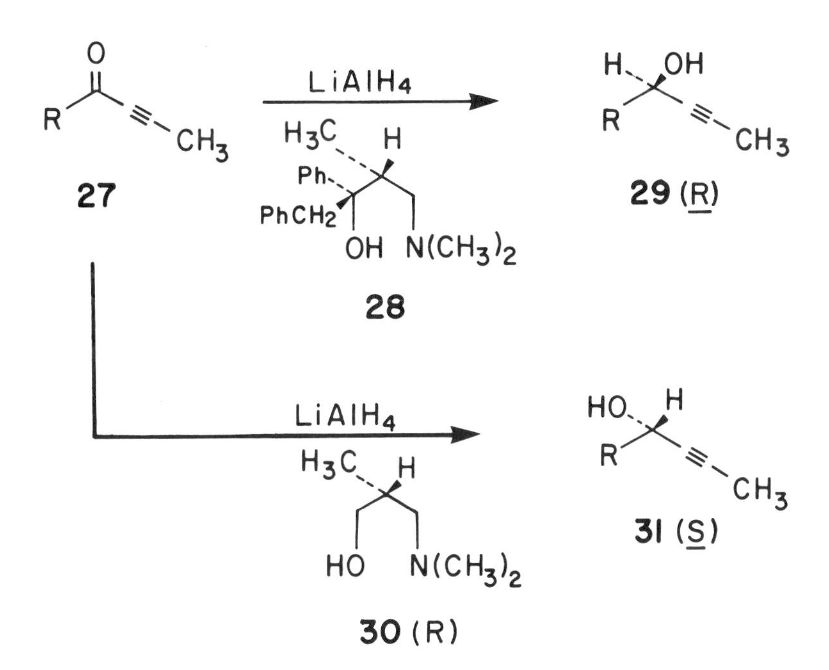

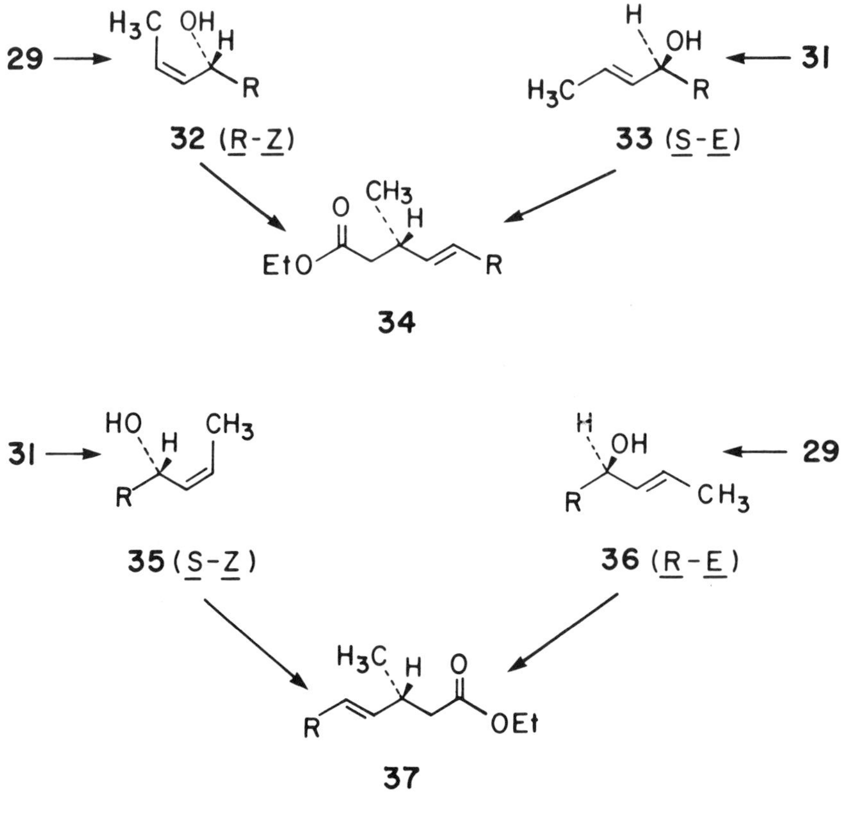

29 → 32 (R-Z) ← 33 (S-E) ← 31

34

31 → 35 (S-Z) ← 36 (R-E) ← 29

37

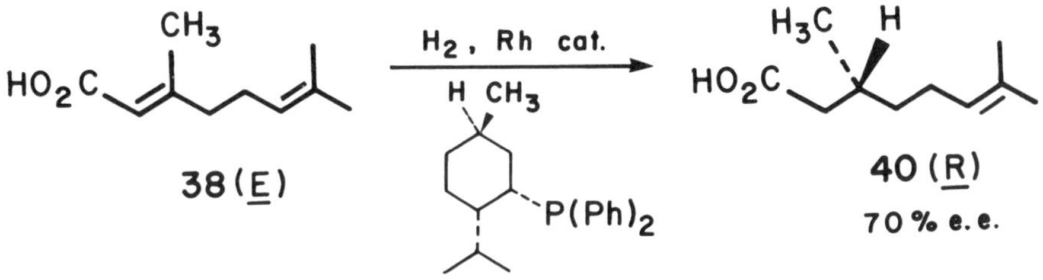

H_2, Rh cat.

38 (*E*) → **40** (*R*), 70% e.e.

39

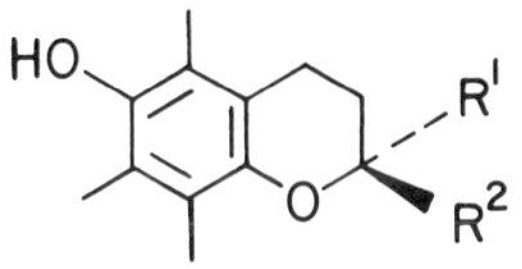

41a; $R^1, R^2 = CH_3, CO_2H$ (±)

41b; $R^1 = CH_3$, $R^2 = CO_2H$ (*S*)

41c; $R^1 = CO_2H$, $R^2 = CH_3$ (*R*)

42a; $R^1, R^2 = CH_3, CH_2CO_2H$ (±)

42b; $R^1 = CH_3$; $R^2 = CH_2CO_2H$ (*S*)

42c; $R^1 = CH_2CO_2H$; $R^2 = CH_3$ (*R*)

Conclusions

For many reasons, the pharmaceutical industry will continue to require facile synthetic routes to diastereoisomerically and enantiomerically pure chiral molecules. In order to achieve these goals, new asymmetric processes, especially catalytic asymmetric reactions, will be needed. Alternatively, there is great potential for the development of industrially useful biotransformations to produce complex optically active compounds. Genetic engineering will probably play an important role in such approaches. Nonetheless, the challenge to the organic chemist will remain.

Acknowledgment

We are grateful to the Research Management of Hoffmann-La Roche Inc. for the opportunity to prepare this review which covers the multidisciplinary efforts of various research groups in the U.S.A. (Nutley, N. J.) and Switzerland (Basle).

Literature Cited

1. Sternbach, L. H. in "Comprehensive Biochemistry"; Vol. 11, Florkin, M.; Stotz, E. H.; Elsevier: New York, 1963; p. 66.
2. Gerecke, M.; Zimmermann, J.-P.; Aschwanden, W. Helv. Chim. Acta 1970, 53, 991.
3. Robinson, F. A. "The Vitamin Co-factors of Enzyme Systems", Pergamon Press: London, 1966, p. 415; Paust, J.; Pfohl, S.; Reif, W.; Schmidt, W. Ann. Chem. 1978, 1024.
4. Reichstein T.; Grüssner, A. Helv. Chim. Acta 1934, 17, 311.
5. Koenig, K. E.; Sabacky, M. J.; Bachman, G. L.; Christopfel, W. C.; Barnstorff, H. D.; Friedman, R. B.; Knowles, W. S.; Stults, B. R.; Vineyard, B. D.; Weinkauff, D. J. Ann. N. Y. Acad. Sci. 1980, 333, 16 and references cited therein.
6. Perry, C.; Saucy, G. unpublished results.
7. Saucy, G.; Borer, R. Helv. Chim. Acta 1971, 54, 2517.
8. Cohen, N. Acc. Chem. Res. 1976, 9, 412 and references cited therein.
9. Rosenberger, M.; Borer, R.; Saucy, G. J. Org. Chem. 1978, 43, 1550.
10. Hajos, Z. G.; Parrish, D. R. J. Org. Chem. 1974, 39, 1615.
11. Eder, U.; Sauer, G.; Wiechert, R. Angew. Chem. Int. Ed. Engl. 1971, 10, 496.
12. Buchschacher, P.; Cassal, J.-M.; Fürst, A.; Meier, W. Helv. Chim. Acta 1977, 60, 2747.
13. Brown, K. L.; Damm, L.; Dunitz, J. D.; Eschenmoser, A.; Hobi, R.; Kratky, C. Helv. Chim. Acta 1978, 61, 3108.
14. Eder, U. J. Steroid Biochem. 1979, 11, 55 and references cited therein.

15. Neef, G.; Eder, U.; Haffer, G., Sauer, G.; Wiechert, R. Chem. Ber. 1977, 110, 3377.
16. Eder, U.; Cleve, G.; Haffer, G.; Neef, G.; Sauer, G.; Wiechert, R.; Fürst, A.; Meier, W. Chem. Ber. 1980, 113, 2249.
17. Pandit, U. K.; Bieräugel, H. Rec. Trav. Chim. 1976, 95, 223.
18. Kametani, T.; Matsumoto, H.; Nemoto, H.; Fukumoto, K. J. Am. Chem. Soc. 1978, 100, 6218.
19. Crabbé, P.; Bieber, L.; Nassim, B. J.C.S. Chem. Comm. 1980, 472.
20. Tsuji, J.; Shimizu, I.; Suzuki, H.; Naito, Y. J. Am. Chem. Soc. 1979, 101, 5070.
21. Shimizu, I.; Naito, Y.; Tsuji, J. Tetrahedron Lett. 1980, 21, 487.
22. Leuenberger, H. G. W.; Boguth, W.; Widmer, E.; Zell, R. Helv. Chim. Acta 1976, 59, 1832.
23. Rütimann, A.; Mayer H. Helv. Chim. Acta 1980, 63, 1456.
24. Partridge, J. J.; Chadha, N. K.; Uskokovic, M. R. J. Am. Chem. Soc. 1973, 95, 7171.
25. Saucy, G.; Weber G. Paper presented at the ACS/CSJ Chemical Congress, Honolulu, Hawaii, April 1-6, 1979; Abstr. ORGN 200.
26. Ojima, I.; Kogure, T.; Terasaki, T.; Achiwa, K. J. Org. Chem. 1978, 43, 3444.
27. Townsend,J; Valentine, D., Jr. unpublished results; Valentine, D., Jr.; Sun, R. C.; Toth, K. J. Org. Chem. 1980, 45, 3703.
28. Aoki, Y.; Suzuki, H.; Akiyama, H.; Okano, S. Chem. Abstr. 1974, 80, 95951z (U. S. Patent no. 3,876,656 - Sumitomo).
29. Confalone, P. N.; Pizzolato, G.; Baggiolini, E. G.; Lollar, D.; Uskokovic, M. R. J. Am. Chem. Soc. 1975, 97, 5936.
30. Vasilevskis, J.; Gualtieri, J. S.; Hutchings, S. D.; West, R. C.; Scott, J. W.; Parrish, D. R.; Bizzarro, F. T.; Field, G. F. J. Am. Chem. Soc. 1978, 100, 7423.
31. Rubel, T., "Vitamin E Manufacture", Noyes Development Corp.: Park Ridge, N. J. 1969.
32. Saucy, G.; Cohen, N. in "New Synthetic Methodology and Biologically Active Substances"; Yoshida, Z. Ed.; Kodansha (Tokyo); Elsevier (New York), 1981; Ch. 9, p. 155.
33. Cohen, N.; Lopresti, R. J.; Neukom, C.; Saucy, G. J. Org. Chem. 1980, 45, 582.
34. Valentine, D. Jr.; Johnson, K. K.; Priester, W.; Sun, R. C.; Toth, K; Saucy, G. J. Org. Chem. 1980, 45, 3698.
35. Takahashi, J.; Mori, K.; Matsui, M. Agric. Biol. Chem. 1979, 43, 1605.
36. Bödeker, C.; de Waard, E. R.; Huisman, H. O. Tetrahedron 1981, 37, 1233.
37. Sato, T.; Kawara, T.; Nishizawa, A.; Fujisawa, T. Tetrahedron Lett. 1980, 3377.

38. Fujisawa, T.; Sato, T.; Kawara, T.; Ohashi, K. Tetrahedron Lett. 1981, 22, 4823.
39. Trost, B. M.; Klun, T. J. Am. Chem. Soc. 1981, 103, 1864.
40. Trost, B. M.; Timko, J. M.; Stanton, J. L. J.C.S. Chem. Commun. 1978, 436.
41. Fischli, A. Chimia 1976, 30, 4.
42. Cohen, N.; Lopresti, R. J.; Neukom, C. J. Org. Chem. 1981, 46, 2445.
43. Cohen, N.; Banner, B. L.; Neukom, C. Synthetic Commun., in press.
44. Cohen, N.; Scott, C. G.; Neukom, C.; Lopresti, R. J.; Weber, G.; Saucy, G. Helv. Chim. Acta 1981, 64, 1158.
45. Slover, H. T.; Thompson, R. H., Jr. Lipids 1981, 16, 268.

RECEIVED December 28, 1981.

11

Stereochemistry of Heterogeneous Asymmetric Catalytic Hydrogenation

KAORU HARADA

University of Tsukuba, Department of Chemistry, Ibaraki 305, Japan

In this paper, the stereochemistry of heterogeneous catalytic hydrogenation of >C=N- and >C=O double bonds of the derivatives of α-keto acids, keto alcohols and diketones is described. The steric course could be explained by the chelation hypothesis.

In 1961, Hiskey et al.(1) reported the successful asymmetric syntheses of α-amino acids. They demonstrated the synthesis of amino acids in 45-70% enantiomeric purity by catalytic hydrogenation of the Schiff bases prepared from α-keto acids and optically active α-methylbenzylamine followed by hydrogenolysis (Scheme 1). When (S)-amine was used, (S)-α-amino acid resulted. This is a highly stereoselective reaction. However, the authors did not discuss the steric course of the asymmetric hydrogenation process.

Scheme 1

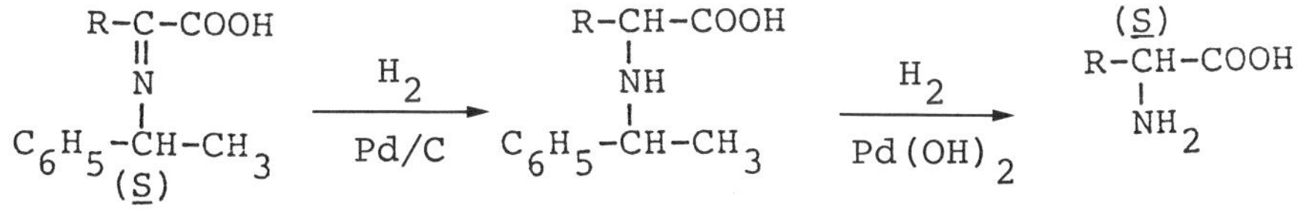

$$\underset{(S)}{\mathrm{C_6H_5\text{-}CH\text{-}CH_3}}\ \text{(N=C(R)-COOH on CH)} \xrightarrow[\mathrm{Pd/C}]{\mathrm{H_2}} \mathrm{R\text{-}CH(COOH)\text{-}NH\text{-}CH(CH_3)\text{-}C_6H_5} \xrightarrow[\mathrm{Pd(OH)_2}]{\mathrm{H_2}} \underset{(S)}{\mathrm{R\text{-}CH(NH_2)\text{-}COOH}}$$

Later Mitsui et al.(2) reported the asymmetric syntheses of phenylglycine by the Hiskey type reaction and proposed a steric course for the asymmetric synthesis as shown in Scheme 2. If it is applicable to all of the Hiskey type reactions, the following may be expected: (a) an increase in optical yield upon substitution of α-methylbenzylamine by α-ethylbenzylamine and (b) a comparable optical yield upon substitution of α-methylbenzylamine by α-(1-naphthyl)ethylamine.

0097-6156/82/0185-0169$05.00/0

Scheme 2

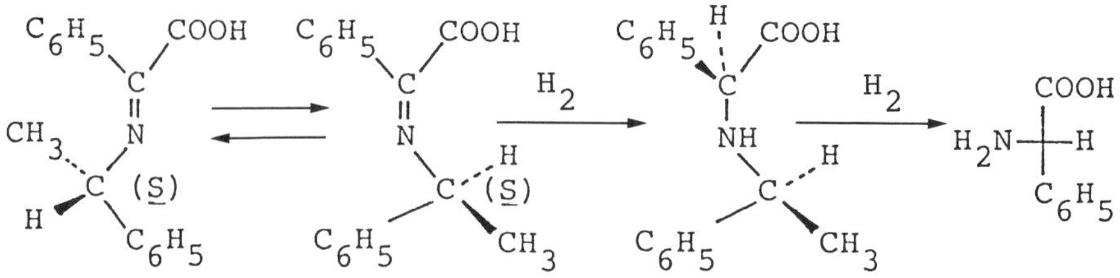

[Scheme 2: imine of C_6H_5-CO-COOH with (S)-α-methylbenzylamine ⇌ rotamer $\xrightarrow{H_2}$ secondary amine $\xrightarrow{H_2}$ H_2N-CH(C_6H_5)-COOH]

In order to examine the steric course proposed by Mitsui et al., we have performed asymmetric syntheses of alanine, α-aminobutyric acid, phenylglycine, phenylalanine and glutamic acid from the corresponding α-keto acids using (S)-α-methylbenzylamine [Me(-)], (S)-α-ethylbenzylamine [Et(-)] and (R)-α-(1-naphthyl)ethylamine [Naph(-)] as the chiral adjuvant. The results are shown in Table I(3,4).

The results indicate that a) the optical purity of amino acids obtained with α-methylbenzylamine is always higher than when α-ethylbenzylamine is used, b) the optical purity of the amino acids decreases steadily as the bulk of the alkyl group of the α-keto acids increases, and c) the optical purity increases when (R)-α-(1-naphthyl)ethylamine is used(4).

These findings show clearly that the steric course proposed previously does not explain any of the experimental results. Based on molecular models we proposed a different steric course consistent with the experiments. Structure I (Scheme 3) represents a conformation of the substrate which satisfies all of the

Table I Asymmetric Synthesis of Amino Acids(3,4)

α-Keto acid R-CO-COOH	Optically active amine	Amino acid	Optical purity(%)
R= CH_3	Me (-)	(S) Alanine	67
	Et (-)	(S) Alanine	52
	Naph(+)	(R) Alanine	83
C_2H_5	Me (-)	(S) Butyrine	44
	Et (-)	(S) Butyrine	33
C_6H_5	Me (-)	(S) Phenylglycine	30
	Et (-)	(S) Phenylglycine	24
CH_2-C_6H_5	Me (-)	(S) Phenylalanine	14
	Et (-)	(S) Phenylalanine	10
$(CH_2)_2COOH$	Me (-)	(S) Glutamic acid	12
	Et (-)	(S) Glutamic acid	6

Solvent:EtOH

Scheme 3

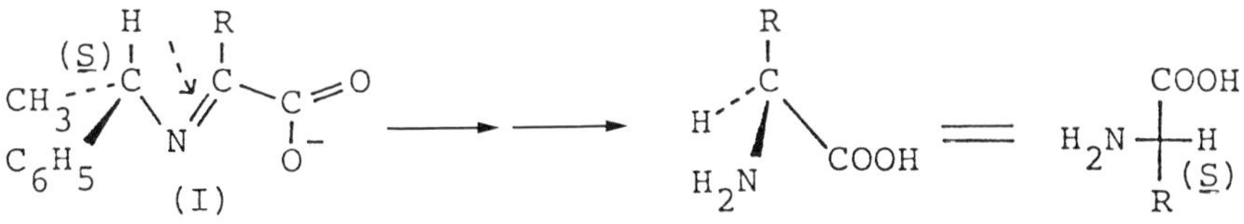

(I)

Scheme 4

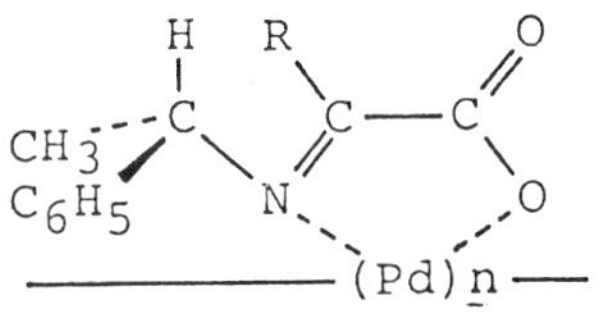

(II)

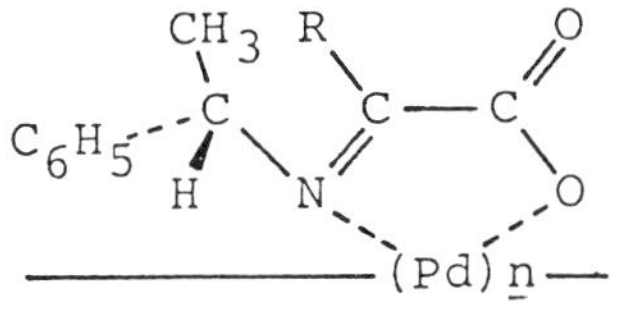

(III)

conditions required by the experimental findings(3). The structure I might be considered to form a substrate-catalyst complex as shown in structure II (Scheme 4). A molecular model of structure II fits very well on the surface of the palladium catalyst. The plane comprising the Schiff base of the α-keto acid is assumed to be perpendicular to the palladium surface, with the phenyl group lying on the palladium surface as shown in structure II. If the phenyl group was placed as shown in structure III (Scheme 4), the alkyl group of the asymmetric moiety and that of the keto acid would interfere with each other, and the structure III would be unstable. Thus we assume (A) the substrate initially interacts with the catalyst, to form a substrate-catalyst complex as shown in structure II before the catalytic hydrogenation takes place, and then (B) the structure II is adsorbed on the catalyst from the less bulky side of the molecule, and catalytic hydrogenation takes place. We have called this hydrogenation process "the chelation hypothesis"(3), and further studies were undertaken to test this hypothesis.

Table II shows solvent effects in the asymmetric synthesis of alanine from pyruvic acid and (*S*)-α-methylbenzylamine(4). The optical purity of alanine decreases with increasing polarity of the solvent. In the case of the asymmetric synthesis of glutamic acid from α-keto-glutaric acid and (*S*)-α-methylbenzylamine, the configuration of the resulting glutamic acid was actually inverted by the use of polar solvents. The substrate appears to interact with the catalyst more strongly in a less polar than in a more polar solvent. Thus, the population of the chelated substrate is

Table II Solvent Effect in the Asymmetric Synthesis of Alanine

Solvent	Yield(%)	Optical purity(%)
Hexane	75	75
AcOEt	49	60
DMFA	47	50
i-PrOH	56	46
MeOH	61	38
MeOH:H_2O(1:2)	75	35
MeOH:H_2O(1:4)	76	29

larger in the less polar solvent than in the more polar one.

The unchelated species, which has a stable conformation IV (Scheme 5) in solution, could be adsorbed on the less bulky front -C=N- face of the molecule (Scheme 5) resulting in an alanine derivative which has the configuration opposite to that obtained from the chelated species.

Asymmetric catalytic hydrogenation of the Schiff base prepared from ethyl pyruvate and an optically active amine in different solvents was carried out and supports the chelation hypothesis. Figure 1 shows solvent effects in the synthesis of alanine and α-aminobutyric acid(5). When (*S*)-benzylic amine was used as the asymmetric moiety, the optical purity of the resulting amino acid increased with a decrease in solvent polarity(6,7). A temperature effect was also observed, and the optical purity of the amino acids increased upon lowering the reaction temperature(8,9,10).

The chelation hypothesis could also be applied to the catalytic hydrogenation of α-keto acid amides carried out initially by Hiskey et al., who explained the steric course assuming intermediate structure V (11) (Scheme 6).

Scheme 5

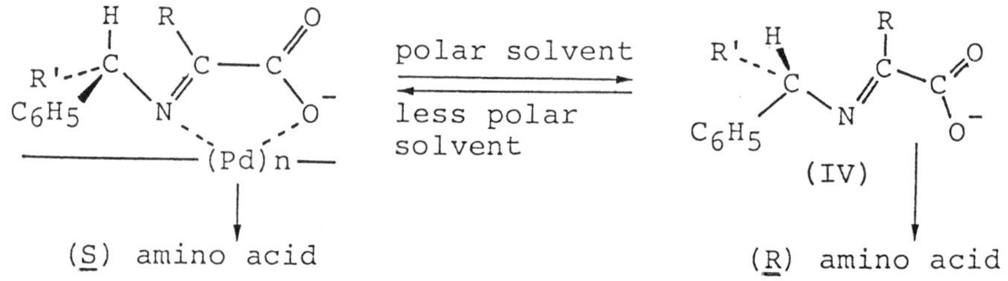

Figure 1 Solvent Effect in the Asymmetric Synthesis of Amino Acids from Ethyl Pyruvate(5)

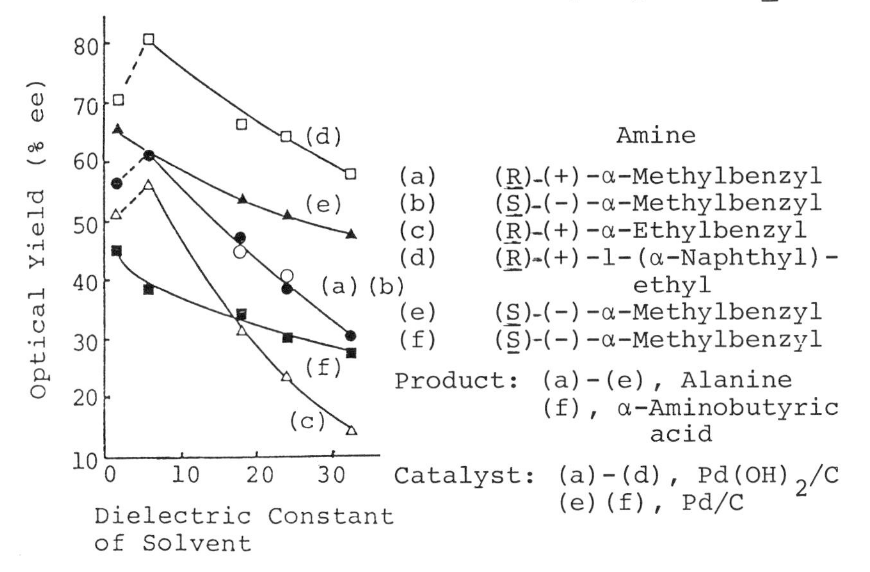

Scheme 6

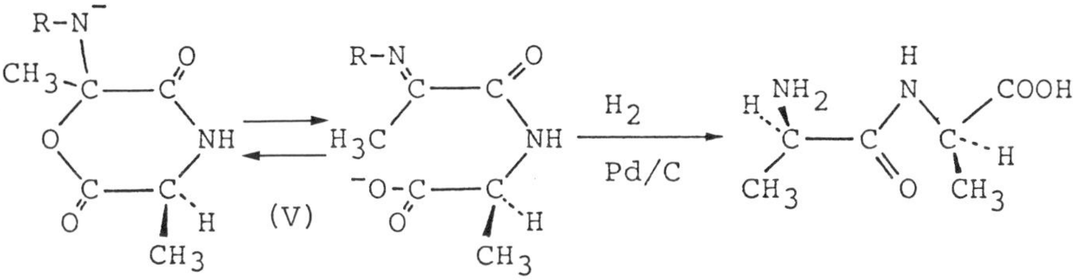

The change in configuration of the resulting amino acid with the use of different asymmetric moieties is also explained by the chelation hypothesis(12) (Scheme 7).

In the sequel, we will discuss a generalization of the chelation hypothesis as it applies to reactions other than hydrogenation of Schiff bases of α-keto acids with chiral amines. The catalytic hydrogenation of pyruvic acid amide resulted in the formation of lactamide in high optical purity (75-99% diastereomeric excess)(13). This might be explained by the chelate conformation of the substrate-catalyst complex shown in Scheme 8.

The catalytic hydrogenation of optically active benzoin oxime resulted in the stereoselective formation of optically active erythro diphenylethanolamine in

Scheme 7

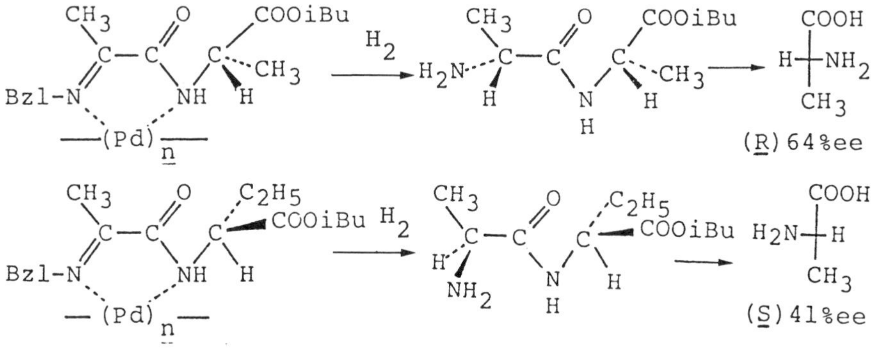

Scheme 8

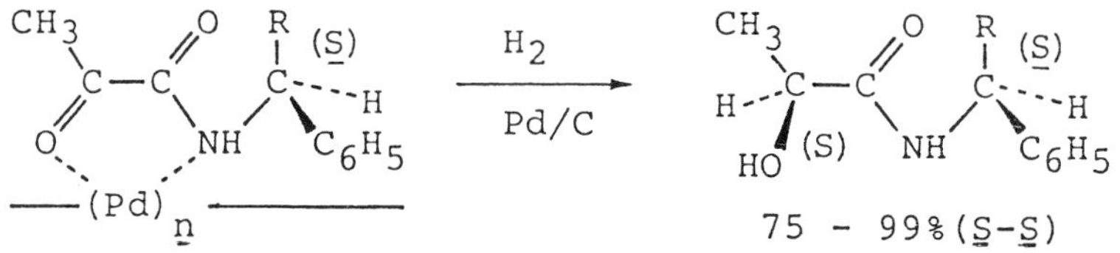

high optical and diastereomeric purity(14) (Scheme 9). Benzil monoxime was similarly hydrogenated using a palladium catalyst to form erythro diphenylethanolamine (15). If the conformation of the substrate in the reaction is planar, as expected from steric and electric considerations, the resulting hydrogenation product should be the threo isomer(16) (Scheme 10). In

Scheme 9

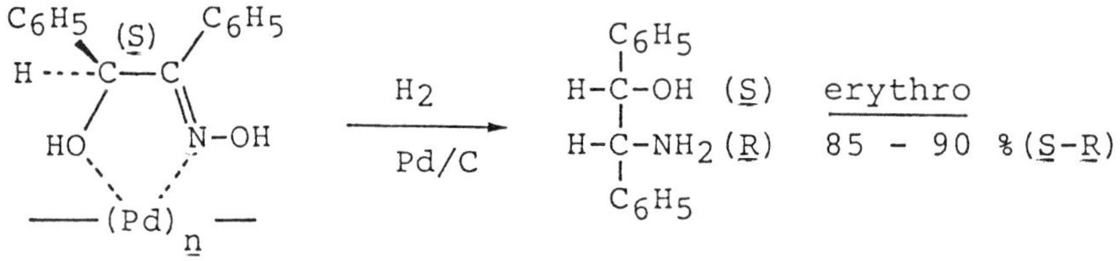

Scheme 10

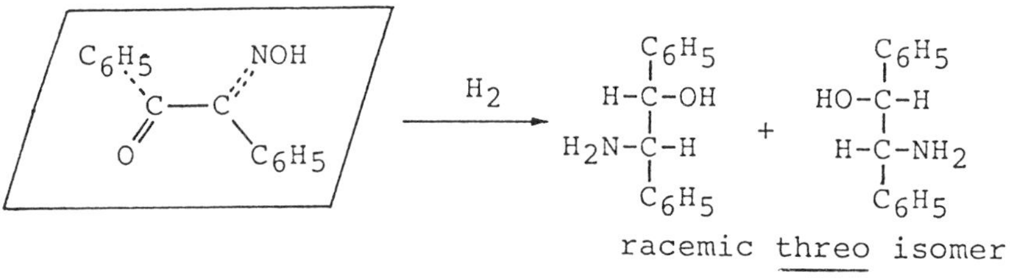

fact, however, the resulting diphenylethanolamine was found to be predominantly the *erythro* form. Similar results have been reported in the synthesis of threonine, phenylserine and ephedrine(17-21) (Scheme 11).

In 1953 Chang and Hartung proposed a mechanism for the hydrogenation of diketone monoxime which explains the formation of a single racemic modification (*erythro* form) by the formation of a rigid ring-like structure with the catalyst(22). The explanation could be regarded as a chelation hypothesis preceding the present generalized chelation hypothesis in catalytic hydrogenation.

Recently, support for the chelation hypothesis was obtained by examining the infrared dichroism of the substrate adsorbed on a metal surface using the high-sensitivity reflection method(23-26). In these investigations the orientation of the substrate on the metal surface is just as assumed by the chelation hypothesis. It was found that the OH, C=O, NH_3, =NOH groups interact with the metal surface and the substrates stand on the catalyst surface vertically. The observation of the infrared dichroism is considered to be a direct physical evidence for the chelation hypothesis (Scheme 12).

Scheme 11

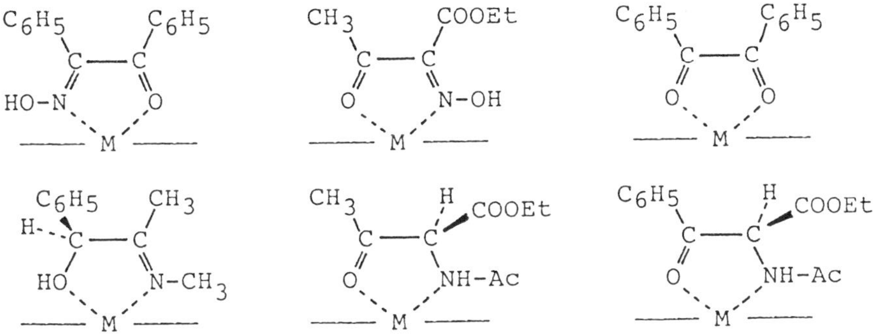

Scheme 12

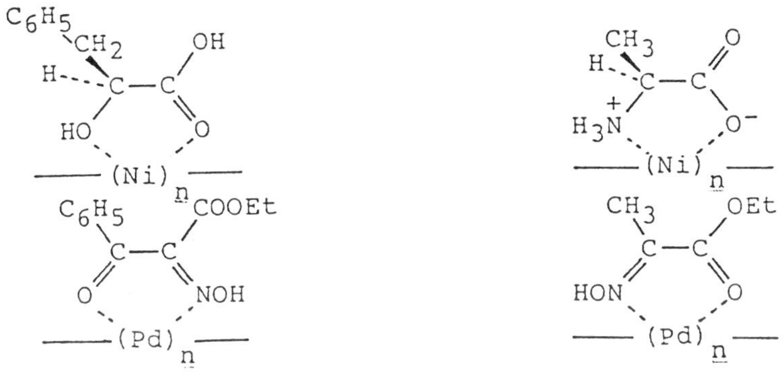

Literature Cited

1. Hiskey,R.G.; Northrop,R.C. J. Am. Chem. Soc. 1961, 83, 4798.
2. Kanai,A.; Mitsui,S. J. Chem. Soc. Jpn. (Pure Chem. Sec.) 1966, 89, 183.
3. Harada,K.; Matsumoto,K. J. Org. Chem. 1967, 32, 1794.
4. Harada,K.; Matsumoto,K. J. Org. Chem. 1968, 33, 4467.
5. Harada,K.; Yoshida,T. Bull. Chem. Soc. Jpn. 1970, 43, 921.
6. Harada,K.; Kataoka,Y. Tetrahedron Lett. 1978, 2103.
7. Harada,K.; Tamura,M. Bull. Chem. Soc. Jpn. 1979, 52, 1227.
8. Harada,K.; Yoshida,T. J. Chem. Soc. Chem. Commun. 1970, 1071.
9. Harada,K.; Yoshida,T. J. Org. Chem. 1972, 37, 4366.
10. Harada,K.; Kataoka,Y. Chem. Lett. 1978, 791.
11. Hiskey,R.G.; Northrop,R.C. J. Am. Chem. Soc. 1965, 87, 1753.
12. Harada,K.; Matsumoto,K. Bull. Chem. Soc. Jpn. 1971, 44, 1068.
13. Harada,K.; Munegumi,T.; Nomoto,S. Tetrahedron Lett. 1981, 22, 111.
14. Harada,K.; Shiono,S.; Nomoto,S. unpublished result.
15. Ishimaru,T. J. Chem. Soc. Jpn. (Pure Chem. Sec.) 1960, 81, 643.
16. Harada,K. In "The Chemistry of the Carbon-nitrogen Double Bond"; Patai,S., Ed.; Interscience Publishers: London, 1970; pp 255-298.
17. Albertson,N.F.; Tullar,B.F.; King,J.A. J. Am. Chem. Soc. 1948, 70, 1150.
18. Pfister,K.; Robinson,C.A.; Schabica,A.C.; Tishler, M. J. Am. Chem. Soc. 1948, 70, 2297.
19. Pfister,K.; Robinson,C.A.; Schabica,A.C.; Tishler, M. J. Am. Chem. Soc. 1949, 71, 1101.
20. Bolhofer,W.A. J. Am. Chem. Soc. 1952, 74, 5459.
21. Freudenberg,K.; Schöffel,E.; Braun,E. J. Am. Chem. Soc. 1932, 54, 234.
22. Chang,Y.; Hartung,W.H. J. Am. Chem. Soc. 1953, 75, 89.
23. Hatta,A.; Suëtaka,W. Bull. Chem. Soc. Jpn. 1975, 48, 2428.
24. Hatta,A.; Moriya,Y.; Suëtaka,W. Bull. Chem. Soc. Jpn. 1975, 48, 3441.
25. Osawa,M.; Hatta,A.; Harada,K.; Suëtaka,W. Bull. Chem. Soc. Jpn. 1976, 49, 1512.
26. Suëtaka,W.; Harada,K. unpublished result.

RECEIVED January 4, 1982.

12

Asymmetric Grignard Cross-Coupling Catalyzed by Chiral Phosphine–Nickel and Phosphine–Palladium Complexes

TAMIO HAYASHI

Kyoto University, Department of Synthetic Chemistry, Faculty of Engineering, Yoshida, Kyoto, Japan 606

A process of kinetic resolution in the coupling of Grignard reagents R*MgX (having a chiral center at the point of attachment to the metal) with various alkenyl halides under the influence of chiral phosphine-nickel or -palladium complexes is described. Enantiomeric excess of the coupling products depends strongly on the phosphine ligand and ranges up to 94% with e.e.'s in the 60-70% range common. Synthetic applications of the procedure are described.

We have prepared various kinds of optically active phosphines,[1,2] e.g., ferrocenylphosphines and β-aminoalkylphosphines, useful for several catalytic asymmetric reactions, viz., hydrogenation of olefins,[3] ketones,[4] and imines[5] catalyzed by rhodium complexes, hydrosilylation of ketones by rhodium complexes[6] and of olefins by a palladium complex,[7] as well as Grignard cross-coupling by nickel complexes.[2,8] Here, we describe the asymmetric cross-coupling of Grignard and organozinc reagents with organic halides catalyzed by chiral phosphine-nickel and -palladium complexes.

Phosphine-nickel and -palladium complexes have been used as catalysts for the reaction of Grignard reagents (RMgX) with vinyl or aryl halides (R'X') to produce, selectively, cross-coupling products (R-R'). The catalytic cycle of the reaction has been proposed to consist of a sequence of steps involving a diorganometal complex ($L_nM(R)R'$) as a key intermediate (Scheme I).

0097-6156/82/0185-0177$05.00/0

Scheme I

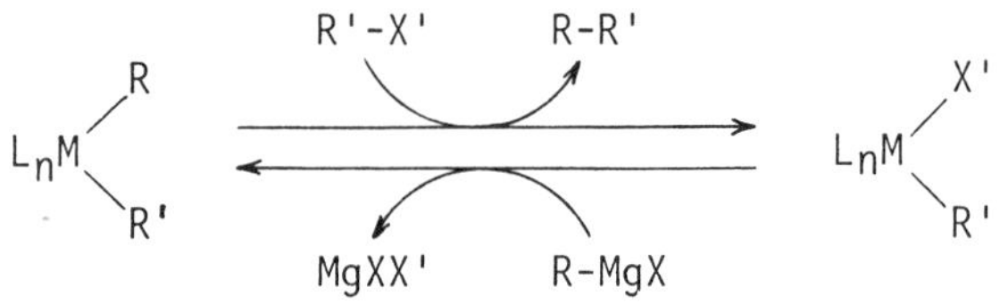

$$L_nM(R)(R') \xrightarrow[\text{R-MgX} \;\to\; \text{MgXX'}]{\text{R'-X'} \;\to\; \text{R-R'}} L_nM(X')(R')$$

Grignard reagents carrying the magnesium atom attached to a chiral carbon center ordinarily undergo racemization because of the stereochemical instability of the magnesium-carbon bond. Should the inversion at this chiral carbon be much faster than the cross-coupling reaction, however, kinetic resolution of the racemic Grignard reagent under the influence of chiral phosphine-metal complexes might occur, leading to a constant optical yield for the coupling product throughout the reaction (Scheme II).

Scheme II

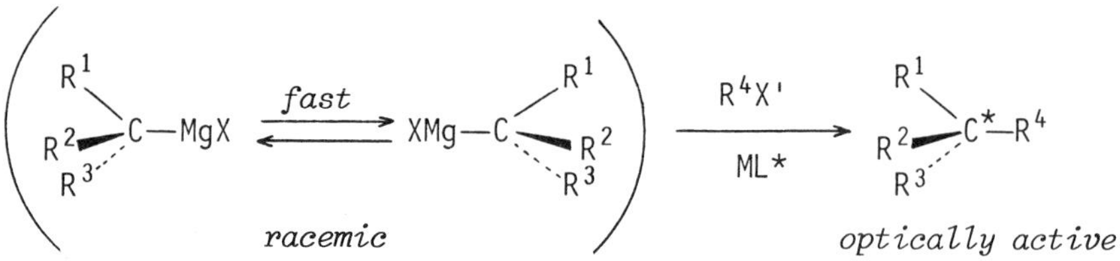

$$\left(R^1R^2R^3C\text{—MgX} \underset{}{\overset{fast}{\rightleftharpoons}} \text{XMg—}CR^1R^2R^3 \right) \xrightarrow[ML^*]{R^4X'} R^1R^2R^3C^*\text{—}R^4$$

racemic *optically active*

The reaction of 1-phenylethyl-, 2-octyl-, and 2-butylmagnesium chloride (**1a,b,c**) with vinyl bromide (**2a**), (*E*)-β-bromostyrene (**2b**), 2-bromopropene (**2c**), and bromobenzene (**2d**) was carried out in the presence of 0.5 mol% of a nickel catalyst prepared in situ from nickel chloride and a chiral ligand, or a chiral palladium-phosphine complex (eq. 1).

The chiral phosphines used are shown in Figure 1 and representative results are summarized in Table I. Among the ferrocenylphosphines, (*S*)-(*R*)-PPFA was one of the most effective ligands giving the coupling product, 3-phenyl-1-butene (**3a**), in up to 68% ee in the reaction of **1a** with **2a**. The ferrocene planar

Table I. Asymmetric Cross-Coupling of 1 with 2 Producing 3[a]

Entry No	Chiral catalyst	Product % ee	Entry No	Chiral catalyst	Product % ee
1	(*S*)-(*R*)-PPFA/$NiCl_2$	3a 63(*R*)	24	(*S*)-(*R*)-PPFA/$NiCl_2$	3b 52(*R*)
2	(*S*)-(*R*)-PPFA/$NiCl_2$[b]	3a 56(*R*)	25	$PdCl_2$[(*S*)-(*R*)-BPPFA]	3b 46(*R*)
3	(*S*)-(*R*)-PPFA/$NiCl_2$[c]	3a 66(*R*)	26	$PdCl_2$[(*S*)-(*R*)-BPPFA]	3c 5(*S*)
4	(*R*)-(*S*)-PPFA/$NiCl_2$[c]	3a 68(*S*)	27	(*S*)-(*R*)-PPFA/$NiCl_2$	3d 37(*S*)
5	$PdCl_2$[(*S*)-(*R*)-PPFA][d]	3a 61(*R*)	28	(*S*)-(*R*)-BPPFA/$NiCl_2$	3d 24(*S*)
6	(*R*)-(*R*)-PPFA/$NiCl_2$	3a 54(*R*)	29	(*R*)-(*S*)-PPFA/$NiCl_2$	3e 30(*R*)
7	(*S*)-FcPN/$NiCl_2$	3a 65(*S*)	30	$PdCl_2$[(*R*)-(*S*)-BPPFA]	3f 12(*R*)
8	(*R*)-PPEF/$NiCl_2$	3a 5(*S*)	31	$PdCl_2$[(*S*)-(*R*)-BPPFA]	3g 22(*R*)
9	(*S*)-(*R*)-4a/$NiCl_2$	3a 33(*R*)	32	(*S*)-Alaphos/$NiCl_2$	3a 38(*S*)
10	(*S*)-(*R*)-4b/$NiCl_2$	3a 65(*R*)	33	(*S*)-Leuphos/$NiCl_2$	3a 57(*S*)
11	(*S*)-(*R*)-4c/$NiCl_2$	3a 65(*R*)	34	(*S*)-Phephos/$NiCl_2$	3a 71(*S*)
12	(*S*)-(*R*)-4d/$NiCl_2$	3a 57(*R*)	35	(*R*)-PhGlyphos/$NiCl_2$	3a 70(*R*)
13	(*S*)-(*R*)-5a/$NiCl_2$	3a 35(*R*)	36	(*S*)-Valphos/$NiCl_2$	3a 81(*S*)
14	(*S*)-(*R*)-5b/$NiCl_2$	3a 7(*S*)	37	(*S*)-Ilephos/$NiCl_2$	3a 81(*S*)
15	(*S*)-(*R*)-5c/$NiCl_2$	3a 15(*S*)	38	(*R*)-ChGlyphos/$NiCl_2$	3a 77(*R*)
16	(*S*)-(*R*)-5d/$NiCl_2$	3a 62(*R*)	39	(*R*)-*t*-Leuphos/$NiCl_2$	3a 83(*R*)
17	(*S*)-(*R*)-5e/$NiCl_2$	3a 42(*S*)			(94)[e]
18	(*S*)-(*R*)-5f/$NiCl_2$	3a 17(*R*)	40	(*S*)-8/$NiCl_2$	3a 25(*R*)[e]
19	(*S*)-(*R*)-5g/$NiCl_2$	3a 65(*R*)	41	$NiCl_2$[(*S*)-prophos]	3a 0
20	(*S*)-(*R*)-6/$NiCl_2$	3a 57(*R*)	42	(*S*)-9/$NiCl_2$	3a 50(*S*)
21	(*S*)-(*R*)-BPPFA/$NiCl_2$	3a 65(*R*)	43	(*S*)-9/$NiCl_2$ (reused)	3a 48(*S*)
22	$PdCl_2$[(*S*)-(*R*)-BPPFA][d]	3a 61(*R*)	44	(*S*)-10/$NiCl_2$	3a 53(*S*)
23	(*R*)-(*S*)-7/$NiCl_2$	3a 17(*R*)			

[a] The reaction was carried out in ether at 0°C for 24 h unless otherwise noted. 1/2 = 2-4. [b] 1/2 = 1. [c] At -20°C. [d] At 25°C for 60 h. [e] Corrected for the optical purity of the phosphine ligand.

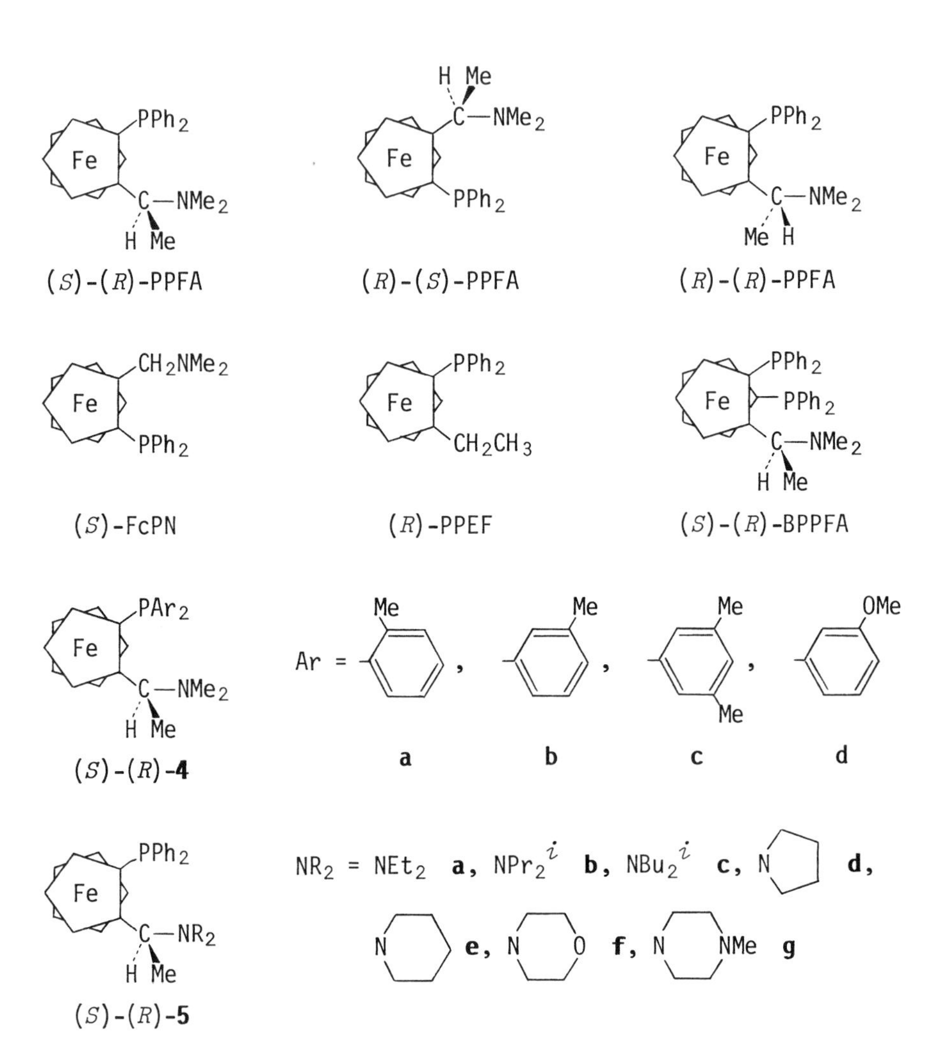

Figure 1. Chiral phosphine ligands and their palladium complexes. Continued on next page.

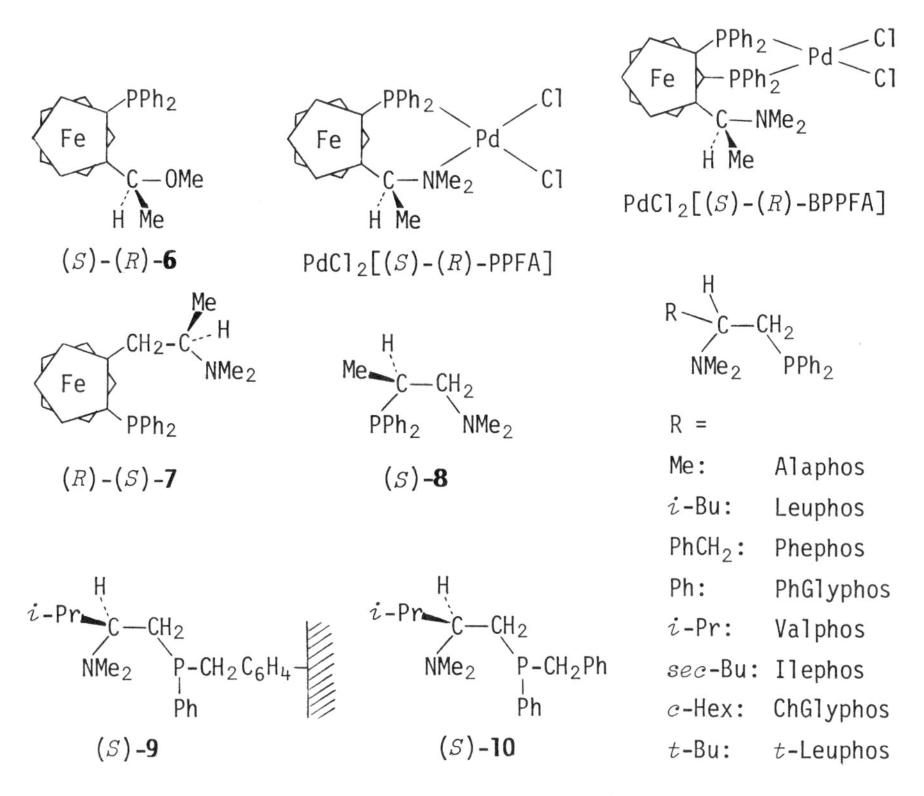

Figure 1 Continued. *Chiral phosphine ligands and their palladium complexes.*

$$\underset{\substack{| \\ \text{Me}}}{\text{R-CH-MgCl}} \quad + \quad \text{R'-Br} \xrightarrow[\text{Et}_2\text{O}]{\text{L*M}} \underset{\substack{| \\ \text{Me}}}{\text{R-}\overset{*}{\text{C}}\text{H-R'}} \qquad (1)$$

1a: R = Ph
1b: R = *n*-HeX
1c: R = Et

2a: R' = $CH{=}CH_2$
2b: R' = $\diagup\!\!\diagdown\!\!\diagup$Ph
2c: R' = $CMe{=}CH_2$
2d: R' = Ph

3a: R = Ph, R' = $CH{=}CH_2$
3b: R = Ph, R' = $\diagup\!\!\diagdown\!\!\diagup$Ph
3c: R = Ph, R' = $CMe{=}CH_2$
3d: R = *n*-HeX, R' = $CH{=}CH_2$
3e: R = Et, R' = $CH{=}CH_2$
3f: R = Et, R' = $CMe{=}CH_2$
3g: R = Et, R' = Ph

chirality was found to be more important than the carbon central chirality and the dimethylamino group also appears to be of primary importance for high stereoselectivity. Thus, (*R*)-(*R*)-PPFA and (*S*)-FcPN showed an asymmetric induction of comparable efficiency to the (*S*)-(*R*)- or (*R*)-(*S*)-PPFA ligand (Table I, entries 1∿7), while (*R*)-PPEF was almost ineffective for the cross-coupling (entry 8). The stereoselectivity was little changed by introduction of substituents onto the diphenylphosphino group of the ligand (Table I, entries 9∿12), but was strongly affected by changing the steric bulk of the secondary amino group on the ferrocenylphosphine side chain (entries 13∿19).

Some of the chiral β-dimethylaminoalkylphosphines derived from amino acids were more effective than the ferrocenylphosphines. (*S*)-Valphos, (*S*)-Ilephos, and (*R*)-*t*-Leuphos gave the coupling product **3a** with over 80% ee. It is clear from the results in Table I that the presence of the dimethylamino group is, again, important for high stereoselectivity in the reaction with the β-aminoalkylphosphine-nickel catalyst. We propose a mechanism involving coordination of the amino group to the magnesium atom as shown in Scheme III.

Scheme III

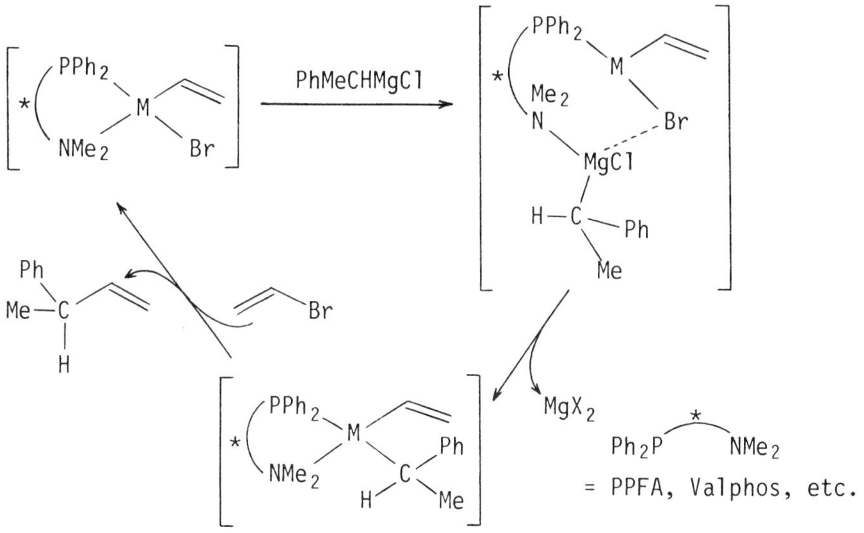
PPh2
NMe2
M
Br
PhMeCHMgCl
PPh2
Me2
N
M
Br
MgCl
H—C—Ph
Me
Ph
Me—C
H
Br
PPh2
NMe2
M
Ph
C
H
Me
MgX2
Ph2P NMe2
= PPFA, Valphos, etc.

The nickel-catalyzed asymmetric cross-coupling between *sec*-alkyl Grignard reagents with vinyl bromide finds many applications in the synthesis of optically and biologically active substances, e.g. α-curcumene (eq. 2) and 2-arylpropionic acids, (anti-inflammatory drugs) (eq. 3).[8b]

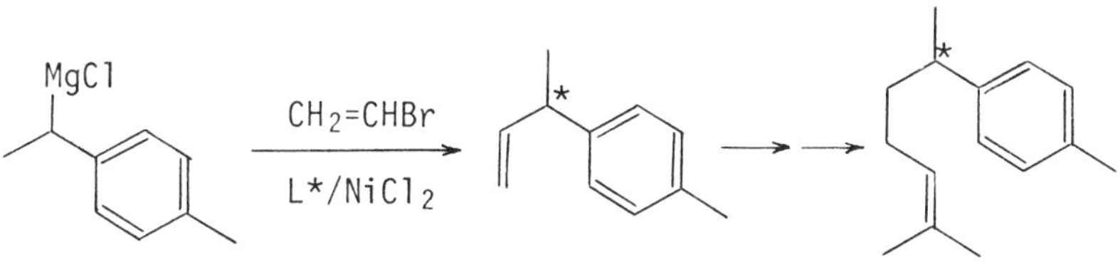
MgCl
CH_2=CHBr
L*/$NiCl_2$

(2)

L* = (*S*)-(*R*)-PPFA: 66% ee (*R*)

L* = (*S*)-Valphos: 83% ee (*S*)

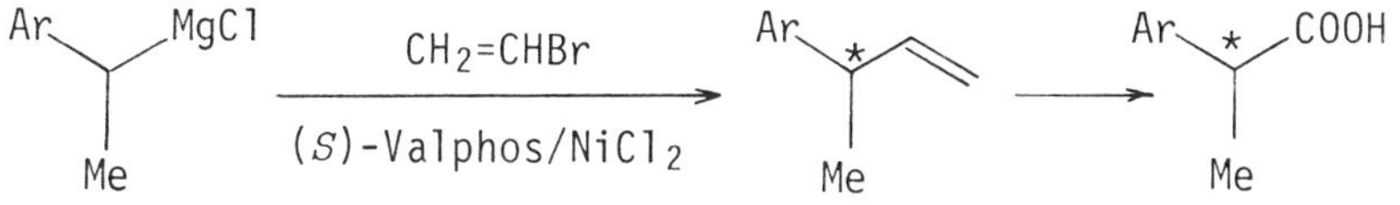
Ar
MgCl
Me
CH_2=CHBr
(*S*)-Valphos/$NiCl_2$
Ar
Me
Ar
COOH
Me

(3)

Ar = *p*-*i*-Bu-C_6H_4: 81% ee (*R*)

Ar = *p*-Ph-C_6H_4: 82% ee (*R*)

The asymmetric cross-coupling of organozinc reagents effected with palladium catalysts in THF was found to proceed with higher stereoselectivity than that of Grignard reagent (eq. 4).[9]

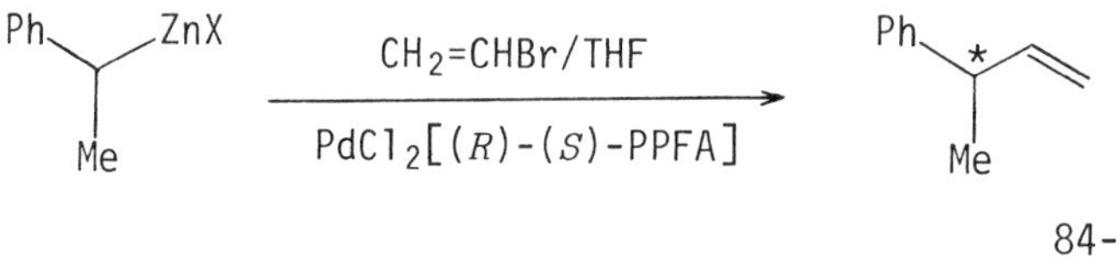

$$\mathrm{PhCH(Me)ZnX} \xrightarrow[\mathrm{PdCl_2[(\mathit{R})\text{-}(\mathit{S})\text{-}PPFA]}]{\mathrm{CH_2{=}CHBr/THF}} \mathrm{PhC^{*}H(Me)CH{=}CH_2} \quad (4)$$

84-86% ee (*S*)

Grignard reagents which do not undergo racemization were kinetically resolved by asymmetric cross-coupling in the presence of (*S*)-Valphos-nickel catalyst. Thus, the reaction of 2-norbornylmagnesium chloride gave olefins of 37% ee (1*S*,2*S*,4*R*), (1*R*,4*S*)-Grignard reagent being recovered (eq. 5).[10]

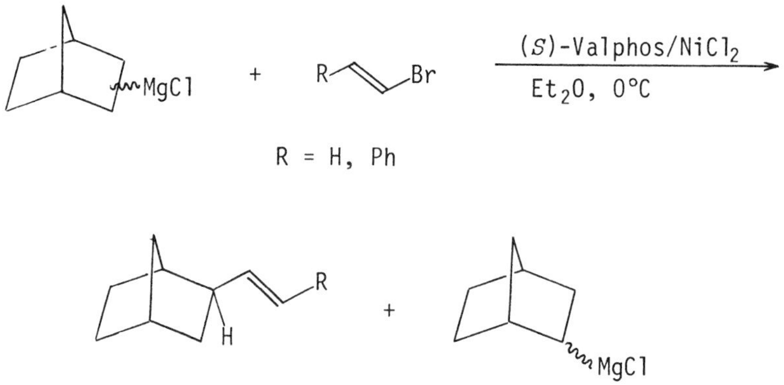

$$\text{2-norbornyl-MgCl} + \mathrm{RCH{=}CHBr} \xrightarrow[\mathrm{Et_2O,\ 0^{\circ}C}]{(\mathit{S})\text{-Valphos/NiCl_2}}$$

R = H, Ph

$$\text{2-norbornyl-CH{=}CHR} + \text{2-norbornyl-MgCl} \quad (5)$$

Cross-coupling of α-trimethylsilylbenzylmagnesium bromide with alkenyl bromides catalyzed by the PPFA-palladium complex gave optically active allylsilanes **11** (eq. 6). Allylsilane **11a** reacted enantioselectively with the prochiral carbonyl compounds **12** in the presence of $TiCl_4$ to produce alcohols **13** of over 90% enantiomeric purity (eq. 7).[11]

$$\text{PhCH(SiMe}_3\text{)MgBr} \xrightarrow[\text{PdCl}_2[(R)\text{-}(S)\text{-PPFA}]]{\text{Br}\diagup\!\!\diagdown\!\!\diagup\text{R}} (R)\text{-}\mathbf{11} \quad (6)$$

$$(R)\text{-}\mathbf{11a} + \underset{\mathbf{12}}{R^1COR^2} \xrightarrow{TiCl_4} \mathbf{13} \quad (7)$$

R^1 = *i*-Pr, R^2 = H: ∿90% ee (*R*)
R^1 = *t*-Bu, R^2 = H: ∿91% ee (*R*)
R^1 = Ph, R^2 = COOMe: ∿92% ee

Literature Cited

1. Hayashi, T.; Mise, T.; Fukushima, M.; Kagotani, M.; Nagashima, N.; Hamada, Y.; Matsumoto, A.; Kawakami, S.; Konishi, M.; Yamamoto, K.; Kumada, M. Bull. Chem. Soc. Jpn. 1980, 53, 1138.
2. (a) Hayashi, T.; Fukushima, M.; Konishi, M.; Kumada, M. Tetrahedron Lett. 1980, 21, 79. (b) Tamao, K.; Yamamoto, H.; Matsumoto, H.; Miyake, N.; Hayashi, T.; Kumada, M. Tetrahedron Lett. 1977, 1389. (c) Hayashi, T.; Nagashima, N.; Kumada, M. Tetrahedron Lett. 1980, 21, 4623. (d) Hayashi, T.; Konishi, M.; Hioki, T.; Kumada, M.; Ratajczak, A.; Niedbała, H. Bull. Chem. Soc. Jpn. in press.
3. Hayashi, T.; Mise, T.; Mitachi, S.; Yamamoto, K.; Kumada, M. Tetrahedron Lett. 1976, 1133.
4. (a) Hayashi, T.; Mise, T.; Kumada, M. Tetrahedron Lett. 1976, 4351. (b) Hayashi, T.; Katsumura, A.; Konishi, M.; Kumada, M. Tetrahedron Lett. 1979, 425.

5. Hayashi, T.; Kumada, M. in "Fundamental Research in Homogeneous Catalysis"; Ishii, Y.; Tsutsui, M., Eds.; Plenum: New York, Vol. 2, 1978; p 159.
6. Hayashi, T.; Yamamoto, K.; Kumada, M. Tetrahedron Lett. 1974, 4405.
7. Hayashi, T.; Tamao, K.; Katsuro, Y.; Nakae, I.; Kumada, M. Tetrahedron Lett. 1980, 21, 1871.
8. (a) Hayashi, T.; Tajika, M.; Tamao, K.; Kumada, M. J. Am. Chem. Soc. 1976, 98, 3718. (b) Tamao, K.; Hayashi, T.; Matsumoto, H.; Yamamoto, H.; Kumada, M. Tetrahedron Lett. 1979, 2155.
9. Hayashi, T.; Hagihara, T.; Katsuro, Y.; Kumada, M. J. Organometal. Chem. to be submitted.
10. Hayashi, T.; Kanehira, K.; Hioki, T.; Kumada, M. Tetrahedron Lett. 1981, 22, 137.
11. Hayashi, T.; Konishi, M.; Ito, H.; Kumada, M. unpublished results.

RECEIVED December 14, 1981.

13

Rhodium(I) Catalyzed Enantioselective Hydrogen Migration of Prochiral Allylamines

K. TANI, T. YAMAGATA, and S. OTSUKA—Osaka University, Department of Chemistry, Faculty of Engineering Science, Toyonaka, Osaka, Japan 560

S. AKUTAGAWA, H. KUMOBAYASHI, and T. TAKETOMI—Takasago Perfumery Co. Ltd., Central Research Laboratory, 31-36, 5-chome, Kamata, Ohta-ku, Tokyo, Japan 144

H. TAKAYA and A. MIYASHITA—Institute for Molecular Science, Chemical Materials Center, Okazaki, Japan 444

R. NOYORI—Nagoya University, Department of Chemistry, Chikusa, Nagoya, Japan 464

Migration of a multi-substituted inner double bond to a less substituted terminal one occurs when the latter gains relative stability, as seen in functionalized allylic systems(eq. 1).[1,2]

$$\begin{matrix} R^1 \\ R^2 \end{matrix}\!\!>\!C\text{=}CH\text{-}CH_2X \longrightarrow \begin{matrix} R^1 \\ R^2 \end{matrix}\!\!>\!CH\text{-}CH\text{=}CHX \qquad (1)$$

Various complexes of Fe[2,3], Ru[2-4], Co[1,2], Rh[2-4], Ir[5], Ni[2,6], and Pt[7] have been proposed as catalysts for such allylic migrations. All of them, of course, lead to racemic product. In a recent

0097-6156/82/0185-0187$05.00/0

communication[1] we have shown that enantioselective hydrogen migration in prochiral allylamines leading to optically active enamines is possible with chiral cobalt catalysts. Both the optical yield and catalytic activity, however, were too low to be of practical use. We have now found better catalysts, namely rhodium(I) chiral diphosphine complexes. In this report we briefly summarize the results of our stereochemical and kinetic studies of such rhodium-catalyzed stereospecific allylamine isomerizations.

As representative Z and E allylamines we chose diethylnerylamine (1)[8] and its geranyl analog (2).[9] The catalyst was a cationic Rh(I) compound, [Rh(diphos)(diene)]ClO_4, prepared from [Rh-(diene)Cl]$_2$ (diene=1,5-cyclooctadiene or norbornadiene). For the chiral diphos ligand we employed (2R, 3R)-DIOP[10], a tetracyclohexyl analog of (2R, 3R)-DIOP [(R)-CyDIOP][11], and (R)-(+)-2,2'-Bis(diphenylphosphino)-1,1'-binaphthyl [(R)-BINAP].[12]

Typically, a THF solution (10 ml) containing the substrate (5 mmol) and a catalytic amount (0.05 mmol) of [Rh(diphos)(COD)]-ClO_4 was stirred under a nitrogen atmosphere at 60° for 17-24h. The isomerization product from 1 and 2 was (3R)-(3) and (3S)-citronellal-*trans*-enamine (4), respectively (eq. 2,3). The chemical yields are virtually quantitative, with 94-96% optical yields. A secondary amine, cyclohexylgeranylamine (5) gave the

(2)

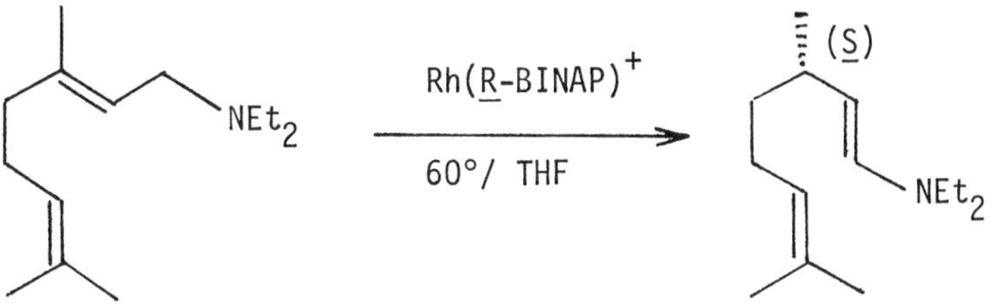

Rh(R-BINAP)⁺ 60°/ THF (3)

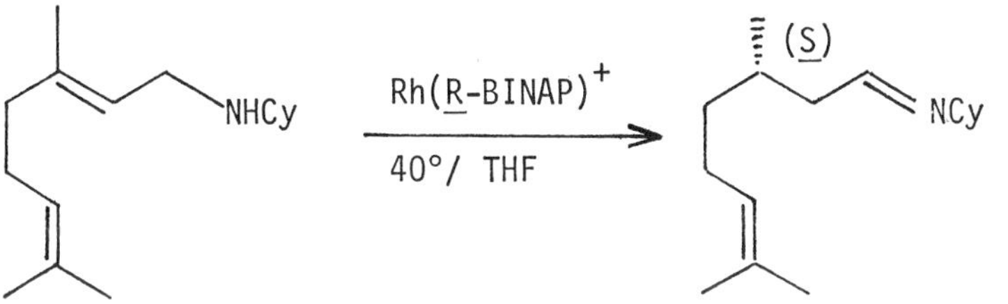

2 → 4 (S) NEt_2 96% ee

Rh(R-BINAP)⁺ 40°/ THF (4)

5 NHCy → 6 (S) NCy 96% ee

corresponding imine (6) in 96% enantiomeric excess (eq. 4). Optical yields were assessed by transforming the hydrolyzed product (citronellal) via isopulegol into menthol which serves as a reliable reference substance for determining optical purity by specific rotation.

A Rh(I) diphos complex with (S)-BINAP was also prepared to examine the stereochemistry of the isomerization product. The E-allylamine (2) was isomerized (60°, 17h, THF) to (3R)-citronellal-trans-enamine (3) with 96% ee. The stereochemical correlation between the configurations of the Rh(I) ligand and product can now be summarized as shown below.

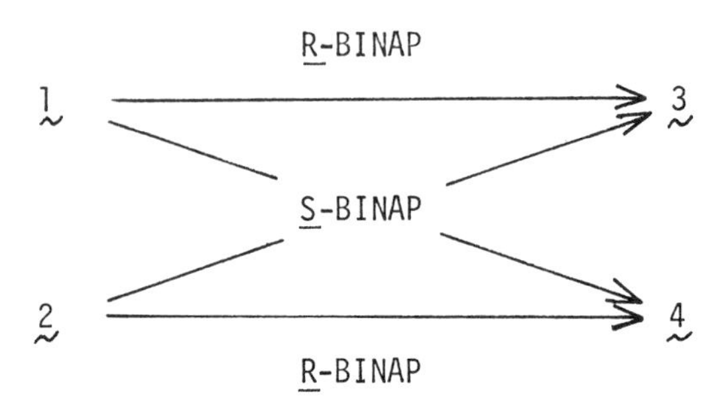

A similar correlation has already been observed for the cobalt catalyst.[1]

Catalysts containing (R)-DIOP and (R)-CyDIOP as the chiral ligand gave lower optical yields. Thus, isomerization of 1 with $[Rh(R\text{-DIOP})(COD)]^+$ and $[Rh(R\text{-CyDIOP})(COD)]^+$ gave 4 in 27% and 68% ee, respectively. The optical yield for the isomerization of 5 with $[Rh(R\text{-CyDIOP})(COD)]^+$ was drastically decreased to only 11%. The reaction rate depends on the nature of the diphos ligand; qualitatively the rate increases in the order of CyDIOP < DIOP < BINAP.

Dimethyl[(E)-3-phenyl-2-butenyl]amine (7) is a slow reacting substrate, presumably because of its styrene type conjugation. Its isomerization, effected with $[Rh(R\text{-BINAP})(COD)]^+$ (60°, 23h, THF), gave the (3R)-trans-enamine (8) in 83% yield with 90% ee (eq. 5). A competitive isomerization of a 9:1 mixture of 7 and its Z-isomer indicated a faster rate for E than for Z.

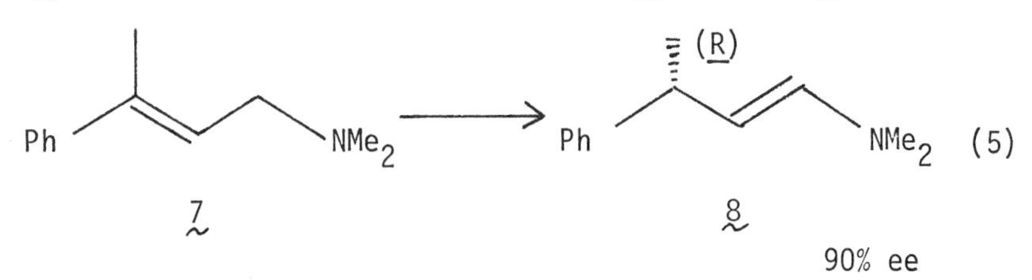

The rate of $[Rh(\underline{rac}\text{-BINAP})(COD)]^+$-catalyzed isomerization of **1** (THF, 40-70°C) was measured by monitoring the ^1H NMR signals of the methylene protons of the N-ethyl substituents in **1** and **3**. The substrate concentration range was 0.05-0.55 mol/l, and that of the Rh(I) catalyst precursor 0.8×10^{-3} - 4.0×10^{-3} mol/l. The reaction follows the following rate law, at least for the initial stage ($\lesssim$ 17% conversion):

$$R = k_{obs} [Rh(I)][Substrate]$$

Compound **2** was isomerized a little faster than **1** by a factor of 1.1. tert-Amines and chelating ligands such as COD retard the rate. 1,3-Dienamines strongly inhibit the reaction.

In order to obtain further mechanistic information, dimethyl[(E)-3-phenyl-2-butenyl]amine-2,2-d_2 was prepared via (E)-3-methylcinnamyl alcohol-1,1-d_2 obtained by $LiAlD_4$ reduction of ethyl (E)-3-methylcinnamate. The $[Rh(\underline{R}\text{-BINAP})(COD)]^+$-catalyzed isomerization (60°, 23h, THF) gave exclusively 1,3-dideuterio-trans-enamine (95% yield), as established unambiguously by ^1H NMR spectra. This conclusively shows that stereospecific 1,3-hydrogen atom migration has occurred and represents the first established example of a transition metal assisted 1,3-suprafacial hydrogen migration for linear allylic systems. A stereoselective 1,3-hydrogen migration has been reported for a cyclic allyl alcohol, namely, endo-β-1-hydroxydihydrocyclopentadiene (**9**) was isomerized by $Fe(CO)_5$ (10 mol%, 130°) to the ketone (**10**), while the isomerization was not observed for the α-isomer (**11**).[13]

Any mechanistic proposal should accommodate the results, (1) 100% 1,3-hydrogen migration and (2) 100% trans-enamine formation. We postulate an η-allyl-hydride Rh(III) complex as the reactive intermediate as it accommodates these features and is consistent with other observations, viz., (1) faster rate for E- than for

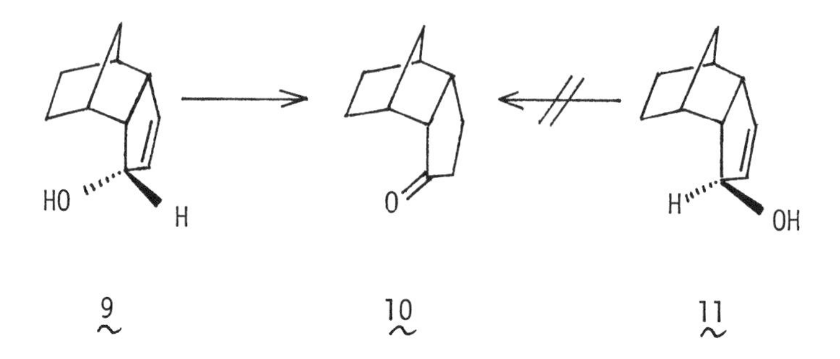

Z-allylamime, (2) inhibition by free amines, diolefins, and conjugate dienamines, and (3) stereochemical correlation between the substrate geometry and product configuration.

The present rhodium catalyzed isomerization reaction provides a convenient access to chiral terpenoid enamines and aldehydes. Its synthetic utility toward such natural products as l-menthol, vitamin E, pheromones etc. is obvious.

Literature Cited

1. Kumobayashi, H.; Akutagawa, S.; Otsuka, S. J. Am. Chem. Soc. 1978, 100, 3949-3950.
2. For a review see: Parshall, G. W. "Homogeneous Catalysis"; Wiley: New York, 1980; Chapter 3 and 4.
3. Stille, J. K.; Becker, Y. J. Org. Chem. 1980, 45, 2139-2145.
4. Suzuki, H.; Koyama, Y.; Moro-oka, Y.; Ikawa, T. Tetrahedron Lett. 1979, 1415-1418.
5. Baudry, D.; Michel, E.; Felkin, H. Nouv. J. Chim. 1978, 2, 355-356.
6. Overman, L. E.; Knoll, F. M. Tetrahedron Lett. 1979, 321-324.

7. Aresta, M.; Greco, R. Synth. React. Inorg. Met. Org. Chem. 1979, 9, 377-390; Chem. Abstr. 1979, 91, 140973y.
8. Takabe, K.; Katagiri, T.; Tanaka, J. Tetrahedron Lett. 1972, 4009-4012.
9. Fujita, T.; Suga, K.; Watanabe, S. Aust. J. Chem. 1974, 531-532.
10. Kagan, H. B.; Dang, T. P. J. Am. Chem. Soc. 1972, 94, 6429-6433.
11. Tani, K.; Suwa, K.; Otsuka, S. this symposium.
12. Miyashita, A.; Yasuda, A.; Takaya, H.; Toriumi, K.; Ito, T.; Souch, T.; Noyori, R. J. Am. Chem. Soc. 1980, 102, 7932-7934.
13. Cowherd, F. G.; von Rosenberg, J. L. J. Am. Chem. Soc. 1969, 91, 2157-2158.

RECEIVED December 14, 1981.

14

Application of Immobilized Enzymes for Asymmetric Reactions

ICHIRO CHIBATA

Tanabe Seiyaku Co., Ltd., Research Laboratory of Applied Biochemistry, 16–89, Kashima-3-chome, Yodogawa-ku, Osaka, Japan

Three immobilized enzyme or microbial cell systems currently used industrially in synthesis of chiral amino acids plus one presently under development are described. L-amino acids are produced by enzymatic hydrolysis of DL-acylamino acid with aminoacylase immobilized by ionic binding to DEAE-Sephadex. *Escherichia coli* cells immobilized by κ-carrageenan crosslinked with glutaraldehyde and hexamethylenediamine are used to convert fumaric acid and ammonia to L-aspartic acid and *Brevibacterium flavum* cells similarly immobilized are used to hydrate fumaric acid to L-malic acid. The decarboxylation of L-aspartic acid by immobilized *Pseudomonas dacunhae* to L-alanine is currently under investigation.

Enzymes are biological catalysts and participate in many chemical reactions occurring in living things. Unlike ordinary chemical catalysts, enzymes have the ability to catalyze a reaction under very mild conditions in neutral aqueous solution at normal temperature and pressure, and with very high substrate specificity. They also have chiral specificity and catalyze asymmetric reactions. However, enzymes are produced by organisms for their own requirements and, though efficient and ef-

0097-6156/82/0185-0195$05.00/0

fective catalysts, are not always ideal for practical applications. Thus, in order to obtain superior catalysts for applications, that is, highly active and stable catalysts having appropriate specificity, modification of enzymes has been carried out. Among such modifications, immobilization has been extensively studied in the past fifteen years. If active and stable immobilized enzymes are prepared, the expected advantages, compared to soluble enzymes, are as follows: (1) stability of the enzymes is improved, (2) enzymes can be tailor-made for specific use, (3) enzymes can be reused, (4) continuous operation becomes practical, (5) reactions require less space, (6) better control of reactions is possible, (7) higher purity and yield of products may be attained, and (8) saving of resources with less attendant pollution may be achieved.

More recently, techniques of direct immobilization of whole microbial cells have been developed, either to avoid the need to extract enzymes from microbial cells or to utilize multi-enzyme systems. Since the early 1960's, we have studied the immobilization of enzymes and microbial cells for industrial applications and have succeeded in the industrialization of three asymmetric reactions. In this report, these already industrialized systems and a potential system for which we are planning industrialization will be described.

I. IMMOBILIZED AMINOACYLASE — Production of L-Amino Acids

For the industrial production of L-amino acids, fermentation and chemical synthetic methods appear to be promising. However, conventional chemical synthesis leads to a racemic mixture. Hence resolution of enantiomers is necessary to obtain optically active L-amino acids. Among many resolution methods, the enzymatic method using mold aminoacylase developed by us proved to be one of the most advantageous procedures. An acyl-DL-amino acid is selectively hydrolyzed by aminoacylase to give L-amino acid and unhydrolyzed acyl-D-amino acid.

$$\underset{\text{N-acyl-DL-amino acid}}{\text{DL-R-CH(NHCOR')-COOH}} + H_2O \xrightarrow{\text{aminoacylase}} \underset{\text{L-amino acid}}{\text{L-R-CH(NH}_2\text{)-COOH}} + \underset{\text{N-acyl-D-amino acid}}{\text{D-R-CH(NHCOR')-COOH}}$$

(N-acyl-D-amino acid → racemization → N-acyl-DL-amino acid)

Between 1954 and 1969, this enzymatic resolution method had been employed by Tanabe Seiyaku Co., Ltd. for the production of several L-amino acids. In the 1960s we extensively studied the immobilization of aminoacylase for continuous optical resolution [1,2]. A variety of immobilization methods were tested for industrial purposes, from which aminoacylase immobilized by ionic binding to DEAE-Sephadex was chosen. Through chemical engineering studies on aminoacylase columns we designed an enzyme reactor for continuous production. Since 1969, we have been operating several series of enzyme reactors for the production of L-methionine, L-valine, L-phenylalanine and so forth. With this immobilized enzyme system, L-amino acids can be produced more economically compared to the conventional batch system using native enzyme as shown in Fig. 1.

II. IMMOBILIZED MICROBIAL CELLS

A. Production of L-Aspartic Acid

We attempted continuous production of L-aspartic acid from fumaric acid and ammonia by immobilized *Escherichia coli* having high aspartase activity [3, 4, 5]. Various methods were tested for the immobilization of microbial cells, and a stable and active enzyme system was obtained by entrapping whole microbial cells in a polyacrylamide gel lattice.

$$\underset{\text{fumaric acid}}{\text{HOOC-CH=CH-COOH}} + NH_3 \underset{\text{aspartase}}{\rightleftharpoons} \underset{\text{L-aspartic acid}}{\text{HOOC-CH}_2\text{-CH(NH}_2\text{)-COOH}}$$

Using a column packed with immobilized *E. coli* cells, conditions for continuous production of L-aspartic acid were investi-

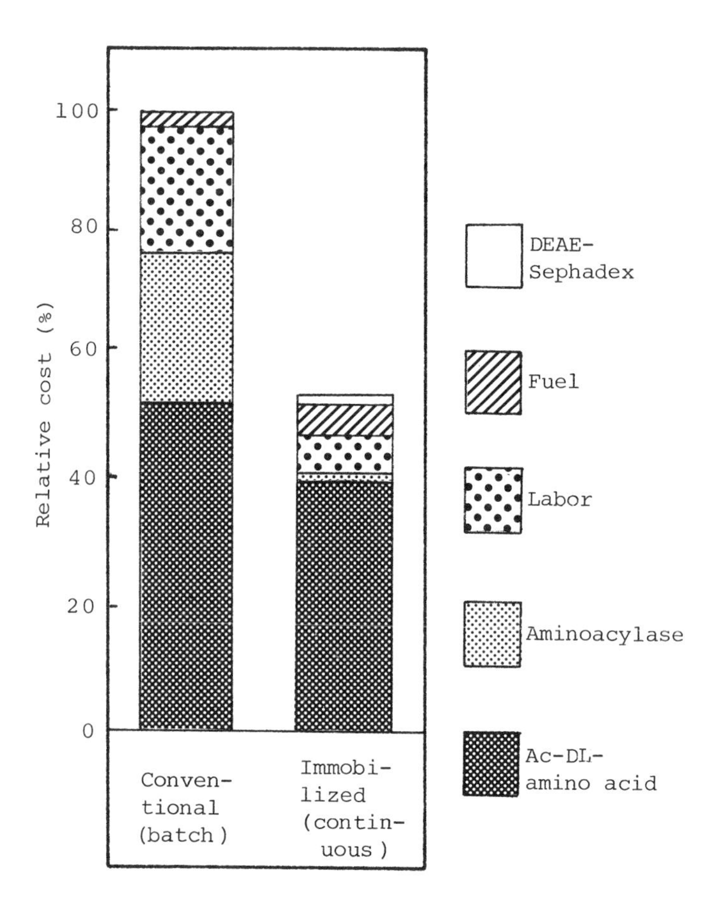

Figure 1. Cost comparison for production of L-*amino acids.*

gated in detail, and an aspartase reactor system was designed. The system is essentially the same as that for the immobilized aminoacylase system and has been operated industrially since 1973.

By way of a further improvement of the process, a new technique using κ-carrageenan was discovered for immobilization of *E. coli* [6, 7, 8]. κ-Carrageenan is a kind of polysaccharide prepared from seaweed and has the characteristic of becoming a gel under mild conditions. We compared the efficiency of *E. coli* cells immobilized with polyacrylamide and κ-carrageenan in production of L-aspartic acid, and found that *E. coli* immobilized with κ-carrageenan and then treated with glutaraldehyde and hexamethylenediamine shows the highest productivity (Table 1). Therefore, we considered that this preparation is most advantageous for continuous production of L-aspartic acid, and in 1978 we replaced the conventional polyacrylamide gel method by the new carrageenan method.

Table 1
COMPARISON OF PRODUCTIVITY OF *E. coli* IMMOBILIZED WITH POLYACRYLAMIDE AND WITH CARRAGEENAN FOR PRODUCTION OF L-ASPARTIC ACID

Immobilization method	Aspartase activity (unit/g cells)	Stability at 37°C (half-life, days)	Relative productivity*
Polyacrylamide	18,850	120	100
Carrageenan	56,340	70	174
Carrageenan (GA)	37,460	240	397
Carrageenan (GA+HMDA)	49,400	680	1,498

GA: glutaraldehyde, HMDA: hexamethylenediamine

* Productivity = $\int_{0}^{t} E_o \exp(-k_d \cdot t)\,dt$

E_o=initial activity, k_d=decay constant, t=operational period

B. Production of L-Malic Acid

In 1974 we succeeded in the industrial production of L-malic acid from fumaric acid by *Brevibacterium ammoniagenes* cells immobilized by the polyacrylamide gel method [9, 10]. The asymmetric reaction catalyzed by the fumarase activity of the cells is shown below.

$$\underset{\text{fumaric acid}}{HOOC\text{-}CH{=}CH\text{-}COOH} + H_2O \underset{\text{fumarase}}{\rightleftharpoons} \underset{\text{L-malic acid}}{HOOC\text{-}CH_2\text{-}\underset{\displaystyle OH}{\underset{|}{CH}}\text{-}COOH}$$

As in the case of L-aspartic acid production, we investigated the carrageenan method to improve the productivity for L-malic acid. After screening various microorganisms for maximal fumarase activity, *Brevibacterium flavum* was found to show a higher enzyme activity after immobilization with κ-carrageenan than the formerly used *B. ammoniagenes*, as shown in Table 2 [11]. Therefore, this polyacrylamide method was also changed to the carrageenan method in 1977. The new method gives satisfactory results for industrial production of L-malic acid.

Table 2
COMPARISON OF PRODUCTIVITIES OF *Brevibacterium ammoniagenes* AND *Brevibacterium flavum* IMMOBILIZED WITH POLYACRYLAMIDE AND WITH CARRAGEENAN FOR PRODUCTION OF L-MALIC ACID

Immobilization method	*B. ammoniagenes*			*B. flavum*		
	Activity (unit/ g cells)	Half-life at 37°C (day)	Relative productivity	Activity (unit/ g cells)	Half-life at 37°C (day)	Relative productivity
Polyacrylamide	5,800	53	100	6,680	94	204
Carrageenan	5,800	75	142	9,920	160	516

* activity after treatment with bile extract

$$\text{Productivity} = \int_0^t E_o \exp(-k_d \cdot t)\,dt$$

E_o=initial activity, k_d=decay constant, t=operational period

C. Production of L-Alanine and D-Aspartic Acid

Besides these two industrial applications of immobilized microbial cells for asymmetric reactions, we are presently studying the continuous production of L-alanine from L-aspartic acid. At present, L-alanine is produced by a batch process. A continuous production system using immobilized *Pseudomonas dacunhae* cells with high L-aspartate β-decarboxylase activity is currently under investigation [12]. The reaction proceeds as shown below.

$$\underset{\text{L-aspartic acid}}{\text{L-HOOC-CH}_2\text{-}\underset{\underset{\text{NH}_2}{|}}{\text{CH}}\text{-COOH}} \xrightarrow[\beta\text{-decarboxylase}]{\text{L-aspartate}} \underset{\text{L-alanine}}{\text{L-CH}_3\text{-}\underset{\underset{\text{NH}_2}{|}}{\text{CH}}\text{-COOH}} + CO_2$$

In this continuous system using immobilized cells, there are problems associated with evolution of CO_2 gas during the reaction. It is difficult to maintain the plug-flow of the substrate solution under normal pressure, and to keep a constant pH of reaction mixture in the reactor because of the CO_2 effervescence. We therefore designed a closed column reactor which performs the enzyme reaction at an elevated pressure such as $10\,Kg/cm^2$. Using this reactor, since liberated CO_2 gas is melded into reaction mixture, the complete plug-flow of the substrate solution is maintained and the pH of reaction mixture is not appreciably changed. Therefore, as shown in Table 3, the efficiency of immobilized cells for production of L-alanine in the closed column system (at high pressure) is much higher than that in the conventional column system at normal pressure.

The decarboxylase enzyme shows high enantiomer selectivity reacting only with L-aspartic acid. Thus, L-alanine and D-aspartic acid can be produced from DL-aspartic acid at the same time.

D-Aspartic acid is used as an important intermediate for synthetic penicillin, whose synthesis has been developed by Tanabe Seiyaku Co.

Table 3
COMPARISON OF EFFICIENCIES AND STABILITIES
OF CONVENTIONAL AND CLOSED COLUMN
REACTORS FOR L-ALANINE PRODUCTION

	Conventional (normal pressure)	Closed (high pressure)
(Efficiency µmole/hr/ml of reactor at 99% convension)	250	360
(Stability half-life, days at 37°C)	46	46

We are planning industrialization of these continuous L-alanine and D-aspartic acid production systems using immobilized *P. dacunhae* in the near future.

Literature Cited

1. Tosa, T.; Mori, T.; Fuse, N.; Chibata, I. Enzymologia 1966, 31, 214-224.
2. Chibata, I.; Tosa, T.; Sato, T.; Mori, T.; Matuo, Y. "Fermentation Technology Today (G. Terui ed.), P.383, Soc. Ferment. Technol., Osaka, Japan (1972).
3. Chibata, I.; Tosa, T.; Sato, T. Appl. Microbiol. 1974, 27, 878-885.
4. Tosa, T.; Sato, T.; Mori, T.; Chibata, I. Appl. Microbiol. 1974, 27, 886-889.
5. Sato, T.; Mori, T.; Tosa, T.; Chibata, I.; Furui, M.; Yamashita, K.; Sumi, A. Biotechnol. Bioeng. 1975, 17, 1797-1804.
6. Tosa, T.; Sato, T.; Mori, T.; Yamamoto, K.; Takata, I.; Nishida, Y.; Chibata, I. Biotechnol. Bioeng. 1979, 21, 133-145.
7. Nishida, Y.; Sato, T.; Tosa, T.; Chibata, I. Enzyme Microb. Technol. 1979, 1, 95-99.

8. Sato, T.; Nishida, Y.; Tosa, T.; Chibata, I. Biochim. Biophys. Acta 1979, 570, 179-186.
9. Yamamoto, K.; Tosa, T.; Yamashita, K.; Chibata, I. Europ. J. Appl. Microbiol. 1976, 3, 169-183.
10. Yamamoto, K.; Tosa, T.; Yamashita, K.; Chibata, I. Biotechnol. Bioeng. 1977, 19, 1101-1114.
11. Takata, I., Yamamoto, K.; Tosa, T.; Chibata, I. Europ. J. Appl. Microbiol. Biotechnol. 1979, 7, 161-172.
12. Yamamoto, K.; Tosa, T.; Chibata, I. Biotechnol. Bioeng. 1980, 22, 2045-2054.

RECEIVED January 4, 1982.

15

Asymmetric Synthesis Using Cofactor-Requiring Enzymes

GEORGE M. WHITESIDES, CHI-HUEY WONG, and ALFRED POLLAK

Massachusetts Institute of Technology, Department of Chemistry, Cambridge, MA 02139

The use of cofactor-requiring enzymes as catalysts for large-scale requires efficient and economical procedures for *in situ* regeneration of these cofactors. This manuscript summarizes the procedures which are now available for cofactor preparation and regeneration. ATP can be effectively regenerated from ADP (and AMP) using acetyl phosphate and acetate kinase (and adenylate kinase), and it can be prepared inexpensively from RNA. Use of ATP-requiring enzymes is now routine (at least as far as the ATP regeneration is concerned). The use of the nicotinamide cofactors is more difficult, because these materials decompose in solution. The best procedure for regenerating NAD(P)H from $NAD(P)^+$ are those based on formate/formate dehydrogenase, glucose 6-phosphate/glucose-6-phosphate dehydrogenase, and ethanol/alcohol dehydrogenase/aldehyde dehydrogenase. The best procedures for regenerating $NAD(P)^+$ from NAD(P)H use dioxygen/methyl viologen or ketoglutarate/glutamic dehydrogenase.

Although enzymes can be effective catalysts for enantioselective reactions, they have been relatively little used for this purpose in practical organic synthesis. The relative indifference of synthetic chemists to the potential of this group of catalysts is a consequence of a number of circumstances. First, enzymes are unfamiliar: they require aqueous environments; they are prepared, characterized, and manipulated using specialized techniques having little in common with techniques used in other areas of synthetic organic chemistry; and they appear to be unstable. Second, certain generally interesting classes of enzymatic reactions (including many reactions which *form* bonds between organic molecules and most reactions which involve oxidation or reduction) involve cofactors; these reactions are expensive. Third, the substrate selectivity

0097-6156/82/0185-0205$05.00/0

of enzyme-catalyzed reactions often limits the generality of their application. Nonetheless, in these reactions in which they are applicable, they can be very efficient catalysts, and their ability to catalyze reactions of naturally-occurring substances (which are, of course, products of and reactants in the reactions which take place in life) makes them of particular interest in pharmaceutical, food, and agricultural chemistry.

The research summarized in this manuscript was directed toward one particular problem in enzymology: that is, the development of techniques which would make possible the use of cofactor-requiring enzymes in organic synthesis. The central problem in this area has been one of expense. ATP costs approximately $800/mole when purchased in mole quantities; the costs of the nicotinamide cofactors range from $1500/mole (for NAD^+) to $250,000/mole (for NADPH). There are few organic reactions which can tolerate costs of this magnitude for stoichiometric reagents. The solution to this problem of cost is, in principle, straightforward, and has been the subject of extensive previous work. The most efficient way of lowering the effective cost of the cofactors is to develop procedures which make possible their regeneration from inexpensive reagents *in situ* (Figure 1)

Among the considerations which determine the usefulness of a synthetic sequence which involves a cofactor-requiring enzymatic step are:

1) The character of the reaction used for regeneration of the cofactor. The reagent A should be readily available, inexpensive, and stable; the product B should not complicate workup; the equilibrium constant for the reaction $A + X \rightleftharpoons B + Y$ should lie far to the right; the enzymes used (if any) should have low cost, high stability, and high specific activity.

2) The intrinsic stabilities of the cofactors X and Y under the conditions of the reaction.

3) The original cost of the cofactor.

4) The operational simplicity of the regeneration scheme.

Here we divide the discussion of approaches to cofactor regeneration into three sections: one each for ATP, oxidized nicotinamide cofactors (NAD^+ and NADP), and reduced nicotinamide cofactors (NADH and NADPH). Most of the other cofactors which appear in biochemistry are either easily regenerated or of little importance, and we shall not discuss their regeneration here.

Although either purely chemical or enzymatic procedures might be used to effect the regeneration reactions, in general enzymatic procedures are superior. To be able to recycle the cofactors a large number of times it is necessary to have high yields for the reactions which regenerate them. Thus, to have 50% of the cofactor remaining after 100 cycles of reaction and regeneration, the yield for each cycle must be 99.3% ($100 \log 0.993 = \log 0.50$), and for 1000 cycles, the corresponding yield must be 99.9%. This type of selectivity is most easily obtained by enzymatic catalysis, and we have therefore used only enzymatic methods in our work.

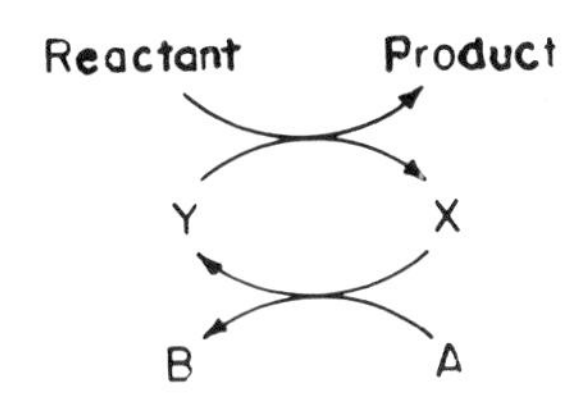

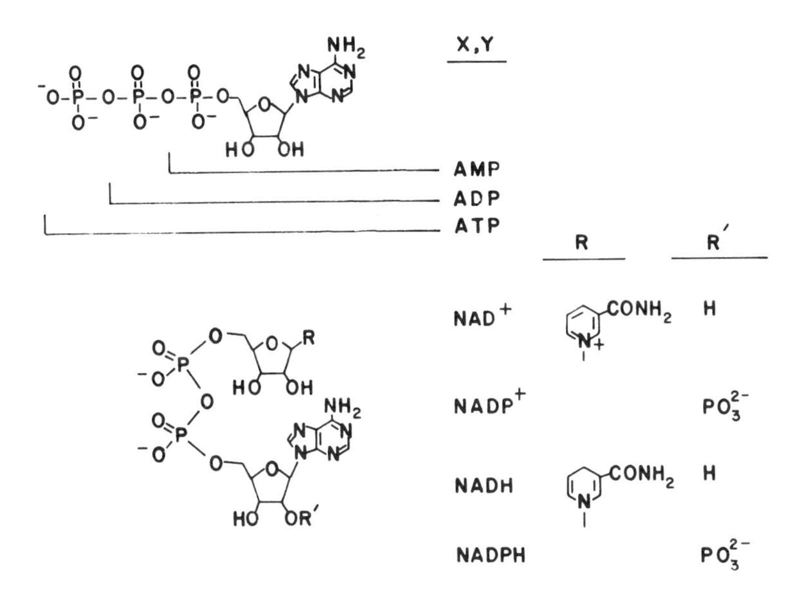

Figure 1. General scheme for cofactor regeneration (top) and structures of adenosine and nicotinamide cofactors (bottom). Key: X, Y, cofactors; A, regenerating agent; and B, product from this reagent.

ATP

Regeneration. Most biochemical reactions which involve ATP as a cofactor convert it to ADP or AMP; adenosine itself is important only as a product of the small group of reactions which proceed through S-adenosyl methionine. Thus, it is necessary to have regeneration procedures which will convert both AMP and ADP to ATP. Chemical methods for these phosphorylation reactions can be rejected out of hand: they are incompatible with the enzymes which would be present in the system as catalysts for reactions which *use* the ATP, and lack the specificity required to give high yields and high total turnover numbers (TTN) for the ATP (TTN = moles of product produced in the reaction per mole of cofactor or enzyme present). The stability of ATP is good: the hydrolysis of ATP at pH 6-8 is slow compared with any synthetic reaction of practical interest.

The choice of phosphorylating agents which might, in principle, be used to convert AMP or ADP to ATP is limited. Table I summarizes values of $\Delta G^{\circ\prime}$ for the reaction XP + ADP $\rightleftharpoons$ X + ATP for those compounds XP which are (relatively) readily available and exergonic with respect to phosphorylation of ADP. Of these, PEP

Table I. Free energy of phosphorylation of ADP to ATP

XP	$\Delta G^{\circ\prime}$ (kcal/mole)
Phosphoenolpyruvate (PEP)	-7.5
Carbamyl phosphate	-5.0
Acetyl phosphate (AcP)	-3.0
Pyrophosphate (PP_i)	-0.7

is relatively expensive (although regeneration systems based on PEP have many advantages in simplicity), carbamyl phosphate has very poor stability in solution, and pyrophosphate is only a weak phosphorylating agent and requires enzymes which are available only with difficulty. Acetyl phosphate (AcP) is a reagent which offers a practical combination in its characteristics: it is prepared easily from inexpensive reagents; the enzymes it requires for use in ATP regeneration are commercially available and acceptably stable; it is a good phosphorylating agent. In this manuscript we focus attention on procedures for ATP regeneration based on AcP. The only other procedure which seems useful for laboratory-scale is that based on PEP; details of this procedure will be published elsewhere.

AcP can be prepared easily by acylation of phosphoric acid with acetic anhydride or with ketene, and isolated as a fairly stable ammonium salt (1,2). Acetate kinase (the enzyme which catalyzes the reaction of acetyl phosphate with ADP) and adenylate kinase (the enzyme which catalyzes phosphate transfer between ATP and AMP) are readily available and inexpensive. The ATP regeneration schemes based on these enzymes are shown in Figure 2 (3,4,5).

These schemes have now been used to prepare organic materials on scales of several moles. An example relevant to asymmetric synthesis is the glycerol kinase-catalyzed phosphorylation of glycerol (equation i) (6).

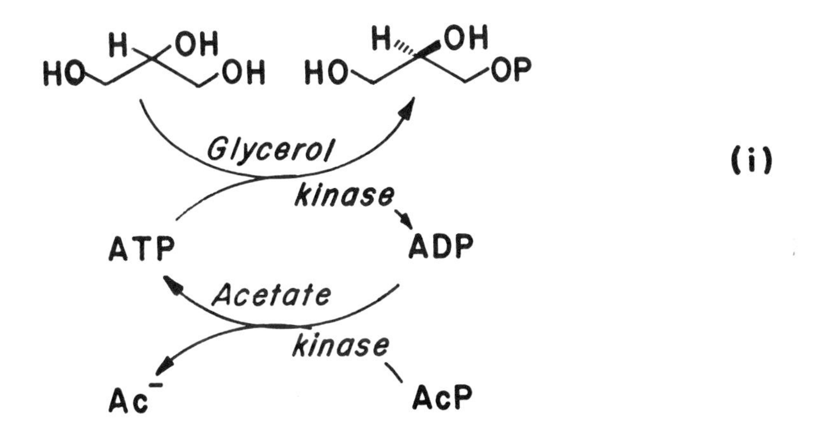

This reaction yields enantiomerically pure sn-glycerol-3-phosphate ((*R*)-glycerol-1-phosphate, a compound having the correct configuration to serve as the basis for the synthesis of phospholipids). The turnover numbers (TTN = moles product per mole cofactor) achieved in these syntheses (TTN $\simeq$ 100) have been limited primarily by convenience: we normally use a relatively large quantity of ATP, to keep reaction rates high. The ATP is, however, essentially all still present at the conclusion of the reaction. For laboratory-scale synthesis of fine chemicals, the methods shown in Figure 2 represent an effective solution to the problem of ATP regeneration.

Synthesis. A final problem related to ATP utilization is that of obtaining the initial quantity of ATP to be used in the reaction. ATP as a pure biochemical is expensive. A material suitable for use in recycling can be obtained from RNA (approximately $80/kg) by the process outlined in equation ii (7).

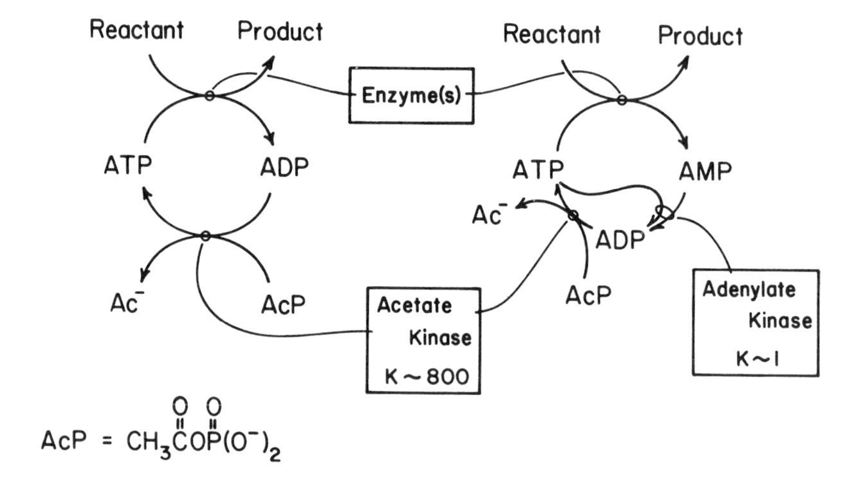

Figure 2. Schemes used for regenerating ATP from ADP (left) and AMP (right).

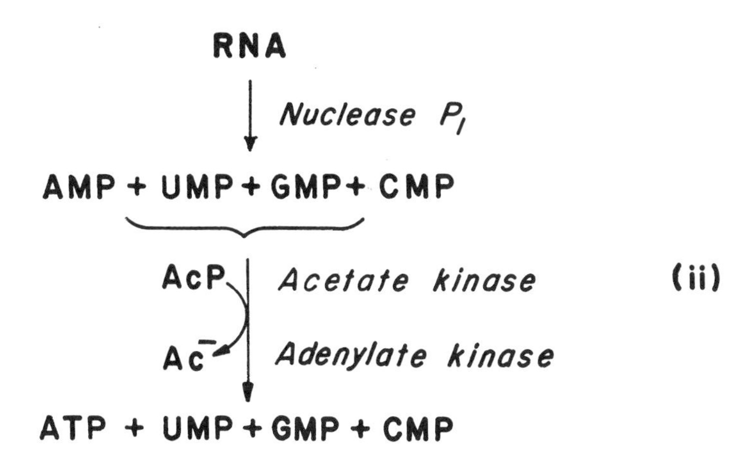

Cleavage of RNA using nuclease P_1 yields a mixture of nucleoside monophosphates, contaminated with oligonucleotides and other materials. The AMP present in this mixture can be converted selectively to ATP by treatment with acetyl phosphate and a mixture of adenylate kinase and acetate kinase. The resulting mixture can be used directly, *without purification*, to supply ATP for use in cofactor recycling.

$NAD(P)^+$

Regeneration. The oxidized nicotinamide cofactors ($NAD(P)^+$) are considerably more difficult to work with than ATP, but are more tractable than the reduced nicotinamide cofactors (NAD(P)H). The oxidized cofactors are sensitive to nucleophiles (8), but are relatively stable at pH 7; the reduced cofactors decompose by acid-catalyzed processes involving protonation at C-5 of the dihydropyridine ring as the rate-limiting step (equation iii) (9,10).

A–H, H, O, NH_2 → H, H, A^-, O, NH_2, N^+ → Other products (iii)

Phosphate, a common component of enzymatic systems and an integral part of $NADP^+$ and NADPH, is a particularly effective acid catalyst for this reaction.

The efficient utilization of the nicotinamide cofactors requires not only a scheme for their recycling but also a method for dealing with their limited lifetime in solution. A number of approaches to recycling have been considered, and some appear to be quite satisfactory. The only obvious approach to the economic problem posed by limited lifetime is to lower the initial cost of the cofactor, and we can only offer suggestions concerning this problem.

The recycling of NAD(P)H to $NAD(P)^+$ can be accomplished by any of several methods. If dioxygen can be used in the system, the most straightforward recycling method involves methyl viologen (MV)-catalyzed oxidation (equation iv). The details of the several

NAD(P)H, MV^{+2} → (Diaphorase) → $NAD(P)^+$, MV^{+}; MV^{+} + O_2 → $[O_2^{\cdot -}, H_2O_2]$ → (Superoxide dismutase, Catalase) → $H_2O + O_2$ (iv)

reactions which may be involved in this recycling scheme are not known, but in practice, it has proved to be a useful synthetic method. When *anaerobic* recycling is required, a procedure based on glutamic dehydrogenase works well (equation v) (11).

NAD(P)H + HO_2C–CH_2CH_2–CO–CO_2H + NH_4^+ → (Glutamic dehydrogenase) → $NAD(P)^+$ + HO_2C–CH_2CH_2–CH(^+H_3N)–CO_2H (v)

We defer discussion of the problem of *synthesizing* $NAD(P)^+$ to the next section.

NAD(P)H

Regeneration. A large number of systems have been tested for utility in regeneration of reduced nicotinamide cofactors. Of these, only three seem likely to be useful in the short term. Here we briefly describe those systems which, in our opinion, have the practicality required for use in mole-scale organic synthesis.

Formate/Formate Dehydrogenase. This system (equation vi) (12) has several advantages: it requires only one enzyme, and this enzyme is readily available in quantity (although it is still moderately expensive when purchased commercially); formate is inexpensive, and removal of CO_2 causes no problem during workup; formate is a strong reducing agent. The principal disadvantages

$$HCO_2^- + NAD^+ \xrightarrow{\textit{Formate dehydrogenase}} CO_2 + NADH \qquad \text{(vi)}$$

of the system are that it is specific for NAD^+ (and thus requires a transhydrogenase for use with $NADP^+$), that the specific activity of the enzyme (3 U/mg; 1 U = 1 μmole of NAD^+ consumed/min) is only modest (13) (thus large reactor volumes are required when immobilized enzymes are being used), and that the enzyme is sensitive to autoxidation.

Glucose-6-Phosphate/Glucose-6-Phosphate Dehydrogenase. The advantage of this system (equation vii) are that it requires a single enzyme (10), that this enzyme is stable and available with high specific activity (700 U/mg for NAD^+; 500 U/mg for $NADP^+$) (14), that the same enzyme catalyzes the reduction of both NAD and NADP , and that the reaction is irreversible because 6-phosphogluconolacetone hydrolyzes rapidly to 6-phosphogluconate. Its disadvantages are that glucose-6-phosphate is not commercially available (although it is relatively readily prepared), that 6-phosphogluconate may cause significant problems in workup, and that glucose-6-phosphate and 6-phosphogluconate both catalyze the decomposition of NAD(P)H (10). In practice, the convenience of having an easily manipulated, stable, active enzyme outweighs the disadvantages of having to prepare glucose-6-phosphate and of suffering a short lifetime for NAD(P)H. Overall, the method is a useful one in laboratory-scale preparations.

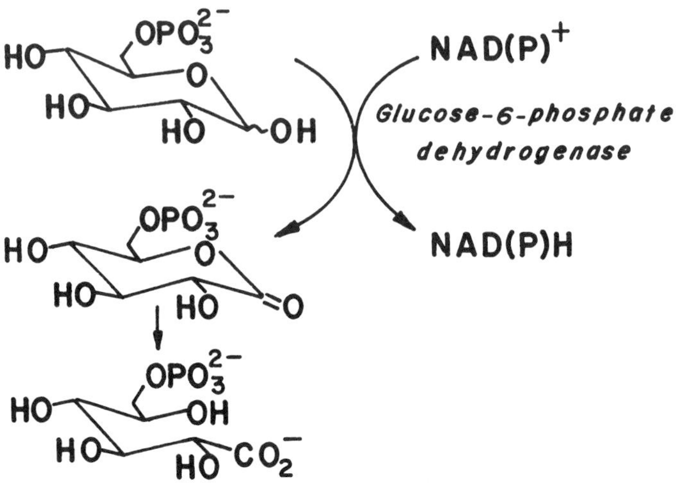

$NAD(P)^+$

Glucose-6-phosphate dehydrogenase

NAD(P)H

(vii)

Ethanol/Alcohol Dehydrogenase/Aldehyde Dehydrogenase. The combination of ethanol and alcohol dehydrogenase has been used extensively to reduce NAD^+ to NADH (15,16). This system is unsatisfactory for two reasons: it is only weakly reducing, and the acetaldehyde produced deactivates many enzymes. By adding an excess of aldehyde dehydrogenase (equation viii), the system becomes a very good one (17,18). Its advantages are that ethanol is

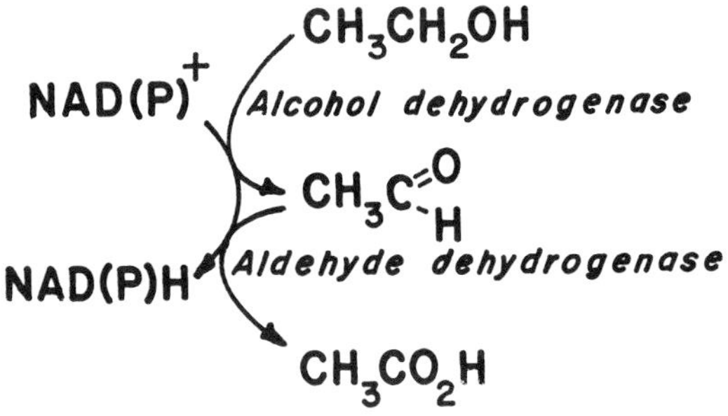

CH_3CH_2OH

$NAD(P)^+$

Alcohol dehydrogenase

CH_3CHO

Aldehyde dehydrogenase

NAD(P)H

CH_3CO_2H

(viii)

inexpensive, the coupled system is strongly reducing, the enzymes have high specific activities (19) (alcohol dehydrogenase: 400 U/mg for ethanol; aldehyde dehydrogenase; 80 U/mg for acetaldehyde), and acetate seldom complicates workup. Its disadvantages are that the system requires two enzymes, that the enzyme which must be in excess (aldehyde dehydrogenase) is the more expensive and the more sensitive, and that it is more reactive toward NAD^+ than $NADP^+$. A similar system which works well for NADH regeneration is based on methanol as substrate and three combined enzymes, including alcohol dehydrogenase, aldehyde dehydrogenase and formate dehydrogenase, as catalysts (18). The side product (CO_2) in this system does not complicate the work-up and methanol is less expensive than ethanol, but the specific activities of the rate-limiting enzymes of this system are less than those based on ethanol as ultimate reducing source.

Others. In addition to these procedures, a number of others have been established to be effective in reducing $NAD(P)^+$. Many of these methods have been reviewed (15,16). Recently a promising procedure based on a secondary alcohol dehydrogenase has been described by Zeikus (20), and methods which use dihydrogen (21) and electrons from a cathode (22,23) have been described. With the exception of the procedure of Zeikus, these methods are not as convenient for laboratory-scale work as those described above.

There are many examples of the use of NAD(P)H to effect organic syntheses in systems in which the reduced nicotinamide cofactor is regenerated *in situ* (15,16,24). Recent examples include syntheses of D-lactic acid (10,12), isocitric acid (10,21) and other α-hydroxy acids (D or L) on scales of 0.1 to 0.5 mole (equation ix). The turnover numbers for NAD(P)(H) in these reactions are TTN ≈ 1000 - 2000.

$RC(O)CO_2^-$ + NADH → (D-Lactic dehydrogenase) → $R(H)(OH)C\text{-}CO_2^-$ + NAD^+

$^-O_2CCH_2C(O)CO_2^-$ + CO_2 + NADPH → (Isocitric dehydrogenase) → $^-O_2CCH_2CH(CO_2^-)C(OH)(H)CO_2^-$ + $NADP^+$ **(ix)**

Synthesis of NAD^+ and $NADP^+$. The nicotinamide cofactors are now isolated from yeast (25,26). A major difficulty in this preparation is simply that of separation of the NAD(P)(H) from the other components in the cell. To reduce the cost of these materials, either the yield must be improved from the yeast preparation, the isolation must be simplified, or some type of synthesis must be developed. We have taken a step toward developing a new synthesis by the combined enzymatic/conventional synthetic procedure summarized in Figure 3 (27). The overall conversion from

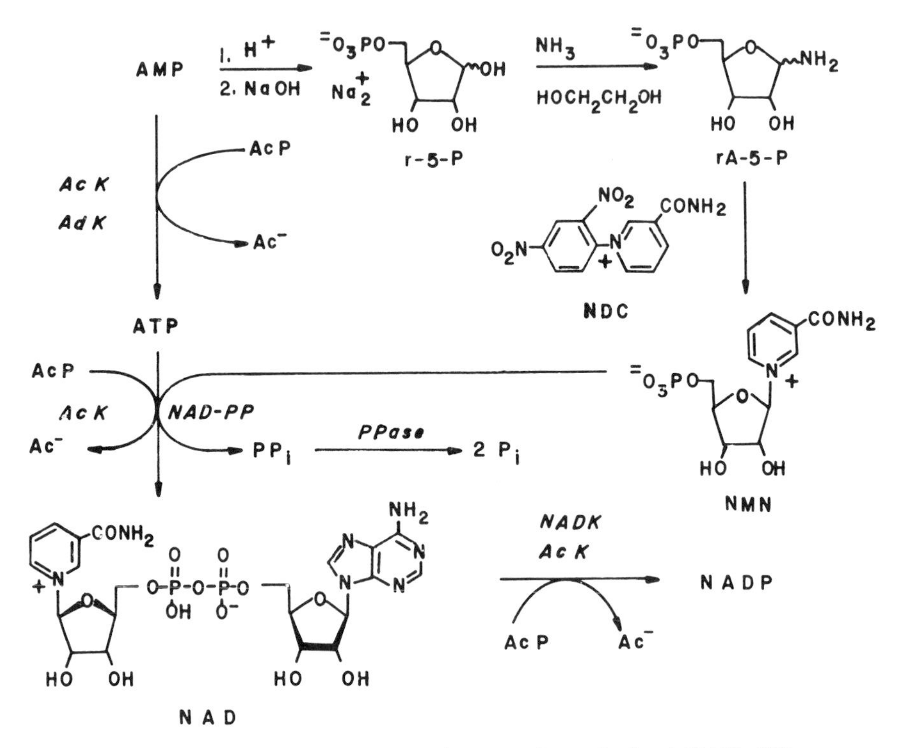

Figure 3. Combined chemical and enzymatic synthesis of NAD^+ (27).
Key: AcK, acetate kinase; AdK, adenylate kinase; NAD–PP, NAD pyrophosphorylase; PPase, pyrophosphorylase; NADK, NAD kinase; r-5-P, ribose-5-phosphate; rA-5-P, ribosylamine-5-phosphate; NMN, nicotinamide mononucleotide; AcP, acetyl phosphate; PP_i, pyrophosphate; and NDC, N_1-(2,4-dinitrophenyl)-3-carbamoylpyridinium chloride.

ribose-5-phosphate to NAD^+ in this procedure is approximately 60% on small scale; that from AMP to NAD^+ via ATP is essentially quantitative. The procedure involves only one isolation (that of ribose-5-phosphate). The solution containing the NAD^+ can be used directly for cofactor recycling: whatever components are present as impurities in this solution apparently do not inactivate or inhibit enzymes. This procedure (after development) or some related procedure may provide the best hope for reducing the cost of the nicotinamide cofactors.

Acknowlegements

Names of many of our coworkers who contributed to this work are listed in the references. The research was supported by the National Institutes of Health, Grant GM 26543, and by the Monsanto Corporation.

Literature Cited

1. Whitesides, G. M.; Siegal, M.; Garrett, P. J. Org. Chem. 1975, 40, 2516-9.
2. Lewis, J. M.; Haynie, S. L.; Whitesides, G. M. J. Org. Chem. 1979, 44, 864-5.
3. Pollak, A.; Baughn, R. L.; Whitesides, G. M. J. Am. Chem. Soc. 1977, 99, 2366-7.
4. Shih, Y. S.; Whitesides, G. M. J. Org. Chem. 1977, 42, 4165-6.
5. Baughn, R. L.; Adalsteinsson, O; Whitesides, G. M. J. Am. Chem. Soc. 1978, 100, 304-6.
6. Rios-Mercadillo, V. M.; Whitesides, G. M. J. Am. Chem. Soc. 1979, 101, 5828-9.
7. Leuchs, H. J.; Lewis, J. M.; Rios-Mercadillo, V. M.; Whitesides, G. M. J. Am. Chem. Soc. 1979, 101, 5829-30.
8. Johnson, S. L.; Morrison, D. L. Biochemistry 1970, 9, 1460-70.
9. Johnson, S. L.; Tuazon, P. T. Biochemistry 1977, 16, 1175-83.
10. Wong, C.-H.; Whitesides, G. M. J. Am. Chem. Soc. 1981, 103, 4890-99.
11. Wong, C.-H.; McCurry, S. D.; Whitesides, G. M. J. Am. Chem. Soc. 1980, 102, 7938-9.
12. Shaked, Z.; Whitesides, G. M. J. Am. Chem. Soc. 1980, 102, 7104.
13. Schutte, H.; Flossdorf, J.; Sahm, H.; Kula, M. R. Eur. J. Biochem. 1976, 62, 151-60.
14. Levy, H. R.; Adv. Enzymol. 1979, 48, 141-3.
15. Jones, J. B.; Beck, J. F. "Application of Biochemical Systems in Organic Chemistry" Jones, J. B.; Perlman, D.; Sih, C. J. Ed.; Wiley-Interscience, New York, 1976, p 107-401.
16. Wang, S. S.; King, C. K. Adv. Biochem. Eng. 1979, 12, 119-46.

17. Chemical and Engineering News, Feb. 25, 1974, p 19.
18. Wong, C. H.; Pollak, A.; Whitesides, G. M. J. Am. Chem. Soc. submitted.
19. Wratten, C. C.; Cleland, W. W. Biochemistry 1963, 2, 935-41.
20. Lamed, R. J.; Keinan, E.; Zeikus, J. G. Enzyme. Microb. Technol. 1981, 3, 144-8.
21. Wong, C. H.; Lacy, D.; Orme-Johnson, W. H.; Whitesides, G. M. J. Am. Chem. Soc. in press.
22. DiCosimo, R.; Wong, C. H.; Lacy, D.; Whitesides, G. M. J. Org. Chem. in press.
23. Shaked, Z.; Barber, J. J.; Whitesides, G. M. J. Org. Chem. in press.
24. Jones, J. B. "Enzymic and Non-enzymic Catalysis", Dunill, P.; Wiseman, A.; Blakebrough, N. Ed.,; Ellis Horwood, Ltd.; Chichester, England, 1980, p 54-81.
25. Nakayama, K.; Sato, Z.; Tanaka, H. Kinoshita, S. Agr. Biol. Chem. 1968, 32, 1331-6.
26. Sakai, T.; Uchida, T.; Chibata, I. Agr. Biol. Chem. 1973, 37, 1041-8.
27. Walt, D. R.; Rios-Mercadillo, V. M.; Auge, J.; Whitesides, G. M. J. Am. Chem. Soc. 1980, 102, 7805-6.

RECEIVED December 14, 1981.

16

Mechanistic Considerations of Biomimetic Asymmetric Reductions

ATSUYOSHI OHNO

Kyoto University, Institute for Chemical Research, Kyoto, Japan

As shown in Table 1, the reduction of ethyl benzoylformate by a 1-substituted-1,4-dihydronicotinamide, a model of NADH or NADPH, in acetonitrile occurs in the presence of a bivalent metal ion such as magnesium(II) or zinc(II)(1)(Scheme 1). When one of the amide-hydrogens is substituted by a chiral group, asymmetric reduction takes place. The stereospecificity of this reduction is also affected by magnesium ion as shown in Table 2 (2,3). Although it is not clear why such a remote chiral center affects the stereochemistry of the reduction, the presence of a nitrogen atom on the side chain appears to play an important role in the stereospecificity, as shown in Table 3(4). The optical yield is still unsatisfactory compared with the enzymic reductions. Expecting that the enantioselectivity would be improved with a model compound having the chiral center and reacting hydrogen at the same position, we synthesized all four possible optical isomers of N-α-methylbenzyl-1-propyl-2,4-dimethyl-1,4-dihydronicotinamide (Me_2PNPH, Scheme 1)[1]. Results for the reduction of various substrates with some of these model compounds are summarized in Table 4.

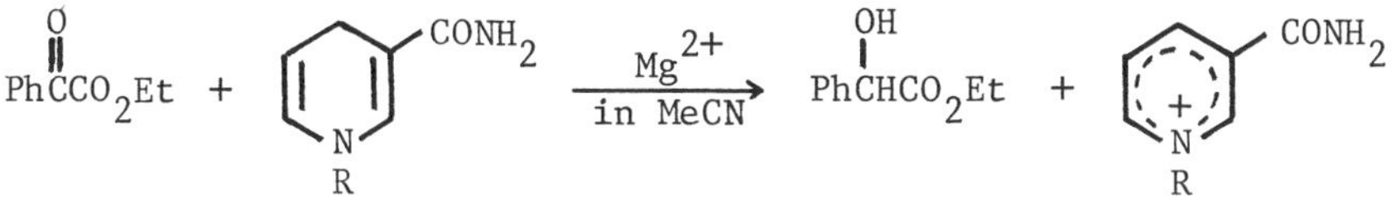

$$PhCOCO_2Et + \text{(1-R-1,4-dihydronicotinamide)} \xrightarrow[\text{in MeCN}]{Mg^{2+}} PhCH(OH)CO_2Et + \text{(1-R-nicotinamide cation)}$$

Scheme 1

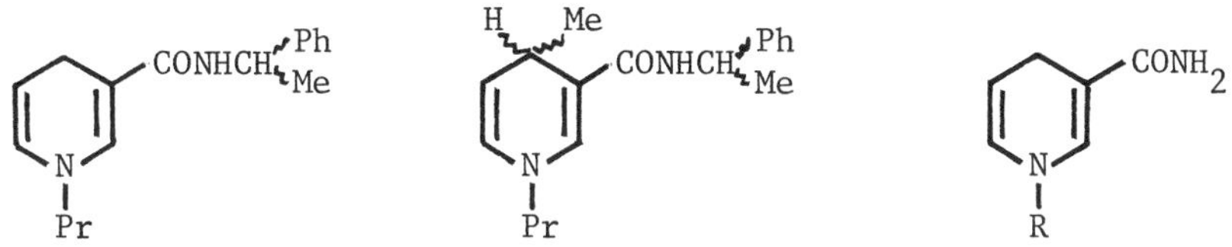

PNPH (1-Pr, 3-CONHCH(Ph)Me) — Me_2PNPH (1-Pr, 4-H/Me, 3-CONHCH(Ph)Me) — R = Pr: PNAH; R = $PhCH_2$: BNAH

[1] Hereafter, the author will denote XY-Me_2PNPH for an isomer of Me_2PNPH which has configurations X at the ring C_4 and Y at the benzylic carbon.

0097-6156/82/0185-0219$05.00/0

Table 1. Reduction of Ethyl Benzoylformate by 1-Benzyl-1,4-dihydronicotinamide (BNAH) [a]

BNAH, mmol	Metal ion,[b] mmol	Isolated Yield,% Recovered Keto ester	Isolated Yield,% Ethyl mandelate
1.06	none	90	0
1.11	Mg^{2+} 1.08[c]	6	86
1.09	Mg^{2+} 1.13	0	100[d]
1.09	Zn^{2+} 1.25[e]	8	66
1.09	Li^{+} 1.25	92	2

[a] The reactions were run with 1 mmol of keto ester in 15 mL of acetonitrile for 17 hr at room temperature in the dark. [b] Perchlorate. [c] Reaction time: 44 hr. [d] Oxidized BNAH (BNA^{+}) was isolated in 90% yield. [e] Hydrated salt was used.

Table 2. Asymmetric Reduction of Ethyl Benzoylformate by Optically Active N-α-Methylbenzyl-1-propyl-1,4-dihydronicotinamide (PNPH) [a]

Config. of PNPH, mmol		Mg^{2+}, mmol[b]	$[Mg^{2+}]/[PNPH]$	Ethyl mandelate Yield,%	Ethyl mandelate e.e.,%
R	1.00	0.26	0.3	86	6.6
	1.02	0.52	0.5	82	8.6
	2.05	1.05	0.5	95	9.9
	0.99	1.04	1.1	94	19.6
	0.98	1.99	2.0	95	18.1
S	1.14	0.96	0.8	96	-18.6[c]

[a] Reactions were run with 1 mmol of the keto ester in 15 mL of acetonitrile for 44 hr at room temperature in the dark. [b] Perchlorate. [c] *S*-Mandelate was obtained in excess.

Table 3. Effect of Substituent on the Stereospecificity of the Reduction of Ethyl Benzoylformate [a)]

X in Model	Ethyl mandelate Isolated,[b)] e.e.,%
NH	26
CH_2	9
O	2

[a)] Reaction conditions are the same as described in Table 2. The structure of the model is:

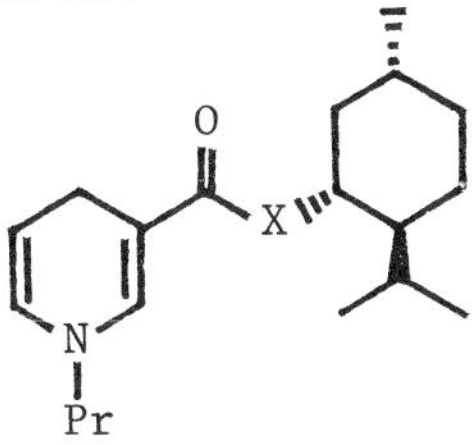

[b)] Chemical yields are quantitative.

Table 4. Reduction by Chiral N-α-methylbenzyl-1-propyl-2,4-dimethyl-1,4-dihydronicotinamide (Me_2PNPH)

Substrate	Config. of Me_2PNPH[a)]	Conv.,%[b)]	Product Structure	Yield,%	e.e.,%
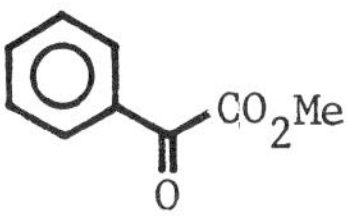 PhCOCO$_2$Me	*RR*	100	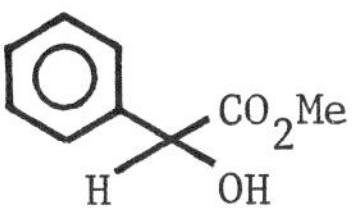 PhCH(OH)CO$_2$Me	100	97.6(*R*)
	SR	100		100	96.5(*S*)
	SS	100		100	94.7(*S*)
	RR[c)]	100		100	52.5(*R*)
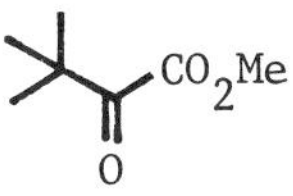 t-BuCOCO$_2$Me	*RR*	95	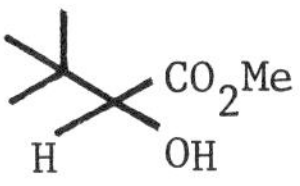 t-BuCH(OH)CO$_2$Me	99	>99 (*R*)
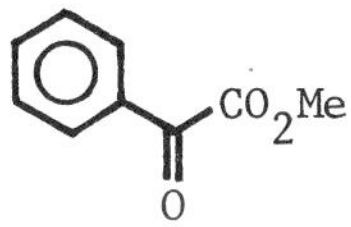 PhCOCO$_2$Me	*RR*	60	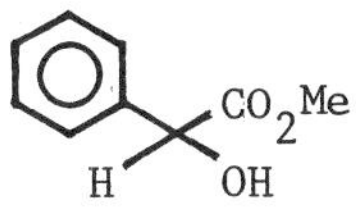 PhCH(OH)CO$_2$Me	100	92.0(*R*)
	SR	56		100	92.0(*S*)
	RR[c)]	68		68	71.3(*R*)
	SR[c)]	79		66	41.4(*S*)

Table 4 Continued

$4\text{-MeOC}_6H_4COCF_3$	*RR*[c)]	97	$4\text{-MeOC}_6H_4CH(OH)CF_3$	78.4	80.2(*R*)
$4\text{-MeC}_6H_4COCF_3$	*RR*	74	$4\text{-MeC}_6H_4CH(OH)CF_3$	100	95 (*R*)
	RR[c)]	95		33.6	76.4(*R*)
$4\text{-ClC}_6H_4COCF_3$	*SS*	75	$4\text{-ClC}_6H_4CH(OH)CF_3$	100	>95 (*S*)
	RR[c)]	99		55.3	82.5(*R*)
$4\text{-BrC}_6H_4COCF_3$	*RR*[c)]	100	$4\text{-BrC}_6H_4CH(OH)CF_3$	52.5	82.2(*R*)
$3\text{-BrC}_6H_4COCF_3$	*RR*	77	$3\text{-BrC}_6H_4CH(OH)CF_3$	100	90 (*R*)
$3\text{-CF}_3C_6H_4COCF_3$	*RR*[c)]	99	$3\text{-CF}_3C_6H_4CH(OH)CF_3$	58.0	85.9(*R*)
$3\text{-O}_2NC_6H_4COCF_3$	*SS*	85	$3\text{-O}_2NC_6H_4CH(OH)CF_3$	100	>99 (*S*)
2-pyridyl-$COCH_3$	*RR*	100	2-pyridyl-$CH(OH)CH_3$	50.1	62.8(*R*)

Table 4 Continued

Substrate			Product	
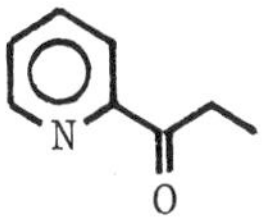 N, O	*RR*	76	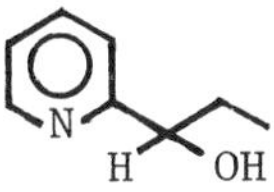 N, H, OH	45.2 52.1(*R*)
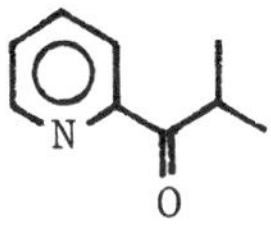 N, O	*RR*	92	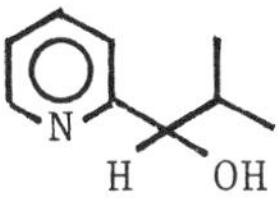 N, H, OH	78.3 ~0
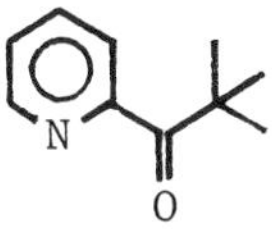 N, O	*RR*	47	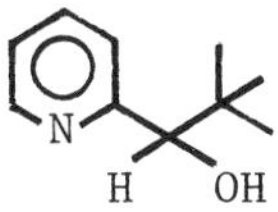 N, H, OH	40.3 43.4(*R*)
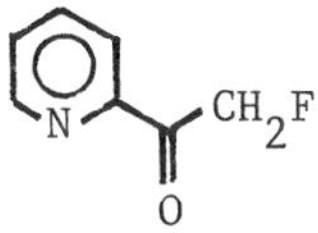 N, O, CH_2F	*RR*	98	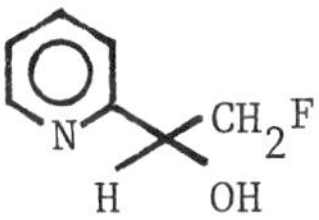 N, H, OH, CH_2F	43.3 53.5(*S*)
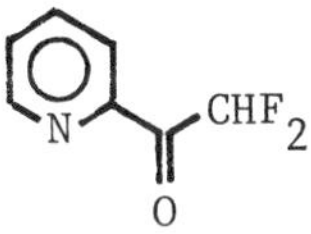 N, O, CHF_2	*RR*	100	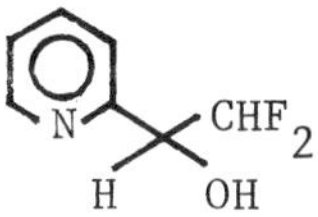 N, H, OH, CHF_2	39.4 30.3(*S*)
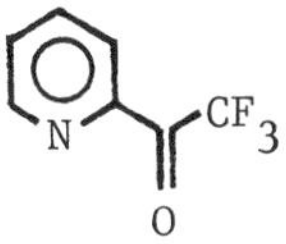 N, O, CF_3	*RR*	100	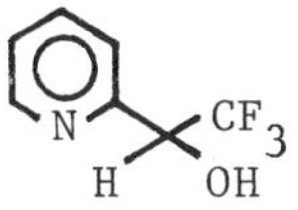 N, H, OH, CF_3	35.7 16.5(*S*)

a) See Note 5 for the notation. b) Amount of substrate consumed.
c) Reaction without magnesium perchlorate.

The results show several interesting characteristics;

1. Introduction of two methyl groups on the dihydropyridine ring (Me_2PNPH) enhances the reactivity compared to PNPH, as exhibited by the reduction of α,α,α-trifluoroacetophenone without magnesium ion.
2. The predominant enantiomer of the product is determined by the configuration at the C_4-position of Me_2PNPH in the presence of Mg^{2+}. However, in the absence of magnesium ion, the configuration at the benzylic carbon exerts a secondary effect on the stereochemistry.

For enzymic reactions, it was proposed that the carbonyl oxygen of the substrate points toward the dihydropyridine ring nitrogen of NAD(P)H in the transition state(6). Based on the same assumption the stereochemistry of the product in the mimetic reduction can be predicted as shown in 1.

R_p: polar substituent

R_n: nonpolar substituent

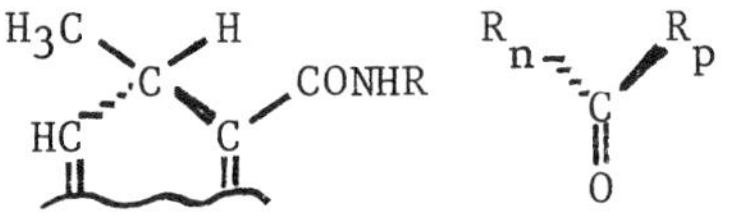

1

The relative bulk of the substituents in the substrate exerts no contribution, at least not a primary one.

In order to obtain information on the transition-state stereochemistry, we studied the reduction of camphoroquinone (CQ) with Me_2PNPH. Products from the reduction of (-)- and (+)-CQ's are shown in Scheme 2 and the results are listed in Table 5.

Scheme 2

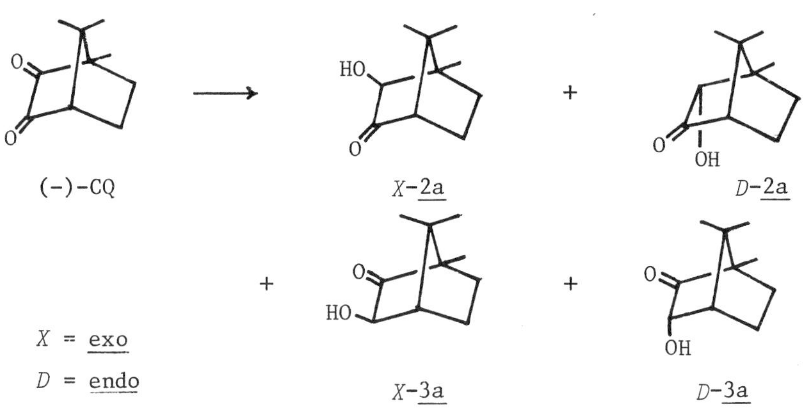

(-)-CQ → *X*-2a + *D*-2a + *X*-3a + *D*-3a

X = exo

D = endo

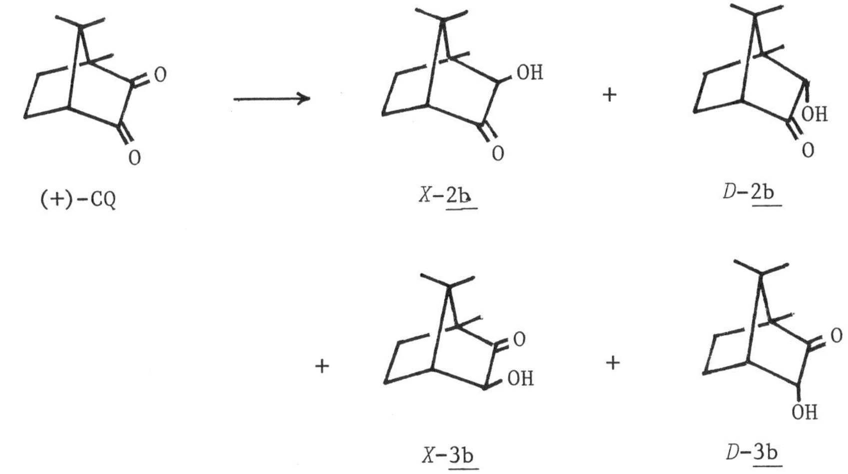

(+)-CQ → *X*-2b + *D*-2b + *X*-3b + *D*-3b

Scheme 3

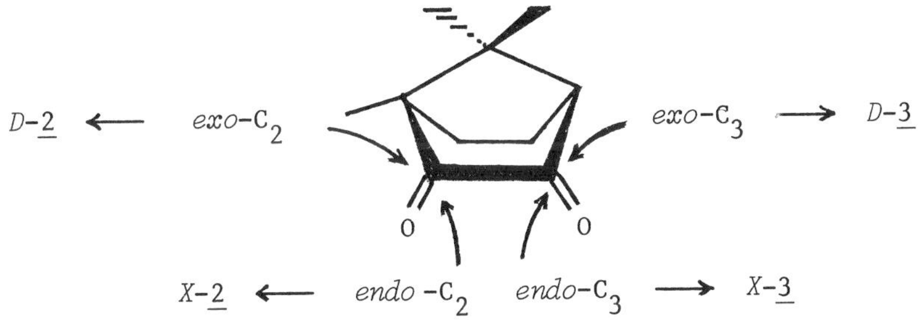

D-2 ← *exo*-C_2 *exo*-C_3 → *D*-3

X-2 ← *endo*-C_2 *endo*-C_3 → *X*-3

Similar results from the reduction with BNAH, PNAH, and *R*-PNPH are summarized in Table 6, which indicates that *exo*-C_3-attack (see Scheme 3) is the most preferential course of the reduction with C_4-achiral dihydronicotinamide derivatives. The chirality in the side-chain of *R*-PNPH plays no important role in determining the stereochemistry of the product. In contrast, the stereochemistry in the reduction with Me_2PNPH reflects mainly the configuration at C_4 of this model compound, and the intrinsic reactivity of each position in CQ, observed above, has only a minor influence now. The result indicates that there is a preference in the orientation of the substituents in CQ with respect to the diastereotopic faces of Me_2PNPH.

The most preferred mode of attack of *SS*-Me_2PNPH is *exo*-C_3 for (-)-CQ (Table 5, entry 2). In this mode the amide moiety in *SS*-Me_2PNPH has to face the carbonyl (electronegative) group of the substrate. Thus, the intermolecular arrangement at the transition state is most likely to be similar to that shown in 1, polar groups facing each other. A magnesium ion probably assists their assembly.

Table 5. Reduction of Camphoroquinone with N-α-methylbenzyl-1-propyl-2,4-dimethyl-1,4-dihydronicotinamide (Me_2PNPH)[a]

Substrate	Config. of Me_2PNPH	Recov'd CQ,%[b]	Product Yield, %[c]	Product Ratio[c]			
				X-2a	*D*-2a	*X*-3a	*D*-3a
(-)-CQ	*RR*	57.7	40.6	8	19	68	5
(-)-CQ	*SS*	36.2	67.6	20	16	6	58
				X-2b	*D*-2b	*X*-3b	*D*-3b
(+)-CQ	*RR*	53.1	47.3	21	14	7	58
(+)-CQ	*SS*	50.9	58.7	7	21	62	10
				X-2[e]	*D*-2[e]	*X*-3[e]	*D*-3[e]
(±)-CQ[d]	*SS*	46.0	54.1	14	16	27	43

a) Reaction was run for 52 hr with each 1 mmol of reagent.
b) Isolated yields. c) Relative intensities of ^{1}H-NMR signals.
d) Racemic camphoroquinone. e) A mixture of a and b.

Table 6. Reduction of Camphoroquinone with NAD(P)H-Models

Substrate	Model	T, hr[a]	Recov'd CQ, %[b]	Product Yield, %[b]	Product Ratio[c]			
					X-2a	*D*-2a	*X*-3a	*D*-3a
(-)-CQ	BNAH	235	42.3	7.3	14	13	16	57
(-)-CQ	PNAH	48	65.6	4.3	13	11	24	52
(-)-CQ	*R*-PNPH	91	50.2	8.6	15	9	14	62
					X-2b	*D*-2b	*X*-3b	*D*-3b
(+)-CQ	*R*-PNPH	91	61.6	6.8	8	10	20	62

a) Reaction time. b) Isolated yields. c) Relative intensities of ^{1}H-NMR signals.

Based on the above results, steric and electronic effects of the substituents of a substrate have been studied further. Results from the reduction of a series of 2-fluoroacylpyridines and 2-acylpyridines indicate that substituent effects are such that the stereospecificity of the reduction is mainly governed by electronic effects. However, the steric bulk of the substituents exerts a certain effect on the conformation of these substrates(7 - 11).

The stereospecificity remains almost constant (>90% e.e.) for the reduction of substituted and unsubstituted α,α,α-trifluoroacetophenones in the presence of magnesium ion. On the other hand, the specificity changes with a change in electronic effect of the substituent for the reduction without magnesium ion. Both electron-releasing and -withdrawing substituents increase the specificity. The results cannot be accounted for by simple steric or electronic substituent effects in a one-step reaction. However, a multi-step mechanism with an initial electron-transfer process(12, 13) explains the variation of the specificity. An electron-releasing substituent reduces the electron-affinity of a substrate and the electron-transfer to a substrate of this sort requires a high activation energy, as illustrated in Scheme 4a. A substrate in this category would form an electron-transfer complex with Me_2PNPH, which is unstable. The subsequent proton-transfer takes place almost spontaneously. The stereochemistry of the net reduction will be defined in the initial electron-transfer step.

The selectivity-reactivity relationship predicts that the less the electron-releasing power of a substituent on the substrate, or the less the activation energy for the electron-transfer process, the less the difference in energy between preferred and other conformations. Consequently, the reduction becomes less stereospecific.

With a strongly electron-withdrawing substituent on a substrate, on the other hand, the electron-transfer takes place quite rapidly and the intermediate electron-transfer complex becomes more stable than the reactant system as shown in Scheme 4c. The preferential course of reduction in this category is, therefore, controlled by the thermodynamic stability of the intermediate, which makes strongly electron-demanding substrates more stereospecific than weakly electron-demanding ones. The stereochemistry of the net reduction is now defined in the second step. Scheme 4b represents the intermediate category, in which both the initial and second steps affect the stereospecificity of the reduction. In Scheme 4, full lines indicate the reduction without magnesium ion and dotted lines represent the reduction with magnesium ion. Since magnesium ion catalyzes the initial electron-transfer process, the stereochemistry of the net reduction in the presence of magnesium ion is controlled by energetics of the second step.

Scheme 4

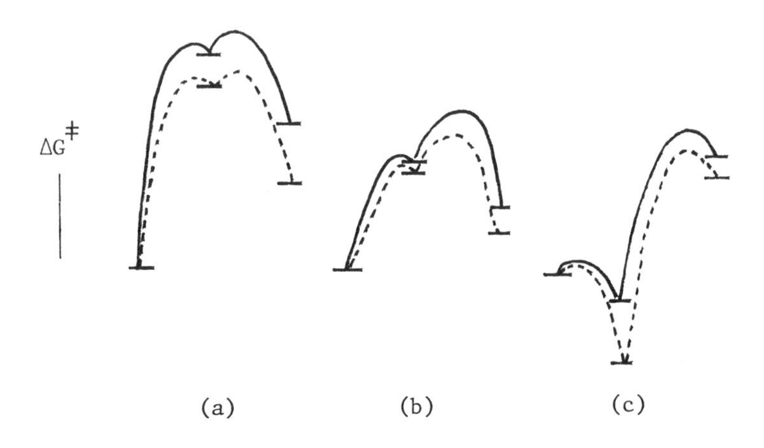

Literature Cited

1. Ohnishi, Y.; Kagami, M.; Ohno, A. *J. Am. Chem. Soc.* 1975, *97*, 4766
2. Ohnishi, Y.; Numakunai, T.; Ohno, A. *Tetrahedron Lett.* 1975, 3813.
3. Ohnishi, Y.; Numakunai, T.; Kimura, T.; Ohno, A. *Tetrahedron Lett.* 1976, 2699.
4. Ohno, A.; Yamamoto, H.; Kimura, T.; Oka, S. *Tetrahedron Lett.* 1977, 4585.
5. Bentley, R. "Molecular Asymmetry in Biology", Vol2, Academic Press, New York, 1970; pp 36 - 39.
6. Ohno, A.; Yasui, S.; Oka, S. *Bull. Chem. Soc. Jpn.* 1980, *53*, 2651.
7. Barassin, J.; Queguiner, G.; Lumbroso, H. *Bull. Soc. Chim. Fr.* 1967, 4707.
8. Osborne, R. R.; McWhinnie, W. R. *J. Chem. Soc., A* 1967, 2075.
9. Kidani, Y.; Noji, M.; Koike, H. *Bull. Chem. Soc. Jpn.* 1975, *48*, 239.
10. Gase, R. A.; Pandit, U. K. *J. Am. Chem. Soc.* 1979, *101*, 7059.
11. Ohno, A.; Yamamoto, H.; Oka, S. *J. Am. Chem. Soc.* 1981, *103*, 2041.
12. Ohno, A.; Shio, T.; Yamamoto, H.; Oka, S. *J. Am. Chem. Soc.* 1981, *103*, 2044.

RECEIVED December 14, 1981.

17

Stereochemistry of One-Carbon Transfer Reactions

HEINZ G. FLOSS

Purdue University, Department of Medicinal Chemistry and Pharmacognosy, School of Pharmacy and Pharmacal Sciences, West Lafayette, IN 47907

The steric course of a number of biological one-carbon transfer reactions has been studied by means of stereospecifically isotope-labeled substrates. These reactions include the transfer of the methylene group of serine to tetrahydrofolate catalyzed by serine transhydroxymethylase, the further utilization of the methylene group of methylene-tetrahydrofolate for the generation of the methyl group of thymidylic acid catalyzed by thymidylate synthetase, the transfer of the S-methyl group of S-adenosylmethionine to various acceptors catalyzed by a number of different methyltransferases, and the transfer of a methyl group from dimethylnitrosamine to DNA or model nucleophiles, a process thought to initiate carcinogenic cell transformation.

As part of a broader interest in stereochemical aspects of biological processes, our laboratory has recently carried out a variety of studies on the stereochemistry of biological one-carbon transfer reactions. Since biologically important single carbon units, like methyl groups, are not *per se* chiral, this work has required the use of one-carbon centers made chiral by virtue of isotopic substitution; for example, methyl groups which are chiral by virtue of the presence of normal hydrogen, deuterium and tritium. The synthesis of such species is not particularly difficult; it can be accomplished essentially by an extension of methods used widely to generate stereospecifically labeled prochiral centers. However, the configurational analysis, i.e., the determination whether an unknown sample represents a methyl group of R- or S- configuration presented a conceptually new problem. This was solved by the pioneering work carried out in the laboratories of Cornforth (1) and Arigoni (2). These authors developed a method which involves conversion of the methyl group in the form of acetic acid into acetyl-CoA followed by condensation with glyoxylate, catalyzed by malate synthase, to give malate, and equilibration with fumarase. Based on an isotope effect in the malate synthase reaction, the percentage tritium retention in the fumarase reaction, called the F value, indicates

0097-6156/82/0185-0229$05.00/0

the configuration and degree of purity of the acetate methyl group. An F value of 79 corresponds to an optically pure R methyl group, an F value of 21 is shown by a chirally pure S methyl group (3). This analytical methodology was employed in most of the studies to be reported here.

Our first study of an enzymatic one-carbon transfer reaction actually involved not the transfer of a methyl group but rather of a methylene group and was carried out in collaboration with the laboratory of Benkovic (4). The pyridoxal phosphate enzyme serine transhydroxymethylase catalyzes the conversion of serine and tetrahydrofolate into glycine and methylene-tetrahydrofolate as shown in Scheme I. Mechanistic considerations suggested that free or enzyme-bound formaldehyde must be a reaction intermediate. To probe this question, we carried out the reaction with serine stereospecifically tritiated in the 3 position and trapped the methylene-tetrahydrofolate generated immediately by further dehydrogenation to methenyl-tetrahydrofolate catalyzed by methylene tetrahydrofolate dehydrogenase. The stereospecific removal of one hydrogen from the methylene group by this enzyme simultaneously served to determine the tritium distribution between the two methylene hydrogens of methylene-tetrahydrofolate. Starting from serine carrying 100% of its tritium in one diastereotopic hydrogen we obtained, under single turnover conditions, methylene-tetrahydrofolate containing 76% of its tritium in one methylene hydrogen and 24% in the other. If the label in serine was in the other diastereotopic hydrogen, the mirror image tritium distribution in methylene-tetrahydrofolate was generated. If the reaction was allowed to go back and forth several times, the methylene group was randomly labeled. This characteristic reaction-dependent scrambling can be explained in either of two ways. Formaldehyde may be a reaction intermediate which remains enzyme bound during its transient existence except for a few molecules which dissociate from the enzyme and rebind before reacting with tetrahydrofolate. Alternatively, the enzyme may bind serine in two conformations around the α,β bond with each conformation reacting stereospecifically as illustrated in Scheme II. It is not possible at the moment to distinguish between these two alternatives, although circumstancial evidence favors the second possibility. The absolute steric course of the reaction was not apparent at the time but can now be written as shown in Scheme III based on the recent determination of the absolute configuration of tetrahydrofolate and additional studies in the laboratory of Benkovic.

Scheme III shows the experimental arrangement to study the second one-carbon transfer reaction we investigated, the formation of thymidylic acid from uridylic acid catalyzed by thymidylate synthetase. In this reaction, the methyl group of thymidylate is derived from the carbon and the two hydrogens of the methylene bridge plus H-6 of methylene-tetrahydrofolate. To study the stereochemistry of this reaction, we (5) synthesized serine stereospecifically labeled with tritium and deuterium at

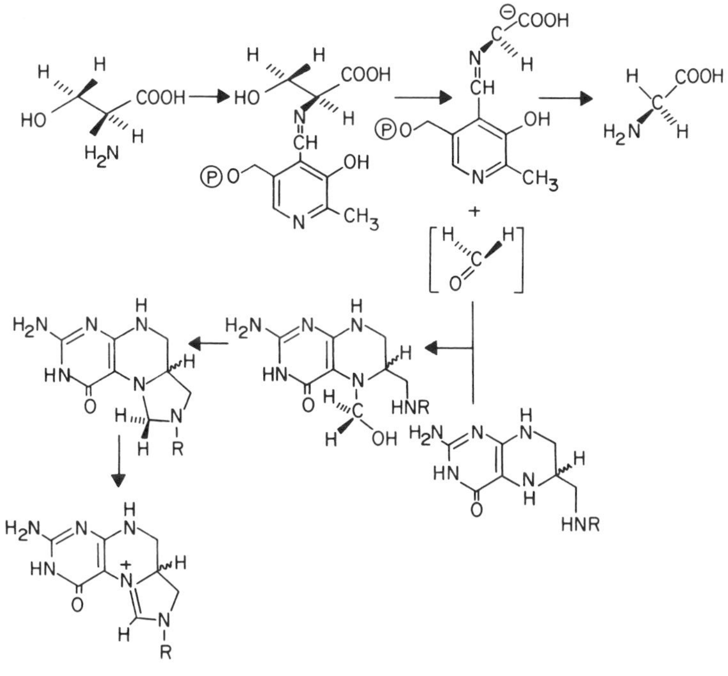

Scheme I. *Serine transhydroxymethylase mechanism and experimental setup for stereochemical analysis.*

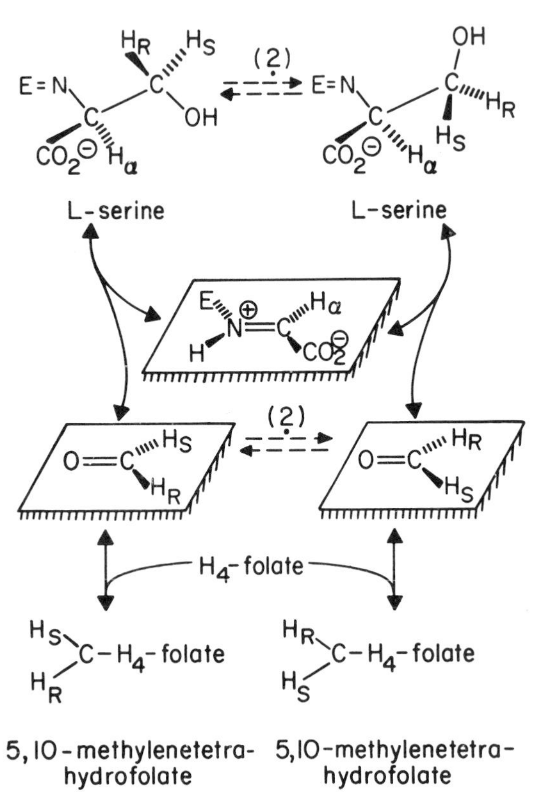

Scheme II. Stereochemical mechanism of serine transhydroxymethylase.

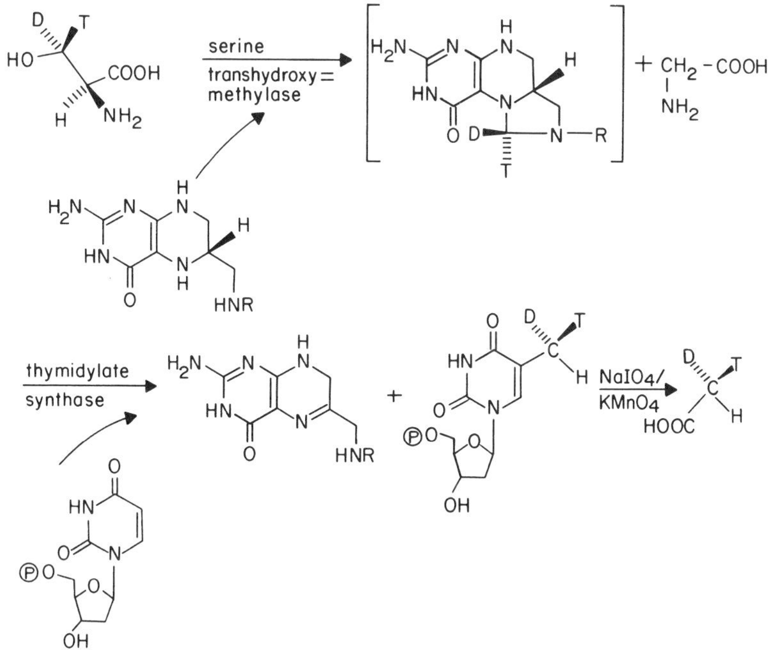

serine transhydroxy=methylase

thymidylate synthase

$NaIO_4$/ $KMnO_4$

Scheme III. Stereochemistry of serine transhydroxymethylase and thymidylate synthetase.

C-3 and converted it in a coupled reaction sequence into thymidylic acid which was then degraded to recover the methyl group as acetic acid. Chirality analysis of the acetic acid showed that it was indeed chiral and had the configuration shown in Scheme III. Generation of the methyl group of thymidylate from methylene-tetrahydrofolate involves four sequential bond breaking and forming steps at the stereospecifically labeled one-carbon unit; the results show that each of these steps occurs in a highly stereospecific manner. However, because of the multitude of steps involved, the result does not yet allow us to describe the steric course of each individual step.

The mode of formation of a methyl group seen in thymidylate is exceptional; most methyl groups in biological molecules arise from the S-methyl group of methionine. Our next goal was to determine the steric course of the transfer of a methyl group from methionine or S-adenosylmethionine (AdoMet) to various C-, N-, or O-atoms in biological molecules catalyzed by methyltransferase enzymes. Pursuit of this goal involved the following tasks:

1) Synthesis of methionine and AdoMet carrying a chiral methyl group of known configuration.

2) Enzymatic transfer of the methyl group to the substrate.

3) Degradation of the product to carve out the chiral methyl group and convert it into a compound suitable for configurational analysis, using only stereospecific reactions of known steric course.

4) Configurational analysis of the methyl group.

The synthesis of methionine and AdoMet carrying a chiral methyl group started from chiral acetate, which had been prepared as shown in Scheme IV (6). The conversion into methionine (Scheme V) involved a Schmidt reaction, known to proceed with retention of configuration, to give methylamine, which, in the form of its ditosylimide, was then used to alkylate the S-anion of homocysteine (6). The latter reaction was expected to proceed with inversion of configuration of the methyl group; the only plausible alternative, racemization due to an S_N1 mechanism, is ruled out by the subsequent finding that the methyl group was indeed still chiral. Enzymatic activation of the two samples of methionine (7) then gave AdoMet.

The first transmethylation studied was that catalyzed by catechol-O-methyltransferase (COMT) using either epinephrine (1a) or 3,4-dihydroxybenzoic acid (1b) as substrate. The products, metanephrine (2a) and 4-hydroxy-3-methoxybenzoic acid (2b), were degraded by the stereospecific reaction sequence shown in Scheme VI to give acetic acid carrying the chiral methyl group. It will be noted that the degradation sequence involves one inversion of the configuration of the methyl group in the cyanide displacement step (8).

Configurational analysis of the various acetic acid samples showed that AdoMet synthesized from acetate of F=28 gave 2a and 2b which, upon degradation, produced acetate of F=68 and 67, respectively. In the other enantiomeric series, the values were

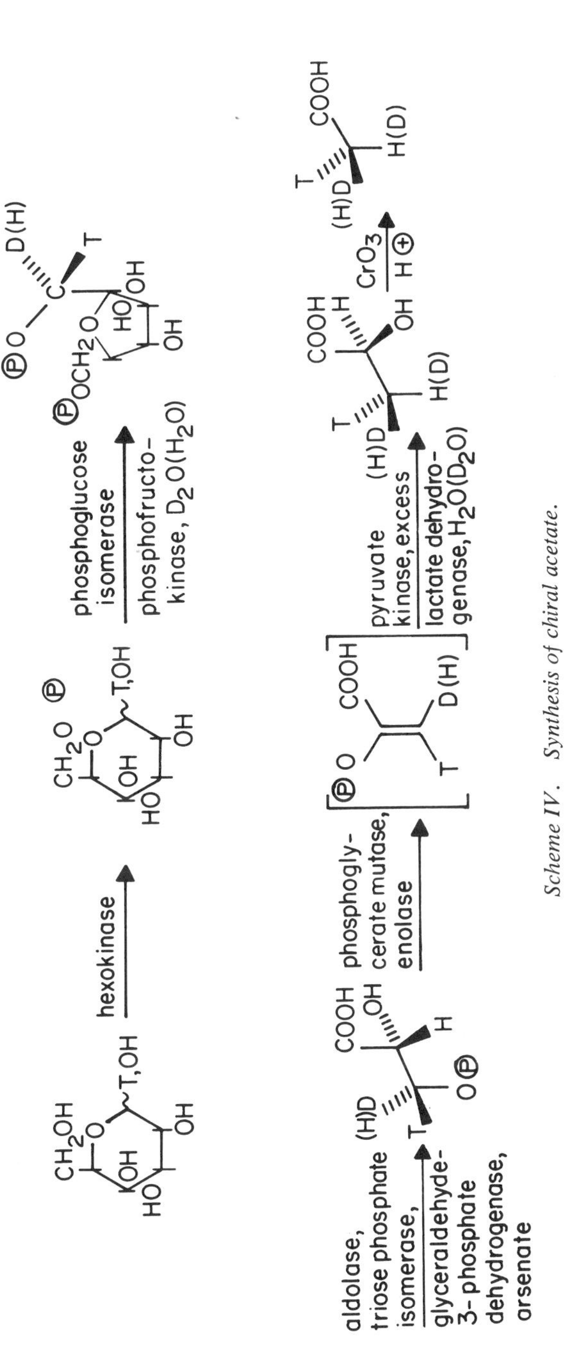

Scheme IV. Synthesis of chiral acetate.

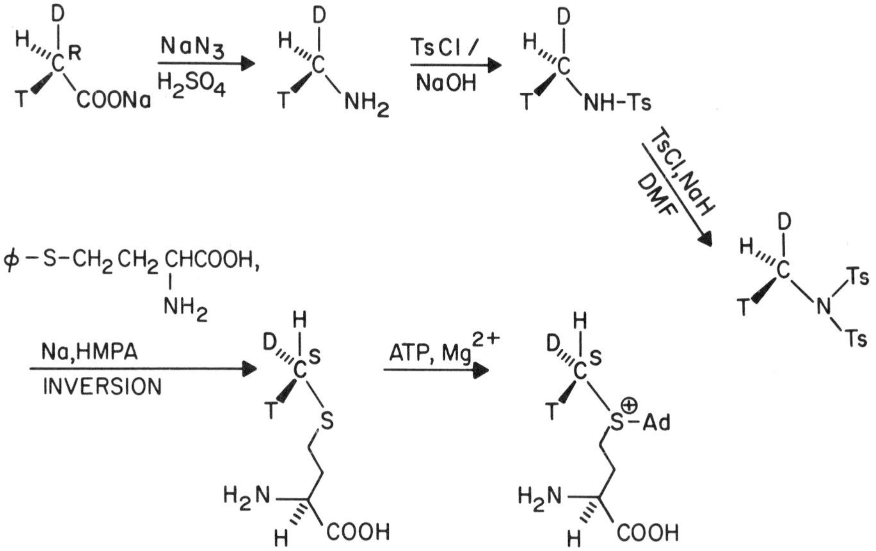

*Scheme V. Synthesis of chiral (*S*)-adenosylmethionine from chiral acetate.*

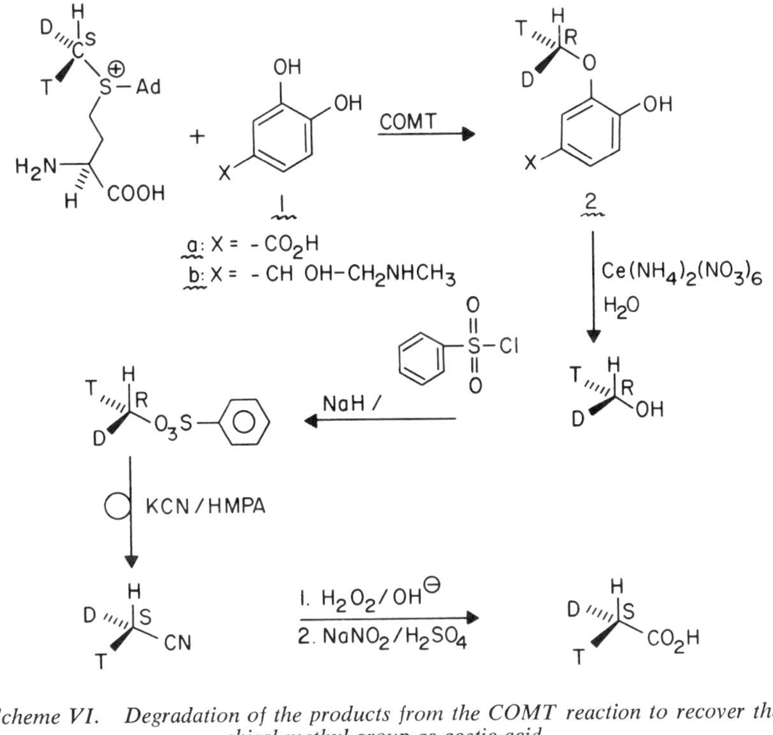

Scheme VI. Degradation of the products from the COMT reaction to recover the chiral methyl group as acetic acid.

F=68 for the starting acetate, and F=39 and 44 for the acetate samples from the degradation of 2a and 2b, respectively. Thus, there is an odd number of inversions in going from the starting acetate (Scheme V, upper left) to the final product (Scheme V, lower right). Since both the synthesis and the degradation each involve one inversion, it follows that the enzymatic transfer of the methyl group catalyzed by COMT must have occurred with inversion of configuration (8).

The same stereochemical course was also observed for another methyl transfer to oxygen, the methylation of the polygalacturonic acid carboxyl groups of pectin catalyzed by an enzyme preparation from mung bean shoots. The methyl group in this case was recovered by direct cyanolysis of the pectin to give acetonitrile (with inversion), which was then converted to acetate for analysis. Again, the starting and the final acetate samples had opposite configurations (F=28 → F=62; F=68 → F=32) (9).

In a microbial system, Streptomyces griseus, we studied simultaneously two methyl transfers, one to carbon and one to nitrogen, which are involved in the biosynthesis of the antibiotic indolmycin (10). In this case, chiral methionine was added directly to the cultures and the resulting indolmycin and indolmycenic acid were degraded as shown in Scheme VII. The results again indicated enzymatic transfer of the methyl group, both to carbon and to nitrogen, with inversion of configuration (6). Earlier work from our laboratory had shown that, in the C-methylation reaction leading to indolmycin, a hydrogen at C-3 of indole-pyruvate is replaced by the methyl group in a retention mode (11). Thus the stereochemistry of indolmycin biosynthesis in Streptomyces griseus can be summarized as shown in Scheme VIII.

In conclusion, all enzymatic methyl transfer reactions studied so far proceed with net inversion of configuration of the methyl group. All these methyl transferases therefore involve an uneven number of transfers of the methyl group, most likely a single, direct transfer from the sulfur of AdoMet to the acceptor atom in an S_N2-type reaction. Ping-pong mechanisms in which a group in the enzyme active site is transiently methylated can be excluded. The two substrates must be oriented in the enzyme active site such that in the transition state the sulfur, the methyl carbon and the acceptor atom form a linear array.

Methyl transferases are not only important in various metabolic processes, but also in the processing of informational macromolecules; for example, DNA by restriction methylases. Aberrations in this processing, as occur in the methylation by carcinogens like dimethylnitrosamine, are probably involved in the transformation of the cell into a tumor cell. This process involves metabolic activation of dimethylnitrosamine, in the manner shown in Scheme IX, to generate the ultimate carcinogen, a methyldiazonium ion which then transfers its methyl group to various nucleophilic sites on DNA. This latter process is presumably not enzyme-controlled and should therefore follow the same rules as the same process in an abiological system. Keeping

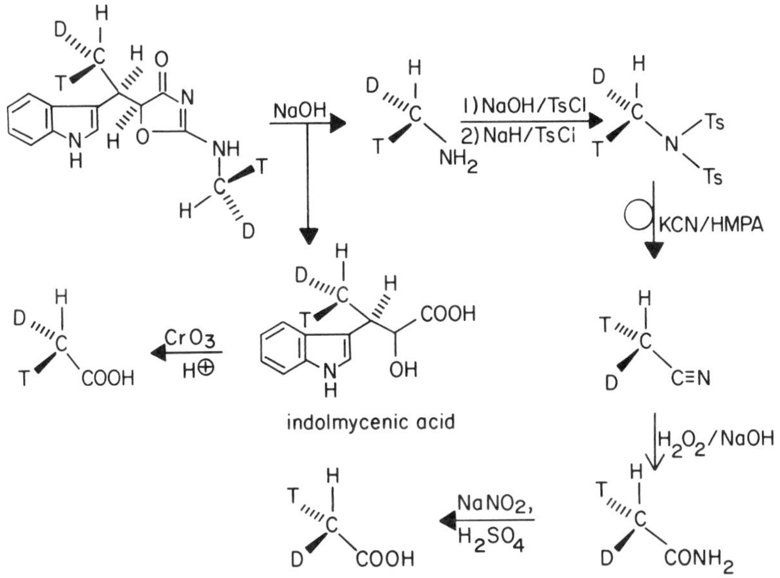

NaOH

1) NaOH/TsCl
2) NaH/TsCl

KCN/HMPA

H_2O_2/NaOH

$NaNO_2$, H_2SO_4

CrO_3 $H^{\oplus}$

indolmycenic acid

Scheme VII. Degradation of indolmycin and indolmycenic acid to recover the chiral methyl groups as acetic acid.

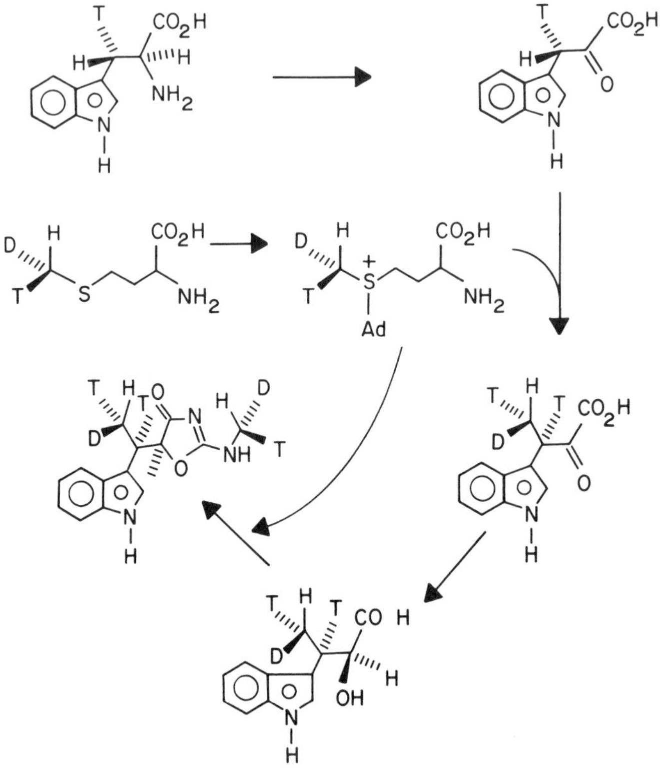

T CO_2H H NH$_2$ H N H → T CO_2H H O N H

D H T S CO_2H NH$_2$ → D H T $\overset{+}{S}$ Ad CO_2H NH$_2$

T H O T N H D D NH T N H

T H T CO_2H D O N H

T H T CO H D H OH N H

Scheme VIII. Steric course of indolmycin biosynthesis.

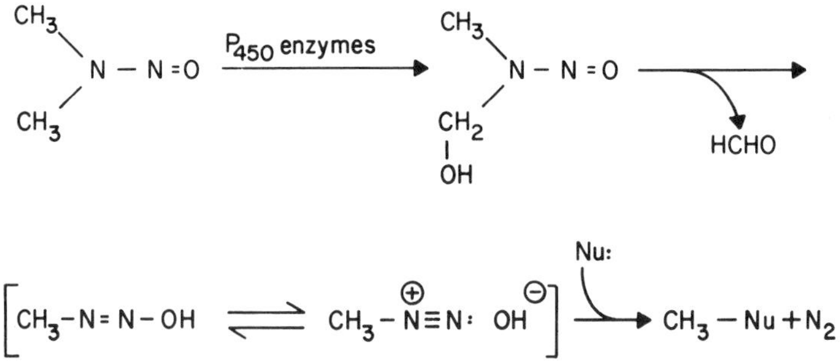

$$(CH_3)_2N-N{=}O \xrightarrow{P_{450}\text{ enzymes}} CH_3(HOCH_2)N-N{=}O \longrightarrow HCHO$$

$$\left[CH_3-N{=}N-OH \rightleftharpoons CH_3-\overset{\oplus}{N}{\equiv}N{:}\ \ OH^{\ominus}\right] \xrightarrow{Nu:} CH_3-Nu+N_2$$

Scheme IX. Metabolic activation of dimethylnitrosamine.

all other factors constant, the stereochemistry of this methyl transfer should be sensitive to the polarity of the reaction environment. Therefore, a comparison of the *in vitro* and the *in vivo* process should enable us to probe whether the reaction in the intact cell takes place in a hydrophobic environment, like the nuclear membrane, or in an aqueous surrounding. To lay the groundwork for experiments probing this question, we synthesized dimethylnitrosamine carrying a chiral methyl group by the reaction sequence shown in Scheme X. Methods for the alkylation of DNA and for the recovery of the methyl group from the most prominent modified base, 7-methylguanine, have been worked out, but results from the stereochemical analysis of these samples are not yet available. We have, however, completed the stereochemical analysis of the alkylation of a model nucleophile, 3,4-dichlorothiophenol, by dimethylnitrosamine activated with rat liver microsomes (12). The reaction sequence is shown in Scheme XI. Based on the structure of the alkyl group in this reaction and the nature of the nucleophile, we expected to see transfer of the methyl group with a high degree of inversion of configuration. However, the first set of analyses of the acetate samples from the degradation of the alkylated material shown in Table I suggests transfer of the methyl group with complete retention of configuration. This surprising result may indicate that even in this *in vitro* system, the reaction takes place entirely in the lipophilic microsomes and therefore proceeds exclusively by an ion pair mechanism with internal return. Alternatively, the mechanism of dimethylnitrosamine activation and alkyl transfer may be more complex than is currently envisioned and may, for example, involve a double displacement process. Further experiments are under way to verify the initial result and to study this problem further.

Table I. Stereochemical analysis of the alkylation of 3,4-dichlorothiophenol by metabolically activated dimethylnitrosamine.

	F-VALUE	CONFIGURATION	F-VALUE	CONFIGURATION
STARTING ACETATE	28	*S*	68	*R*
DIMETHYLNITROSAMINE	--	*S*	--	*R*
3,4-DICHLOROTHIOPHENOL METHYL ETHER	--	*S*	--	*R*
ACETATE FROM DEGRADATION	71	*R*	33	*S*

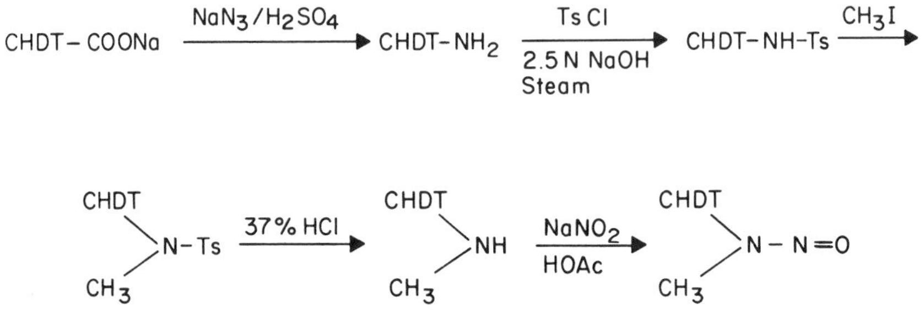

$$CHDT\text{-}COONa \xrightarrow{NaN_3/H_2SO_4} CHDT\text{-}NH_2 \xrightarrow[\text{2.5 N NaOH, Steam}]{TsCl} CHDT\text{-}NH\text{-}Ts \xrightarrow{CH_3I}$$

$$(CHDT)(CH_3)N\text{-}Ts \xrightarrow{37\%\ HCl} (CHDT)(CH_3)NH \xrightarrow[HOAc]{NaNO_2} (CHDT)(CH_3)N-N{=}O$$

Scheme X. Synthesis of dimethylnitrosamine carrying a chiral methyl group.

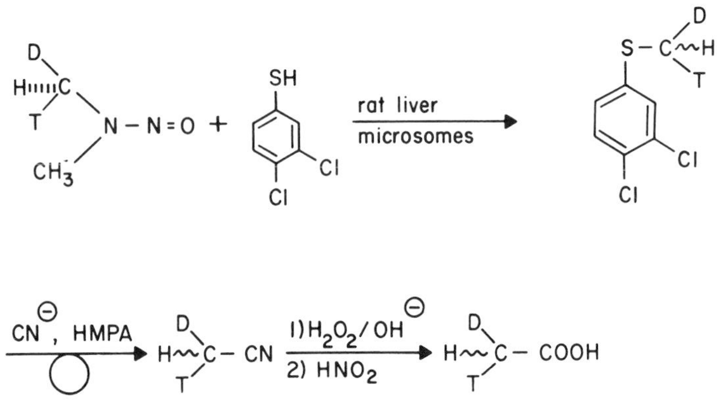

$$(HDTC)(CH_3)N-N{=}O + \text{3,4-dichlorothiophenol (SH, Cl, Cl)} \xrightarrow[\text{microsomes}]{\text{rat liver}} \text{3,4-}Cl_2C_6H_3\text{-}S-C(D)(H)(T)$$

$$\xrightarrow{CN^{\ominus},\ HMPA} H\text{-}C(D)(T)\text{-}CN \xrightarrow[2)\ HNO_2]{1)\ H_2O_2/OH^{\ominus}} H\text{-}C(D)(T)\text{-}COOH$$

Scheme XI. Alkylation of 3,4-dichlorothiophenol by chiral dimethylnitrosamine.

Acknowledgements

This work was supported by the National Institutes of Health. I wish to acknowledge with gratitude the enthusiastic contributions of numerous coworkers and collaborators whose names appear on the publications listed.

Literature Cited

1 Cornforth, J.W.; Redmond, J.W.; Eggerer, H.; Buckel, W.; Gutschow, C. Eur. J. Biochem., 1970, 14, 1.
2 Lüthy, J.; Rétey, J.; Arigoni, D. Nature (Lond.), 1969, 221, 1213.
3 For a review see: Floss, H.G.,; Tsai, M.D. Adv. Enzymol., 1979, 50, 243.
4 Tatum, C.M.; Benkovic, P.A.; Benkovic, S.J.; Potts, R.; Schleicher, E.; Floss, H.G. Biochemistry, 1977, 16, 1093.
5 Tatum, C.; Vederas, J.; Schleicher, E.; Benkovic, S.J.; Floss, H.G. J. Chem. Soc. Chem. Commun., 1977, 218.
6 Woodard, R.W.; Mascaro, L.; Hörhammer, R.; Eisenstein, S.; Floss, H.G. J. Amer. Chem. Soc., 1980, 102, 6314.
7 Cantoni, G.L. Biochem. Prep., 1957, 5, 58.
8 Woodard, R.W.; Tsai, M.D.; Floss, H.G.; Crooks, P.A.; Coward, J.K. J. Biol. Chem., 1980, 255, 9124.
9 Woodard, R.W.; Weaver, J.; Floss, H.G. Arch. Biochem. Biophys., 1981, 207, 51.
10 Hornemann, U.; Hurley, L.H.; Speedie, M.K.; Floss, H.G. J. Amer. Chem. Soc., 1971, 93, 3029.
11 Zee, L.; Hornemann, U.; Floss, H.G. Biochem. Physiol. Pflanzen (Jena), 1975, 168, 19.
12 Shen, S.J.; Tsai, M.D.; Floss, H.G., unpublished results.

RECEIVED December 14, 1981.

A Useful and Conveniently Accessible Chiral Stationary Phase for the Liquid Chromatographic Separation of Enantiomers

WILLIAM H. PIRKLE, JOHN M. FINN, BRUCE C. HAMPER, JAMES SCHREINER, and JAMES R. PRIBISH

University of Illinois, School of Chemical Sciences, Urbana, IL 61801

The use of a simply prepared high efficiency chiral HPLC column capable of separating the enantiomers of thousands of compounds is described, documentation being provided by more than 120 specific examples covering 19 classes of compounds. Chiral recognition models are presented to account for elution orders of the enantiomers. Practical applications of the chiral column, including preparative separations, are described.

It has long been perceived that chromatography of enantiomers upon a chiral stationary phase (CSP) might, in principle, result in separation of the enantiomers. Owing to the potential utility of such a resolution procedure, a great many workers have attempted to so effect resolutions. Most early attempts involve empirically chosen, readily accessible CSP's (e.g., starches, modified celluloses, wool) with varying degrees of success. More recently, synthetic CSP's (coupled with modern HPLC technology) have begun to afford impressive examples of chromatographically effected resolutions. Several recent reviews of this work are available (1-4).

The development of CSP's for the direct chromatographic separation of enantiomers is revolutionizing stereochemical analysis and will considerably alter future synthetic approaches to chiral compounds. From the analytical standpoint, an effective chiral HPLC column makes possible the accurate determination of enantiomeric purity upon as little as nanogram quantities of sample, absolute configurations being obtained simultaneously. From the preparative standpoint, multigram quantities of racemate can be resolved per pass through large chiral columns. Automation of this process will enable hundreds of grams of racemate to be resolved daily. Hence, a wide variety of chiral precursors will be available for the synthesis

0097-6156/82/0185-0245$05.00/0

of more complex chiral materials. In many cases, the final products will themselves be chromatographically resolvable. Finally, chromatographic resolution conveniently provides both enantiomers for biological evaluation.

The critical point in the preceding Utopian prediction is whether or not chiral columns can be devised which will indeed efficiently and predictably separate the enantiomers of a wide array of solutes. Work conducted in our laboratory in Urbana leads us to believe that such "Broad Spectrum" CSP's are clearly possible, that their chiral recognition mechanisms can be discerned, and that an understanding of these mechanisms can be used for the rational design of still more effective CSP's (5-10). To support this belief, let us describe a simply prepared chiral chromatography column capable of separating the enantiomers of thousands of compounds of diverse functional types.

Results and Discussion

Treatment of γ-aminopropylsilanized silica with a THF solution of R-N-3,5-dinitrobenzoylphenylglycine affords CSP 1, a CSP in which the chiral moiety is ionically bonded to the achiral support. This treatment may be performed either upon a prepacked HPLC column or upon bulk material (8, 9).

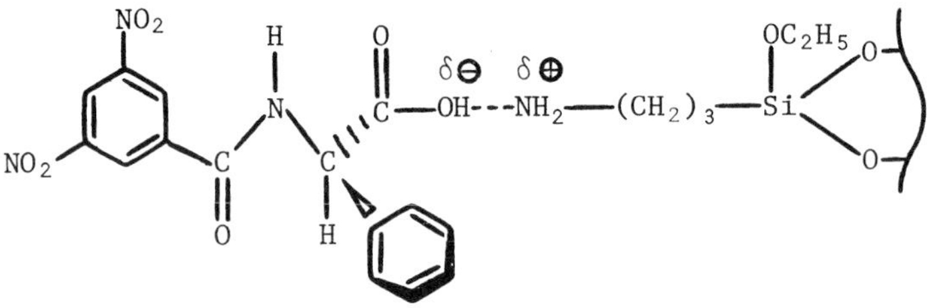

CSP 1

For analytical (and small scale preparative) applications, we modified a Regis 4.6 mm x 250 mm 5 μ Spherisorb NH_2 column. Using hexane-isopropyl alcohol as a mobile phase, we have been able to resolve the enantiomers of the types of compounds indicated in Tables I-IV. It can be seen from Figure 1 that this column is of high efficiency, enabling one to accurately determine enantiomeric purity (by peak area comparison) for compounds having an enantiomeric separability factor of 1.05 or greater (the enantiomeric separability factor, α, is simply the ratio of retention times of the enantiomers measured, not from injection, but from the elution point of a non-retained compound.) For compounds having an α value between unity and 1.05, multiple column or recycle techniques may have to be

Table I. Resolution of Some Alcohols Upon CSP 1 Using 1-10% 2-Propanol in Hexane.

BENZYL ALCOHOLS

Ar	R	α
Ph	Me	1.05*
Ph	t-Butyl	1.08*
α-Naph	Me	1.14*
α-Naph	CF_3	1.08*
9-Anth	CF_3	1.33*
9-Anth	n-Butyl	1.48
9-Anth	$CH_2COOC_2H_5$	1.27

PROPANOLOL ANALOGS (AS N-LAUROYL DERIVATIVE)

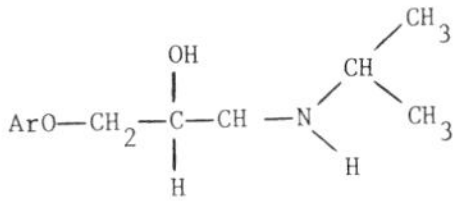

Ar	α
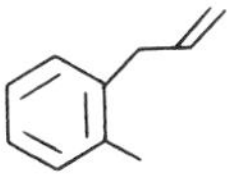 2-allylphenyl	1.08
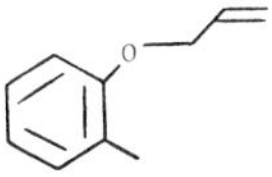 2-(allyloxy)phenyl	1.08
α-Naph	1.07

CYCLIC ALCOHOLS

R	α
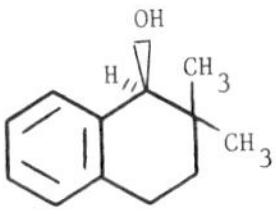	1.04*
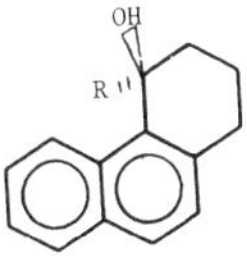 H	1.17
Me	1.37
n-Butyl	1.73
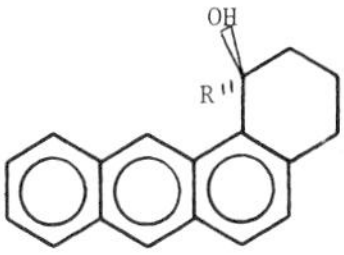 H	1.17
Me	1.39
n-Butyl	1.50
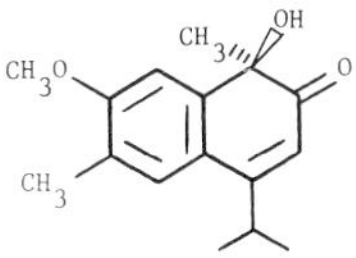	1.11

Table I. (continued).

HYDROXY SULFIDES

R_1	R_2	α
H	Me	1.03
H	n-$C_{12}H_{25}$	1.04
Me	Me	1.03
n-C_6H_{13}	n-$C_{10}H_{21}$	1.07/1.10
n-$C_{12}H_{25}$	H	1.06
H	Ph	1.06 (first?)

ARYL-SUBSTITUTED HYDROXY PHOSPHONATES

Ar	R	α
Ph	Ph	1.07
α-Naph	Et	1.19
p-Anisyl	Et	1.12

Table II. Resolution of Some Bi-β-naphthols Upon CSP 1 Using 10-20% 2-Propanol in Hexane.

BI-β-NAPHTHOLS

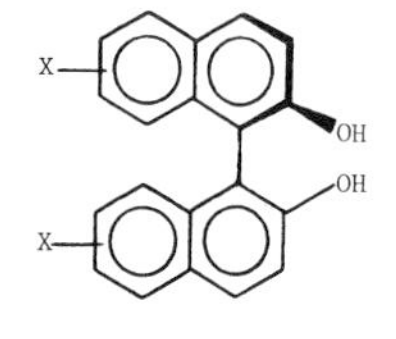

X	α
H	1.45*
6,7-$(Me)_2$	2.40*
6-Br	1.41*
7-OCH_3	1.51

X	R	α
H	CH_3	1.75*
H	C_2H_5	1.23*
6,7-$(Me)_2$	CH_3	3.67*

Table III. Resolution of Some Sulfoxides Upon CSP 1 Using 5-20% 2-Propanol in Hexane.

ARYL ALKYL SULFOXIDES

Ar	R	α
Ph	Me	1.05*
Ph	t-Butyl	1.09*
p-Tolyl	i-Propyl	1.10*
α-Naph	Me	1.09*
9-Anth	Me	1.19*
9-Anth	i-Propyl	1.22*
9-Anth	$CH_2COOC_2H_5$	1.20

DIARYL SULFOXIDES

Ar	Ar'	α
p-Tolyl	Ph	1.09
p-Tolyl	o-Tolyl	1.04
9-Anth	Ph	1.59
9-Anth	p-Anisyl	1.57
9-Anth	p-NO_2-C_6H_4	1.40
9-Anth	2,4-$(NO_2)_2C_6H_3$	1.20

SPIRO-2,2-DITHIOLANE-1-OXIDES

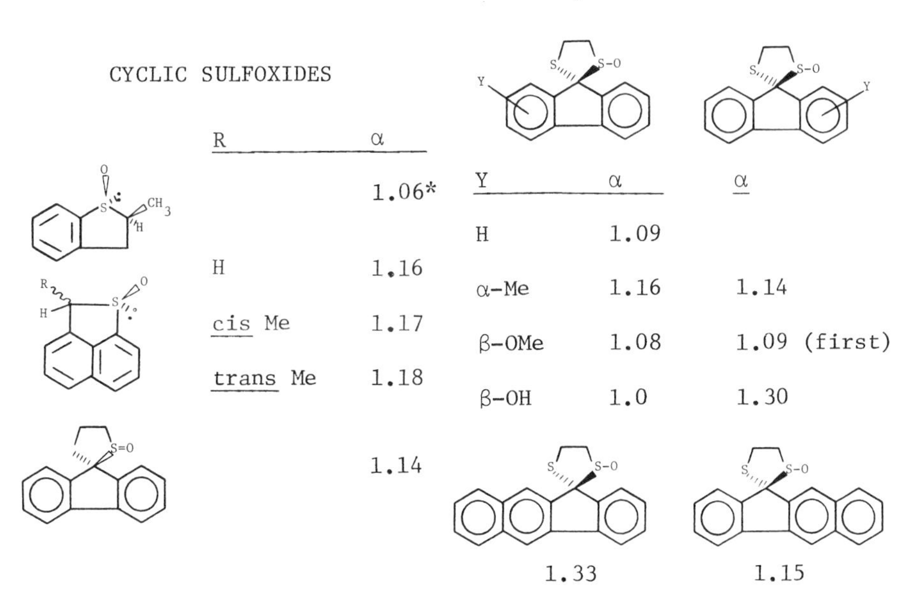

CYCLIC SULFOXIDES

R	α
	1.06*
H	1.16
cis Me	1.17
trans Me	1.18
	1.14

Y	α	α
H	1.09	
α-Me	1.16	1.14
β-OMe	1.08	1.09 (first)
β-OH	1.0	1.30

1.33 1.15

Table III. (continued)

β-HYDROXY SULFOXIDES

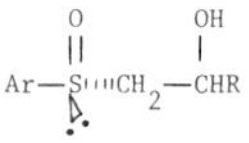

Ar—S(O)—CH$_2$—CHR(OH)

Ar	R	α
9-Anth	H	1.31
9-Anth	Me	1.38/1.22
9-Anth	$\underline{n}$-$C_{18}H_{37}$	1.52/1.28
9-Anth	Ph	1.18/1.04

β-HYDROXY SULFOXIDES CONTINUED

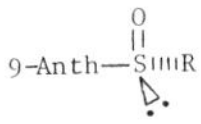

9-Anth—S(O)—R

R	α
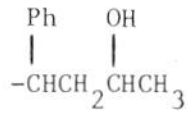 -CH(Ph)CH$_2$CH(OH)CH$_3$	1.42, 1.22, 1.13
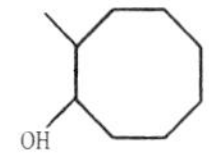 2-hydroxycyclooctyl	1.15/?
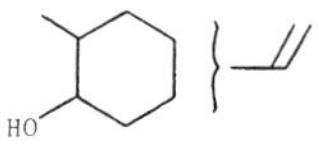 2-hydroxycyclohexyl; allyl	1.22, 1.26, 1.07, 1.11

1-ARYL-1-ALKYL-2,2-DITHIOLANE-1-OXIDES

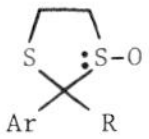

Ar	R	α
Ph	Et	1.06/NS
Ph	$\underline{n}$-$C_{11}H_{23}$	1.10/NS
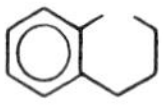 tetralin		1.13/1.04
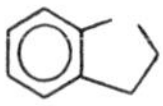 indan		1.10/≈1.01
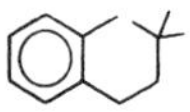 dimethyltetralin		1.24/NS

Note: Multiple entries for α are those of diastereomers. If in parentheses, α values refer to other positional isomers.

Table IV. Resolution of Some Amides Upon CSP 1.

ARYL-SUBSTITUTED SUCCINIMIDES

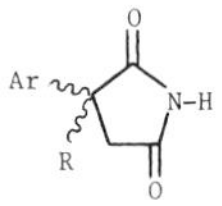

Ar	R	α
Ph	H	1.13
Ph	Me	1.07
Ph	Et	1.13
$\underline{p}$-Anisyl	H	1.24

ARYL ACETAMIDES

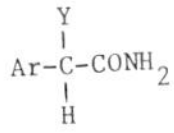

Ar-C(Y)(H)-CONH$_2$

Ar	Y	α
Ph	$\underline{i}$-Propyl	1.08
Ph	Methoxy	1.13
Ph	$\underline{i}$-Propoxy	1.30
Ph	SPh	1.03

Table IV. (continued)

ARYL-SUBSTITUTED LACTAMS

Ar	R	n	α
Ph	H	1	1.18* (1.02) (1.03)
p-Anisyl	H	1	1.33*
α-Naph	H	1	1.14
p-Anisyl	H	2	1.30
α-Naph	H	2	1.23

PHTHALIDES

Ar	R	α
Ph	H	1.03
Ph	Et	1.03*
p-Anisyl	H	1.07
α-Naph	H	1.13
α-Naph	Me	1.37
9-Anth	CF_3	1.20

OXAZOLIDONES

A	B	α
H	Ph	1.05*
H	α-Naph	1.04
Ph	Ph	1.02*

DIELS-ALDER ADDUCTS OF ACRYLAMIDES AND ANTHRACENES

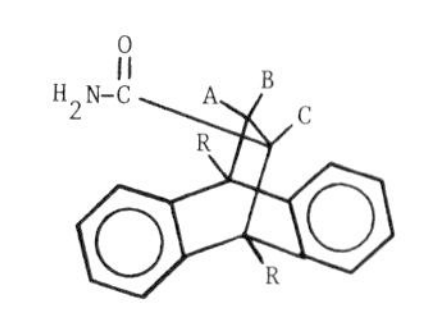

R	A	B	C	α
H	H	H	H	1.13
Me	H	H	H	1.40
Me	H	Ph	H	1.42
Me	H	Cl	H	1.56
Me	H	Br	H	1.80
Me	H	Br	Br	1.96

ARYL-SUBSTITUTED HYDANTOINS

Ar	R	Y	X	α
Ph	H	H	O	1.13
Ph	Et	H	O	1.26
p-Anisyl	i-Propyl	H	O	1.50
α-Naph	Me	H	O	1.35
α-Naph	$(CH_2)_2CH=CH_2$	H	O	1.48
β-Naph	CH_3	H	O	1.39
β-Naph	$(CH_2)_2CH=CH_2$	H	O	1.33
Ph	H	$COCH_3$	S	1.27
β-Naph	CH_3	CH_3	O	1.53

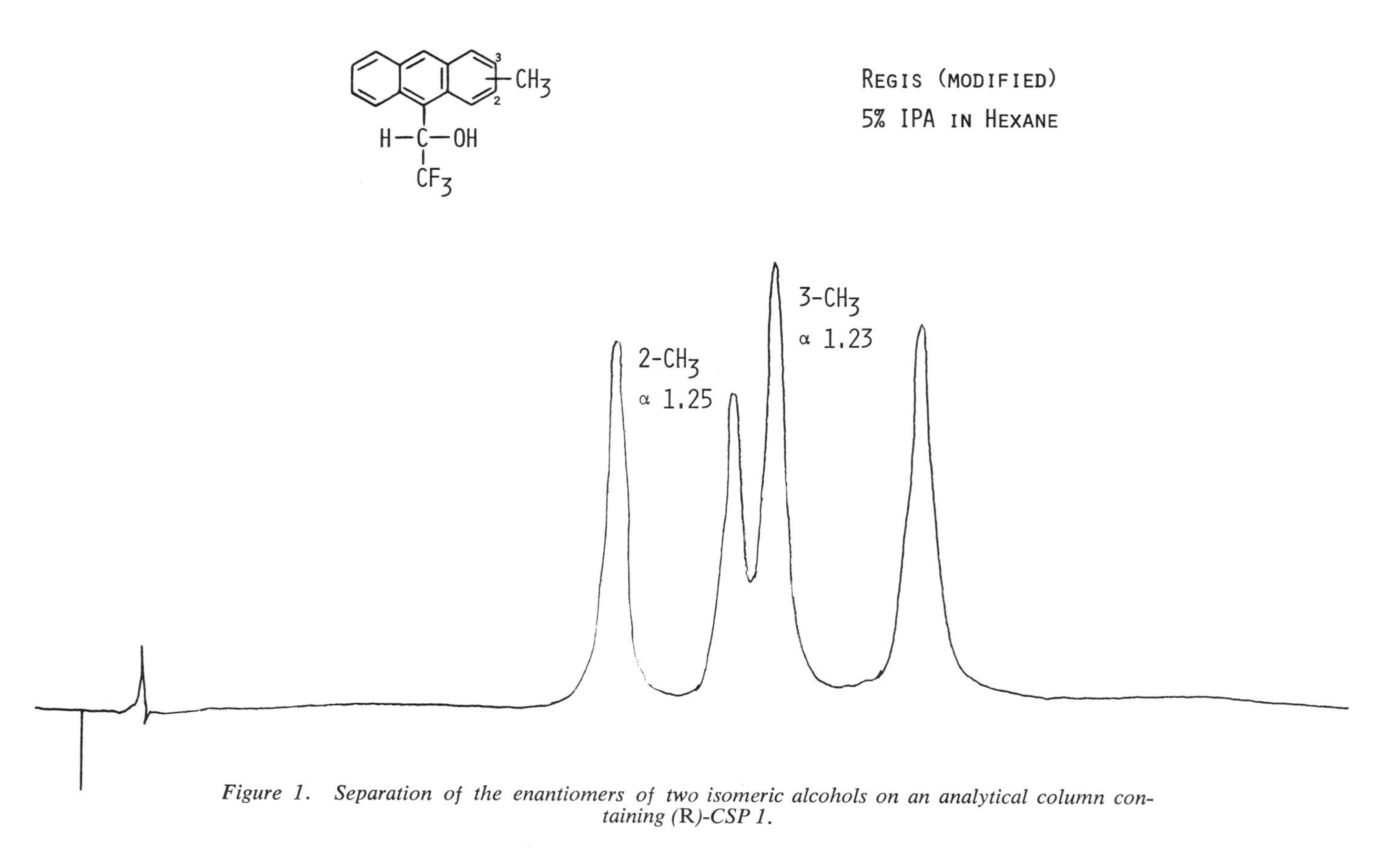

Figure 1. Separation of the enantiomers of two isomeric alcohols on an analytical column containing (R)-CSP 1.

employed. Each Table indicates the magnitude of α noted upon the chiral analytical column for the indicated solute. As a rule, many more compounds have been resolved within each class than are presented, great generality being encountered.

Elution order from the column is diagnostic of the absolute configuration of the enantiomers. When a stereochemical representation is shown for a generalized solute type, that enantiomer is either known (indicated by an asterisk) or believed (for mechanistic reasons) to be last eluted. Elution orders were determined either by chromatographing enriched samples of known configuration or through use of a polarimetric detector (Figure 2), the sign of rotation afforded by each enantiomer being noted as it eluted. These rotational signs were then compared to literature assignments of stereochemistry.

Chiral Recognition Mechanisms

The conformation depicted in CSP 1 is analogous to the solution conformations preferentially populated by amides of primary amines having a single α-hydrogen. This conformation is used in our present chiral recognition mechanisms. A minimum of three simultaneous interactions, at least one of which must be stereochemically dependent, is required for chiral recognition. CSP 1 uses the following types of interactions. The DNB group is used to π-complex to a π-base (usually an aryl group) in the solute, the amide hydrogen bonds to a basic site in the solute, the third stereochemically dependent interaction being either hydrogen bonding of the carboxylate group by an acidic solute site or steric repulsion between the phenyl of CSP 1 with a "steric barrier" contained within the solute. Steric repulsions are notably less effective for chiral recognition than are bonding interactions involving the carboxylate group. Solutes containing two acidic sites (and a π-base) but no suitable basic site can substitute a hydrogen bond to the DNB carbonyl oxygen for the hydrogen bond from the amide hydrogen. Figures 3-8 illustrate these multiple interactions between R-CSP 1 and the most strongly retained enantiomer for several solute types.

For benzyl alcohols, one can see from Figure 3 the simultaneous occurrence of a π-π interaction, a conventional hydrogen bond, and a weaker "carbinyl hydrogen bond" (8, 11) to the carbonyl oxygen of the DNB group. Altering the configuration of either chiral center breaks one of these bonding interactions, hence a stability difference occurs for the diastereomeric solvates. Figure 4 shows recognizably similar interactions between bi-β-naphthol and CSP 1. Increasing the π-basicity of the naphthol system or increasing the basicity of one of the oxygens by methylation increases the magnitude of α (Table II). Two of the interactions shown in Figure 5 are analogous to those just discussed. The third stereochemically dependent interaction seems to be repulsion between the steric

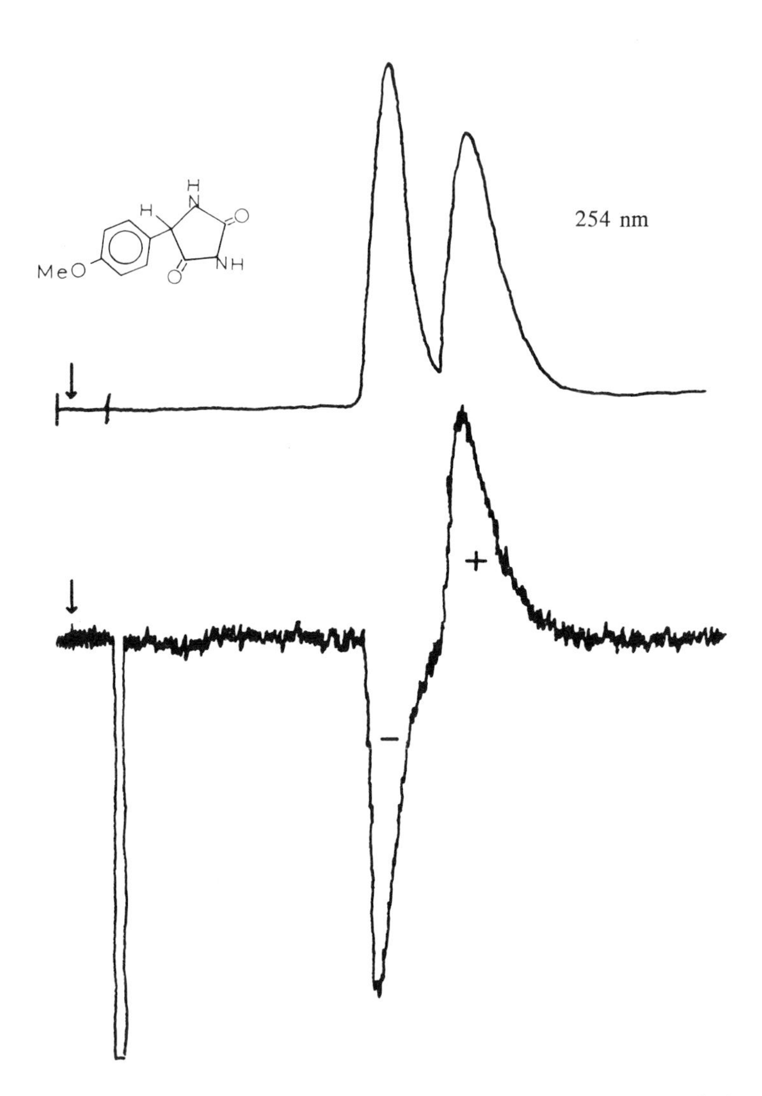

Figure 2. Separation on (R)-CSP 1 of the enantiomers of 5-anisylhydantoin employing UV (254 nm) and polarimetric (589 nm) detectors.

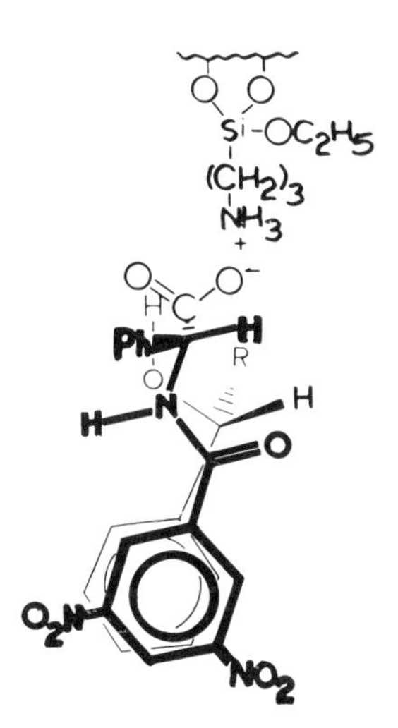

Figure 3. Chiral recognition model showing the relative arrangement for three simultaneous bonding interactions between (R)-CSP 1 and the most retained enantiomer of an alkyl aryl carbinol.

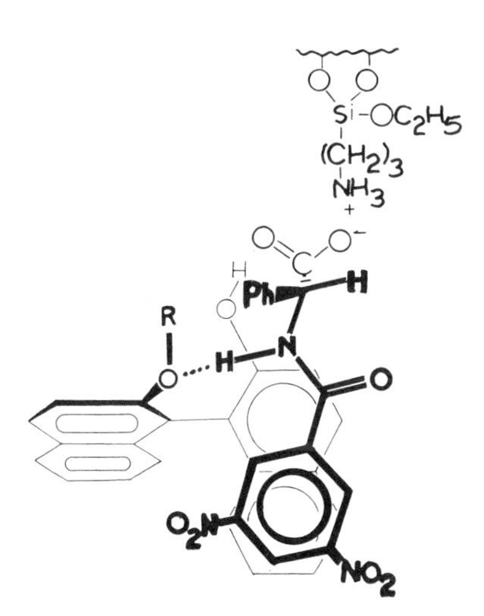

Figure 4. Chiral recognition model showing the relative arrangement for three simultaneous bonding interactions between (R)-CSP 1 and the most retained enantiomer of bi-β-naphthol.

barrier of the alicyclic ring and either the carboxylate or phenyl group of CSP 1, depending upon relative configuration. The interaction with phenyl is the most severe. Figure 6 shows rather similar interactions with aryl alkyl sulfoxides, the stereochemically dependent interaction being, at least in part, steric repulsion with the alkyl group. A similar model accounts for the resolution of diaryl sulfoxides, the π-π interaction occurring at the most π-basic aryl group, the steric barrier being the least π-basic aryl group. Note from the data in Table III that incorporation of a hydroxyl group into the alkyl group of 9-anthryl alkyl sulfoxides can enhance α by allowing hydrogen bonding to the carboxylate group to augment steric repulsion with the phenyl group. Consequently, SN_2 ring opening of an epoxide with 9-anthryl thiol followed by oxidation to the sulfoxide appears to offer promise in terms of HPLC enantiomeric purity determinations of epoxides. Figure 9 shows separation of the eight entities so derived from racemic disparlure, the Gypsy moth sex attractant.

Figures 7 and 8 show that chiral recognition of hydantoins and lactams by CSP 1 utilizes the same, now familiar, three bonding interactions. Table IV shows resolution data for these and other amide-like compounds.

The foregoing discussion makes clear that CSP 1 requires that solutes contain structural subunits capable of undergoing the required multiple simultaneous interactions employed by CSP 1 in effecting enantiomer separation. It should be obvious that not only must appropriate "complementary functionality" be present but that it must be arrayed so that it can effectively contribute to the chiral recognition process. Because HPLC can effectively reveal quite small stability differences between diastereomeric "solvates", the conformational behavior of both solute and stationary phase must be considered in advancing chiral recognition models to account for observed chromatographic behavior. One is seldom in a position to fully describe the conformations assumed by conformationally mobile molecules. Nevertheless, our work indicates that chiral recognition rationales of rather broad scope can be formulated and can be used in assigning absolute configurations to a variety of compounds. Moreover, the rationales can be used to formulate still more effective chiral chromatography columns.

Preparative Resolutions

Larger scale resolutions have been accomplished using a 2" x 30" column filled with CSP 1 derived from J. T. Baker 40 μ irregular "amino" silica. This column is considerably less efficient (in terms of total plates) than the analytical column but affords somewhat larger α values owing to the use of a different type of silica. Used in conjunction with a homemade automated prep chromatography system, we have been able to

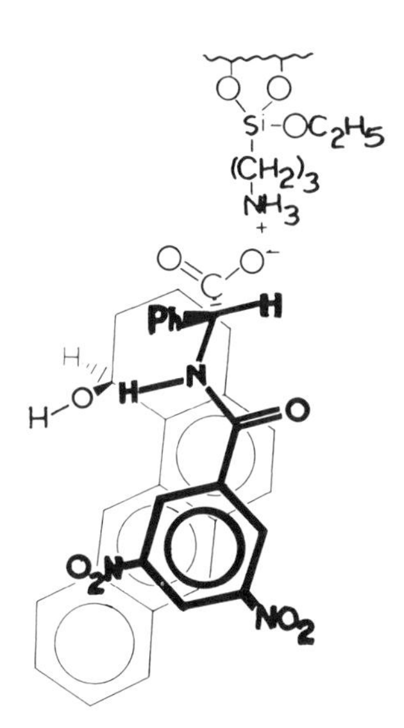

*Figure 5. Chiral recognition model showing the relative arrangement for two simultaneous bonding and one (least) repulsive interaction between (*R*)-CSP 1 and the most retained enantiomer of 1,2,3,4-tetrahydrobenz[*a*]anthracen-1-ol.*

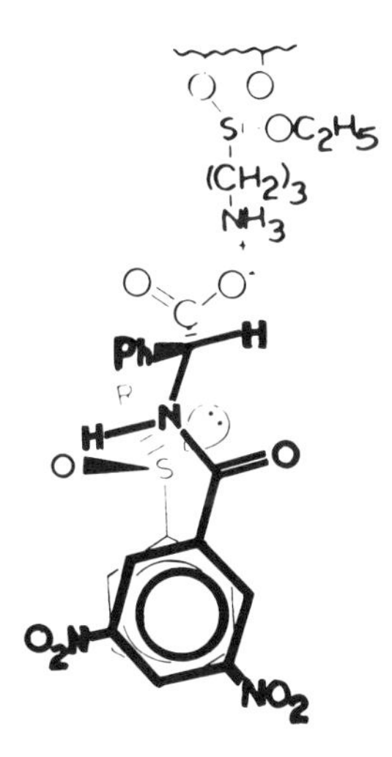

*Figure 6. Chiral recognition model showing the relative arrangement between (*R*)-CSP 1 and the most retained enantiomer of an alkyl aryl sulfoxide.*

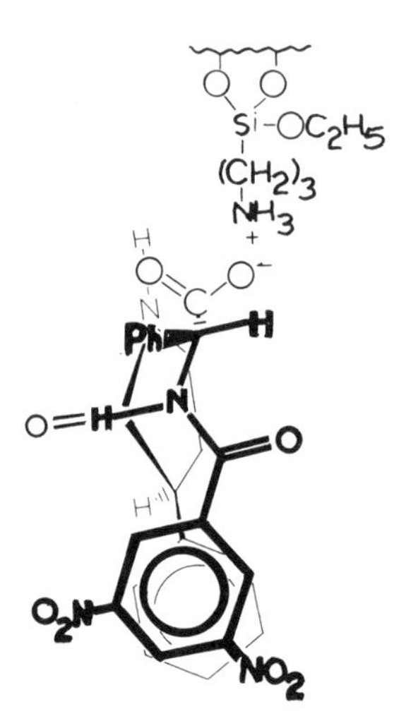

*Figure 7. Chiral recognition model showing the relative arrangement for three simultaneous bonding interactions between (*R*)-CSP 1 and the most retained enantiomer of a 3-aryllactam.*

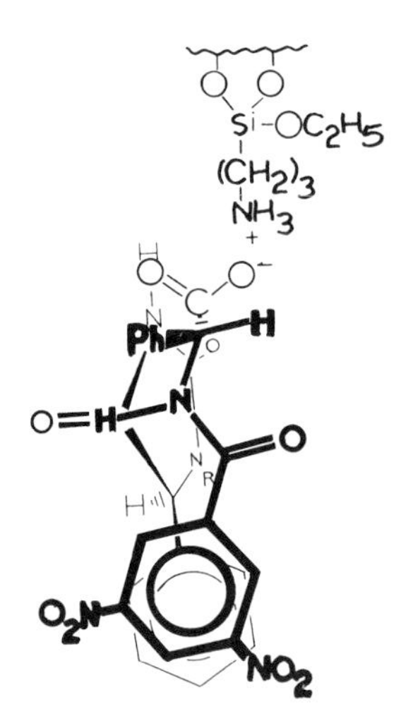

*Figure 8. Chiral recognition model showing the relative arrangement for three simultaneous bonding interactions between (*R*)-CSP 1 and a 5-arylhydantoin.*

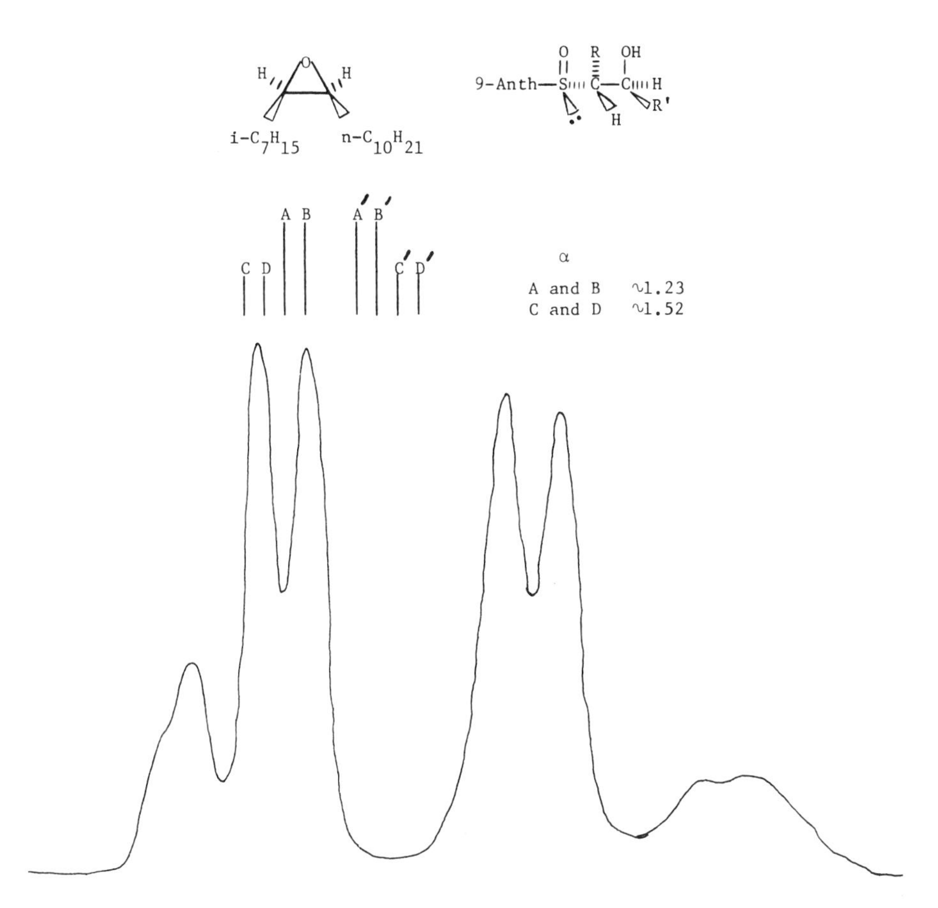

Figure 9. Chromatogram showing the separation on CSP 1 of the eight possible β-hydroxy sulfoxides derived from racemic disparlure. Bands bearing the same letter designation arise from enantiomers. A and B differ in relative stereochemistry from C and D. A and B (and C and D) are identical stereochemically but are regioisomers.

effect essentially total resolution of multigram samples of racemates having an α value of 1.4 or greater. For example, 4.0 g samples of racemic 2,2,2-trifluoro-1-[9-(10-methyl)-anthryl]ethanol [the precursor of another type of CSP (6, 7)] have been so separated into "first" and "second" chromatographic bands, the enantiomeric purities being 99 and 86%, respectively. Eight-gram samples of this racemate have been similarly resolved with only slightly poorer results. For samples having α values of 1.05 to 1.25, a more efficient prep column derived from 5 μ particles would be desirable.

Acknowledgement

This work has been supported by a grant from the National Science Foundation. A number of the samples used in these studies have been provided by colleagues throughout the world.

Literature Cited

1. Krull, I. S. Advan. Chromatogr. 1978, 16, 175.
2. Audebert, R. J. Liq. Chromatogr. 1979, 2, 1063.
3. Blaschke, G. Angew. Chem., Int. Ed. Engl. 1980, 19, 13.
4. Davankov, . Advan. Chromatogr. 1980, 18, 139.
5. Pirkle, W. H.; Sikkenga, D. L. J. Chromatogr. 1976, 123, 440.
6. Pirkle, W. H.; House, D. W. J. Org. Chem. 1979, 44, 1957.
7. Pirkle, W. H.; House, D. W.; Finn, J. M. J. Chromatogr. 1980, 192, 143.
8. Pirkle, W. H.; Finn, J. M. J. Org. Chem. 1981, 46, 2935.
9. Pirkle, W. H.; Finn, J. M.; Schreiner, J. L.; Hamper, B. C. J. Am. Chem. Soc. 1981, 103, 3964.
10. Pirkle, W. H.; Schreiner, J. L. J. Org. Chem. 1981, 46, 4988.
11. Pirkle, W. H.; Hauske, J. R. J. Org. Chem. 1976, 41, 801.

RECEIVED December 21, 1981.

SHORT COMMUNICATIONS

New Asymmetric Reactions Using (*S*)-2-Aminomethylpyrrolidine Derivatives

MASATOSHI ASAMI

University of Tokyo, Department of Chemistry, Faculty of Science, Tokyo, Japan 113

A new chiral auxiliary reagent, (S)-2-substituted-aminomethylpyrrolidine 1, has been designed based on the fundamental assumption that a conformationally restricted cis-fused five-membered bicyclic structure would be effective for asymmetric induction. The effectiveness of the new reagent was realized in the following highly stereoselective reaction:

Asymmetric reduction of prochiral ketones. A chiral hydride reagent formed by treating the chiral diamine 1 with $LiAlH_4$ was postulated to assume a cis-fused five-membered bicyclic ring structure 2. Highly enantiomerically pure alcohols were obtained when the reaction was carried out in ether at low temperature (-100°C) by employing diamines 1 having 2,6-xylyl or phenyl substituents on nitrogen.[1]

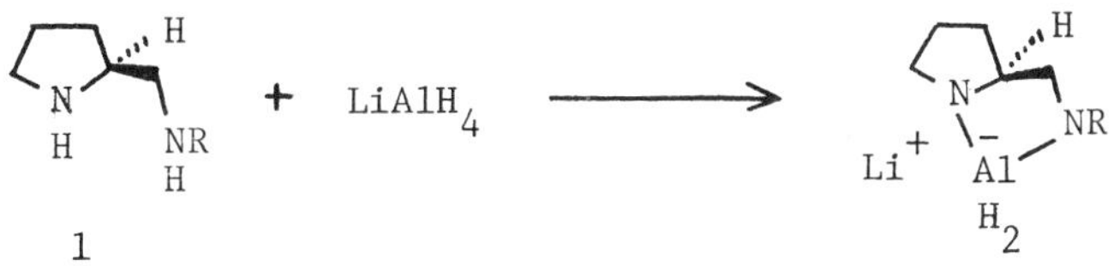

1 + $LiAlH_4$ ⟶ 2 (Li^+, N–Al^-H_2–NR)

$R^1C(=O)R^2$ —i) 2; ii) H_3O^+→ R^1*CH(OH)R^2

R= 2,6-dimethylphenyl (Me, Me)

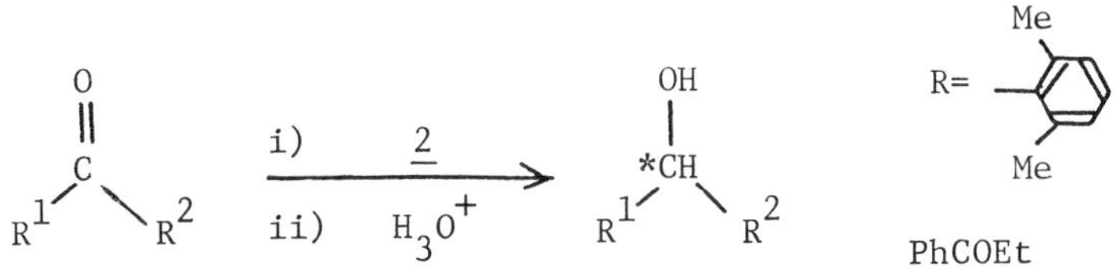

PhCOEt	96% ee
α-Tetralone	86% ee

0097-6156/82/0185-0263$05.00/0

Asymmetric synthesis of optically active aldehydes. The idea was extended to the synthsis of various synthetically useful optically active aldehydes utilizing aminals having a similar rigid structure as 2. The first example is an asymmetric 1,2-addition of Grignard reagents to a chiral keto aminal leading to various α-hydroxyaldehydes.[2] The utility of the aminal structure was also shown in an asymmetric 1,4-addition of Grignard reagents to an aminal 3, prepared from the diamine 1 (R=Ph) and fumaraldehydic acid methyl ester. Various 3-alkylsuccinaldehydic acid methyl esters were thus obtained with high optical yields.[3]

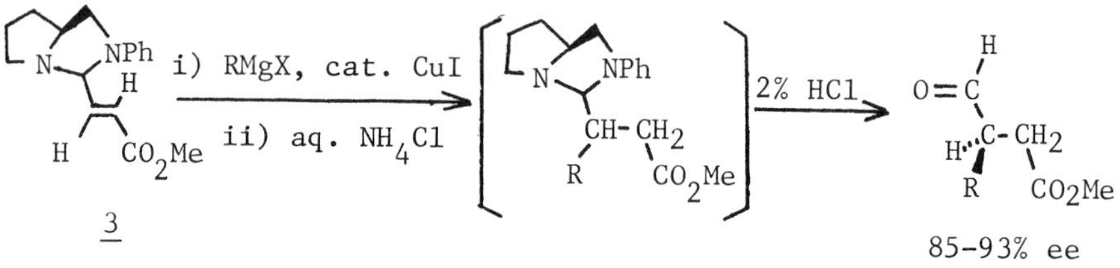

3 → i) RMgX, cat. CuI; ii) aq. NH_4Cl → [intermediate] → 2% HCl → $O{=}C(H)$–C(H)(R)–CH_2–CO_2Me

85-93% ee

Another highly enantioselective addition was achieved by using the chiral aryllithium derived from 4. Presumably the lithium compound assumes structure 5. Highly optically pure lactols 6 were obtained by its reaction with aldehydes. The resulting lactols 6 were successfully converted to optically active 3-alkylphthalide, e.g., (S)-3-butylphthalide, an essential oil of celery.[4]

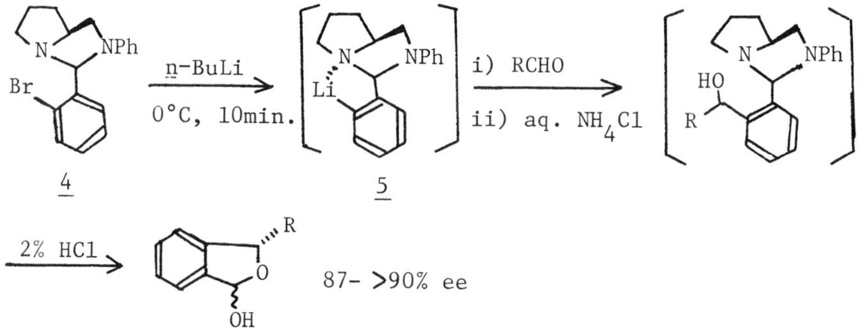

4 → n-BuLi, 0°C, 10min. → 5 → i) RCHO; ii) aq. NH_4Cl → [intermediate] → 2% HCl → 6

87- >90% ee

Bu OH $\xrightarrow[88\%]{Ag_2O}$ Bu O

(S)-3-Butylphthalide
88% ee

Literature Cited

1. a) Mukaiyama, T., Asami, M., Hanna, J., Kobayashi, S. *Chem. Lett.*, **1977**, 783. b) Asami, M., Ohno, H., Kobayashi, S., Mukaiyama, T. *Bull. Chem. Soc. Jpn.*, **1978**, *51*, 1864. c) Asami, M., Mukaiyama, T. *Heterocycles*, **1979**, *12*, 499.
2. See Mukaiyama, T., article in this volume.
3. Asami, M., Mukaiyama, T. *Chem. Lett.*, **1979**, 569.
4. Asami, M., Mukaiyama, T. *ibid.*, **1980**, 17.

RECEIVED December 21, 1981.

Liquid Chromatographic Resolution of Enantiomeric α-Amino Acid Derivatives Employing a Chiral Diamide Phase

SHOJI HARA, AKIRA DOBASHI, and MASAKATZU EGUCHI

Tokyo College of Pharmacy, Horinouchi, Hachioji, Tokyo 192-03, Japan

Liquid chromatographic resolutions based on highly selective host-guest, metal chelate and charge-transfer complexations have been described (1,2). Recently, a chiral diamide-bonded stationary phase (I) has been prepared, which relies entirely on hydrogen bond associations for the material to be resolved. Despite the weak and flexible interaction in this system, direct resolution of enantiomeric N-acyl-α-amino acid esters (II) was accomplished with the advent of a highly efficient column technology (2-4).

Amide derivatives of α-amino acid solutes (III) were tested for resolution. An increase in the bulkiness of the *N*-alkyl moiety R' improved the separation factors (α), i.e., the enantioselectivity. The highest α value (1.43; 2(v/v)% 2-propanol in *n*-hexane) was obtained for *N*-tert-butylamides. Thus, enantiomeric *N*-tert-butylamide derivatives of *N*-acyl-α-amino acids were separated with larger α values than corresponding *O*-alkyl ester derivatives.

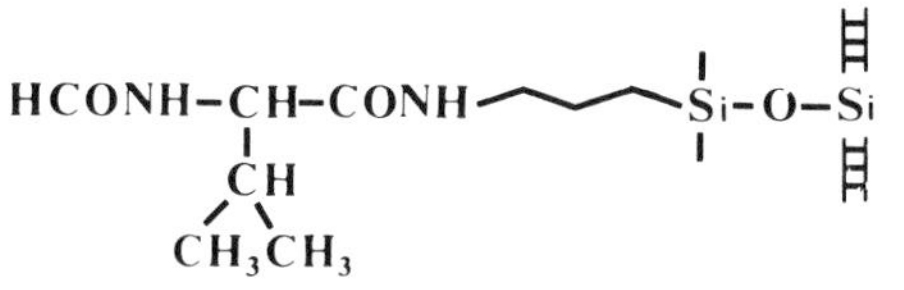

HCONH–CH(CH(CH_3)CH_3)–CONH–$(CH_2)_3$–Si–O–Si

(I)

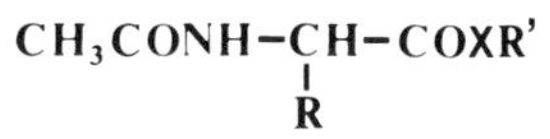

CH_3CONH–CH(R)–COXR'

X=O, (II) X=NH (III)

0097-6156/82/0185-0266$05.00/0

For characterization and exploitation of the diamide-phase system, a chiral diamide, e.g., (III) was examined as a modifier in the mobile phase (solvent) in conjunction with a non-bonded (bare) silica. Such a chiral carrier separated enantiomeric N-acyl-α-amino acid esters and amides with separation factors comparable to those for bonded stationary phase systems. The resolution can be ascribed to diastereomeric complexation through amide-amide hydrogen bonding between the amide additive and enantiomeric solute molecules in the carrier solvent, followed by separation of the diastereomeric complexes by the (achiral) silica phase. This process should be applicable as widely as that involving chiral diamide-bonded stationary phase systems.

Analytical high resolution of enantiomers was achieved with high sensitivity by using glass capillary micro-column technology based on these diamide-phase systems.

The amide phase systems are also applicable to preparative scale separations. A semi-preparative bonded column (10 mm i.d. x 25 cm) was prepared, yielding a loading capacity of ca. 1 mg per 1 g packing material. Enantiomeric and diastereomeric pairs of benzyloxycarbonyl and tert-butyloxycarbonyl protected di- and tri-peptides were resolved successfully using this chiral amide-bonded column system.

Literature Cited

1. Audebert,R. J. Liq. Chromatogr., Special Issues on Liquid Chromatographic Separation of Enantiomers, Diastereomers, and Configurational Isomers, Marcel Dekker, New York 1979, 2, 1063.

2. Dobashi, A.; Hara, S. Kagaku no Ryoiki, Special Issue on Biomedical Chromatography, Nanko-do, Tokyo 1981, 132, 171.

3. Hara, S.; Dobashi, A. J. Chromatogr. 1979, 186, 543.

4. Dobashi, A.; Oka, K.; Hara, S. J. Am. Chem. Soc., 1980, 102, 7122.

RECEIVED December 14, 1981.

Asymmetric Reduction with Chiral NADH Model Compounds

YUZO INOUYE
Kyoto University, Institute for Chemical Research, Uji, Kyoto, Japan 611

Novel chiral bis(1,4-dihydronicotinamide) derivatives bearing an *S*-prolinamide moiety were prepared and used to reduce ethyl phenylglyoxylate and other substrates. High optical yields of the reduction product *R*-mandelate (95.6-98.1%) were obtained with the *p*-xylylene- and hexamethylene-bridged bis(NAH) reductants.[1]

The e.e. was unaffected by an excess of Mg and also did not change at all during the course of reduction. Upon addition of Mg to bis(NAH), the carbonyl absorption band at 1680^{-cm} shifted to lower frequency,whereas the C-N band at 1605^{-cm} moved to a higher value. This shows[2,3] that Mg ion complexes to the primary amide carbonyl oxygen of prolinamide. The mole ratio method[4] showed the formation of a 1:1 complex between the bis(NAH) and Mg.

The Table shows the outcome of this reaction and the related ones. The spectral evidence, when combined with the stereochemical outcome that the e.e. was at a maximum when equimolar quantities of bis(NAH) and Mg were employed, shows that the present reduction with bis(NAH) is a single kinetically controlled process in contrast to that with the mono-derivatives.[5,6,7] The stereochemical requirements are well accommodated in a stoichiometric intramolecular chelation complex which assumes a C_2-conformation (I) with one specific diastereotopic face of the dihydropyridine moiety disposed toward the outside for the attack on substrates.

The chiral bis(NAH) reductants were easily regenerated by reduction of the resulting oxidized forms with aqueous solutions of

0097-6156/82/0185-0268$05.00/0

Table. Asymmetric Reductions with Chiral NADH Model Compounds

bis-NAH

-X-	Substrate	Reaction		Product			
		Temp. (°C)	Period (hr)	% Yield	$[\alpha]_D^{25}$ (°)	Config.	% ee
o-xylylene	PhCOCOOEt	25	23	69.5	-38.0	*R*	36.4
m-xylylene	PhCOCOOEt	25	23	61.5	-35.4	*R*	34.0
p-xylylene	PhCOCOOEt	25	1	66.6	-102.4	*R*	98.1
	PhCOCOOEt	50	2	79.8	-97.8	*R*	93.5
	2-pyridyl-COMe	60	16	66.9	+50.8	*R*	89.7
	2-pyridyl-COPh	50	100	71.5	-123.3	-	99.7
	Ph(Me)C=C(CN)$_2$	25	192	16.8	+4.1	*R*	23.2
	dihydro-4,4-dimethylfuran-2,3-dione	25	67	100	-19.3	*R*	38.1
-(CH$_2$)$_4$-	PhCOCOOEt	20	19	50.4	-41.7	*R*	39.9
-(CH$_2$)$_5$-	PhCOCOOEt	20	17	73.4	-44.9	*R*	43.0
-(CH$_2$)$_6$-	PhCOCOOEt	20	17	63.5	-99.8	*R*	95.6
	2-pyridyl-COMe	50	23	84.3	-37.8	*R*	66.7
	2-pyridyl-COPh	50	23	67.0	-114.6	-	92.7
	Ph(Me)C=C(CN)$_2$	50	96	59.2	+4.3	*R*	24.5
	dihydro-4,4-dimethylfuran-2,3-dione	50	23	65.7	-17.2	*R*	33.9
-(CH$_2$)$_7$-	PhCOCOOEt	20	15	57.9	-61.3	*R*	58.7
-(CH$_2$)$_8$-	PhCOCOOEt	20	15	63.2	-84.8	*R*	81.2
mesitylylene (tris-NAH)	PhCOCOOEt	20	16	16.5	+18.3	*S*	17.6

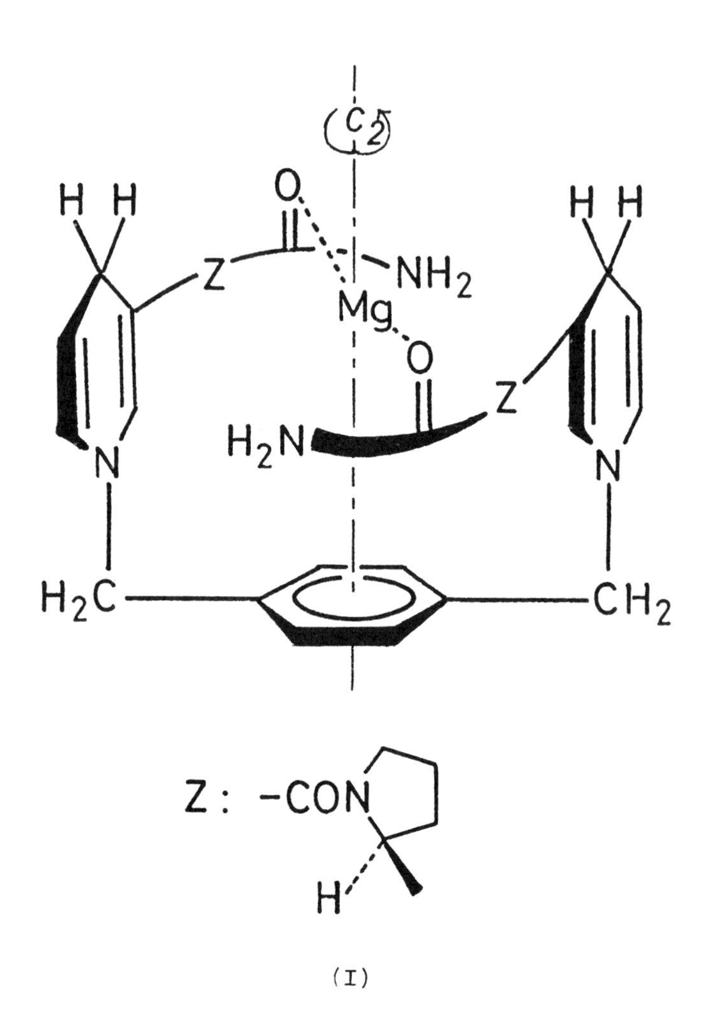

(I)

sodium hydrosulfite in 42-60% recovery and can be recycled with the optical yields remaining unchanged.

The mesitylylene-bridged tris(NAH) derivative of *S*-prolinamide and the *p*-xylylene-bridged bis(NAH) derivative of *S*-prolinol switched the steric course of reduction so as to give the enantiomeric *S*-mandelate in lower e.e.

Literature Cited

1. Seki,M;Baba,N.;Oda,J.;Inouye,Y.*J.Amer.Chem.Soc.*, 1981, 103, 4613.
2. Hughes,M;Prince,R.H.;Weyth,P. *J.Inorg.Nucl.Chem.*, 1978,40,713.
3. Kanzaki,M.;Nonoyama,M.;Yamazaki,K. *Kagaku*,1971, 27, 1182.
4. Yoe,J.H.;Jones,A.L. *Ind.Eng.Chem. Anal.Ed.*,1944, 16, 111.
5. Makino,T.;Nunozawa,T.;Baba,N.;Oda,J.;Inouye,Y. *J.Chem.Soc. Perkin I*, 1980,7.
6. Makino,T.;Nunozawa,T.;Baba,N.;Oda,J.;Inouye,Y.*Tetrahedron Lett.* 1979, 1683.
7. Baba,N;Oda,J.;Inouye,Y. *J.Chem.Soc., Chem.Commun.*,1980,815.

RECEIVED January 4, 1982.

Asymmetric Hydrogenation of Cyclic Dipeptides Containing α,β-Dehydroamino Acid Residues and Subsequent Preparation of Optically Pure α-Amino Acids

NOBUO IZUMIYA

Kyushu University, Laboratory of Biochemistry, Faculty of Science, Higashi-ku, Fukuoka 812, Japan

AM-Toxin I (I, Scheme 1) is a host specific phytotoxin. To elucidate the role of the double bond in the ΔAla^2 residue, we planned to prepare [L-Ala^2]- or [D-Ala^2]-AM-toxin I (II) by hydrogenation of AM-toxin I. By way of a preliminary study we hydrogenated cyclo(ΔAla-L-Leu) (III) and observed unexpectedly high asymmetric induction, affording pure cyclo(L-Ala-L-Leu) (IV).

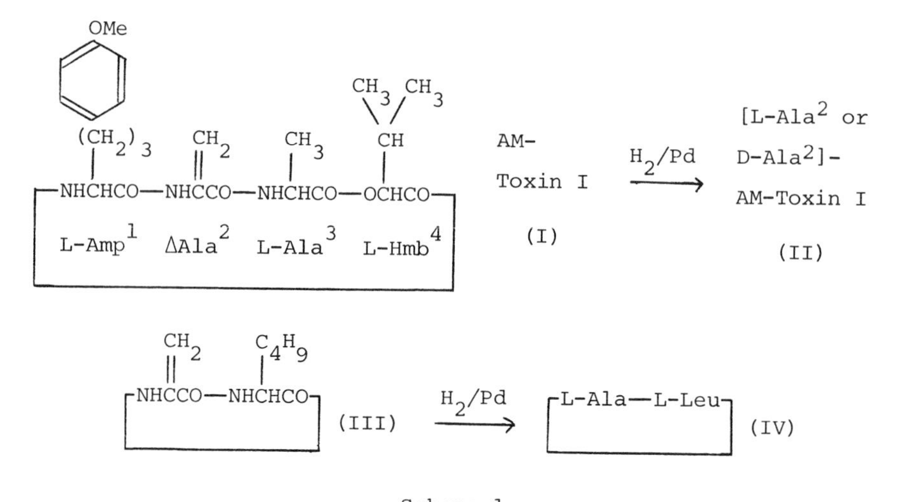

Scheme 1

Cyclo(L-Ser-L-AA) (AA=Ala, Val, Phe or Lys(ε-Ac)) was converted by the Photaki method (1) into the corresponding cyclo (ΔAla-L-AA) and subsequently hydrogenated with Pd black in methanol at 25°C affording cyclo(Ala-L-AA) (2). Generally high chiral inductions defined as %L-Ala minus %D-Ala in the cyclo(Ala-L-AA) were observed, ranging from 92 to 98% with yields of 63-75%. Similarly high chiral inductions (96-99%) were observed for

0097-6156/82/0185-0272$05.00/0

hydrogenation of a series of cyclo(ΔAA'-L-AA) (ΔAA'=ΔAba, ΔVal or ΔLeu) prepared from cyclo(Gly-L-AA) and appropriate aldehydes (3). Hydrogenation of ΔPhe or ΔTrp in cyclo(ΔPhe or ΔTrp-L-AA) resulted in a slightly lower asymmetric induction at 25°C (3). Hydrogenation of a series of its higher homologs (e.g. cyclo(ΔHomophe-L-Ala)), however, afforded high induction (4). Cyclo(ΔPhe-L-Ala) was hydrogenated with high chiral induction at a low temperature, 0°C (4).

Optically pure α-amino acids can be prepared by this route. For example, pure cyclo(L-Aba-L-Lys(ε-Ac)) obtained from cyclo (ΔAba-L-Lys(ε-Ac)) was hydrolyzed by 6 M HCl to give pure L-Aba (L-2-aminobutanoic acid) (3). Pure L-App (L-2-amino-5-phenyl-pentanoic acid) was prepared from cyclo(ΔApp-L-Ala) and used for the synthesis of AM-toxin II (5). $^{2}H_2$-D-Phe was prepared from cyclo(ΔPhe-D-Lys(ε-Ac)) and deuterium at 0°C, and synthesis of [$^{2}H_2$-D-Phe$^{4,4'}$]-gramicidin S (cyclic decapeptide) for NMR investigation is under study.

The mechanism of the chiral inductions has been discussed (2-4). In cyclo(ΔAla or ΔLeu-L-AA), the rigid and planar structure of the diketopiperazine ring and the side chain containing the double bond is an important factor inducing high asymmetry. In cyclo(ΔPhe or ΔTrp-L-AA), however, the diketopiperazine ring and the aromatic ring cannot be coplanar; a somewhat poorer stereoselectivity in the adsorption of the diketopiperazine ring on Pd is assumed to lower the degree of asymmetric hydrogenation.

Acknowledgments

The author thanks Drs. S. Lee, T. Kanmera, H. Aoyagi, and Y. Hashimoto in his laboratory for their experimental assistance.

Literature cited

1. Photaki, I. J. Am. Chem. Soc. 1963, 85, 1123.
2. Lee, S.; Kanmera, T.; Aoyagi, H.; Izumiya, N. Int. J. Pept. Protein Res. 1979, 13, 207.
3. Kanmera, T.; Lee, S.; Aoyagi, H.; Izumiya, N. Int. J. Pept. Protein Res. 1980, 16, 280.
4. Hashimoto, Y.; Aoyagi, H.; Izumiya, N. Int. J. Pept. Protein Res. in preparation.
5. Shimohigashi, Y.; Izumiya, N. Int. J. Pept. Protein Res. 1978, 12, 7.

RECEIVED December 14, 1981.

2,2'-Bis(diphenylphosphino)-1,1'-binaphthyl: A New Axially Dissymmetric Bis(triaryl)phosphine

A. MIYASHITA and H. TAKAYA

Okazaki National Research Institutes, Chemical Materials Center, Institute for Molecular Science, Okazaki 444, Japan

R. NOYORI

Nagoya University, Department of Chemistry, Chikusa, Nagoya 464, Japan

The title diphosphine (BINAP, 1) was synthesized in optically pure form according to the procedure outlined in Scheme I.[1] The absolute configuration of BINAP was determined by the X-ray crystallographic analysis of [Rh((R)-(+)-1)(norbornadiene)]ClO_4 [(R)-6].

Scheme I

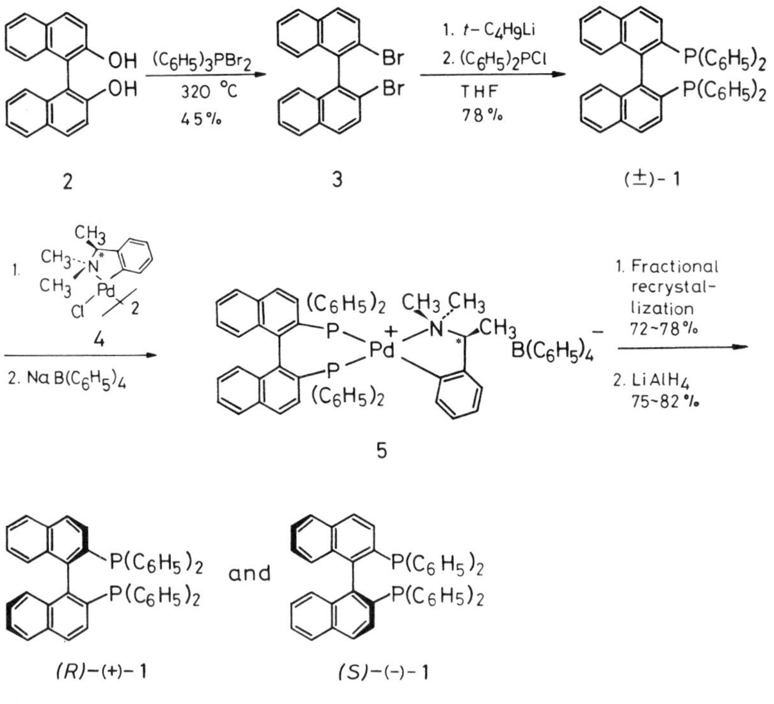

2 —($(C_6H_5)_3PBr_2$, 320 °C, 45%)→ 3 —(1. t-C_4H_9Li; 2. $(C_6H_5)_2PCl$; THF, 78%)→ (±)-1

—(1. 4; 2. $NaB(C_6H_5)_4$)→ 5 —(1. Fractional recrystallization, 72~78%; 2. $LiAlH_4$, 75~82%)→ (R)-(+)-1 and (S)-(−)-1

0097-6156/82/0185-0274$05.00/0

Two kinds of Rh complexes, (*S*)-**7** and (*S*)-**8**, were formed in a ratio of 9:1 when (*S*)-**6** in methanol was exposed to hydrogen. The major complex (*S*)-**7** serves as a highly efficient catalyst for the asymmetric hydrogenation of α-acylaminoacrylic acids and esters as shown in Table I.

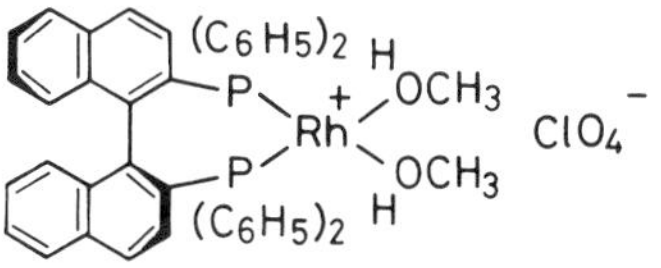

(*S*)-7

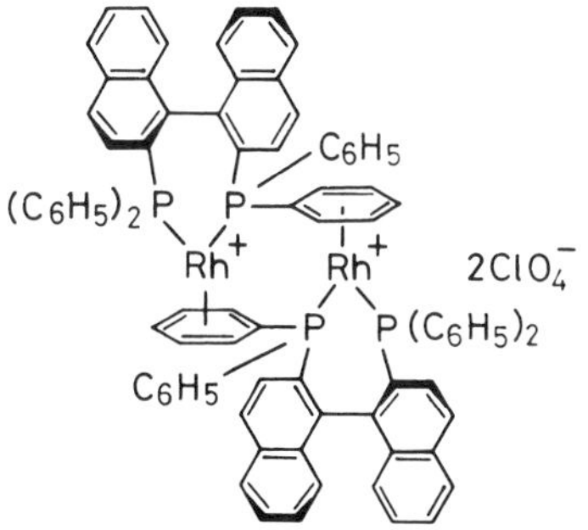

(*S*)-8

The reaction of the chiral Rh complex (*S*)-**7** and (*Z*)-α-acetamido- or (*Z*)-α-benzamidocinnamic acid, or their esters in methanol gave only one of the two possible diastereomeric complexes (within the limits of ^{31}P NMR detection). This indicates that the BINAP-coordinated Rh(I) complex has a very high ability of enantioface differentiation, although this might not necessarily be the criterion for the efficiency of the asymmetric reaction.[2] Other factors controlling the enantioselectivity were also examined on the basis of the ^{31}P NMR measurement.

Table I. Asymmetric Hydrogenation of α-(acylamino)acrylic Acids Catalyzed by the Rh–BINAP complex [a]

substrate	config. of BINAP	% yield	% ee (config.)	substrate	config. of BINAP	% yield	% ee (config.)
H(Ph)C=C(COOH)NHCOPh	(S)	96	96 (R)	H(3-MeO-4-HO-C_6H_3)C=C(COOH)NHCOPh	(S)	97	79 (R)
	(R)	97	100 (S)				
	(S)	98	71 (R)[b]				
Ph(H)C=C(COOH)NHCOPh	(R)	93	87 (R)[c]	$H_2C=C(COOH)NHCOPh$	(S)	97	98 (R)
H(Ph)C=C(COOH)NHCOMe	(S)	99	84 (R)	$H_2C=C(COOH)NHCOMe$	(S)	97	67 (R)
H(Ph)C=C(COOMe)NHCOPh	(S)	98	93 (R)				
	(R)	97	92 (S)				

[a] H_2, 3—4 atm; substrate/cat. = 100—150; solvent, C_2H_5OH. [b] H_2, 50 atm. [c] Solvent, THF.

References

1) A. Miyashita, A. Yasuda, H. Takaya, K. Toriumi, T. Ito, T. Souchi, and R. Noyori, J. Am. Chem. Soc., 102, 7932 (1980).

2) A. S. C. Chan, J. J. Pluth, and J. Halpern, J. Am. Chem. Soc., 102, 5952 (1980).

RECEIVED December 14, 1981.

Asymmetric Reduction of Ketones with Metal Hydride Reagents Modified with Chiral Aminodiols and Aminotriols

JAMES D. MORRISON, EDWARD R. GRANDBOIS, and GARY R. WEISMAN
University of New Hampshire, Department of Chemistry, Durham, NH 03824

A chiral aminotriol, tris-[(S)-2-hydroxypropyl]amine (1) was synthesized from ammonia and three equivalents of (S)-propylene oxide. The triol was allowed to condense with boric acid to form a bicyclic borate [(S,S,S,)-triisopropanolamine borate, 2] which was allowed to react with a suspension of potassium hydride in tetrahydrofuran to form a borate hydride (3). Propiophenone was reduced to optically active alcohol [14% ee (R)] by 3 in THF.

By reaction of n-butylamine, t-butylamine and (R)- and (S)-1-phenylethylamine with two equivalents of (S)-propylene oxide or ethylene oxide five chiral aminodiols were prepared: N-n-butyl-bis-[(S)-2-hydroxypropyl]-amine (4), N-[(R)-1-phenylethyl]-bis-(2-hydroxyethyl)amine (5), N-[(R)-1-phenylethyl]-bis[(S)-2 hydroxypropyl]amine (6), N-[(S)-1-phenylethyl]-bis-[(S)-2-hydroxypropyl]amine (7) and N-[tert-butyl]-bis-[(S)-2-hydroxyl-propyl]amine (8).

The aminodiols were added to ethereal lithium aluminum hydride (LAH) to produce chiral, modified LAH reagents which were used to reduce prochiral ketones. The modified LAH reagents from ligands 4, 5, 6 and 7 quantitatively reduced acetophenone [44%ee (R), 10%ee (R), 35%ee (R), and 82%ee (R), respectively] and propiophenone [57%ee (R), 10%ee (S), 19% (R) and 77%ee (R), respectively]. A general additivity of asymmetric induction due to the chiral centers in the carbinol and noncarbinol arms of the ligands was perceived, with 7 optimizing the inductive influences. The direction of the asymmetric induction was rationalized by means of a stereocorrelation model in which the ketone is coordinated with a lithium cation that is simultaneously coordinated with the nitrogen and two oxygens of the aminodiol ligand (in the form of a dialkoxydihydridoaluminate).

Unique and unpredictable behavior was observed with the LAH reagent from 8. The enantiomeric composition was observed to vary wildly [83%ee (R) to 32%ee (S)] in roughly comparable reductions of propiophenone. This strange behavior has not been fully rationalized.

0097-6156/82/0185-0278$05.00/0

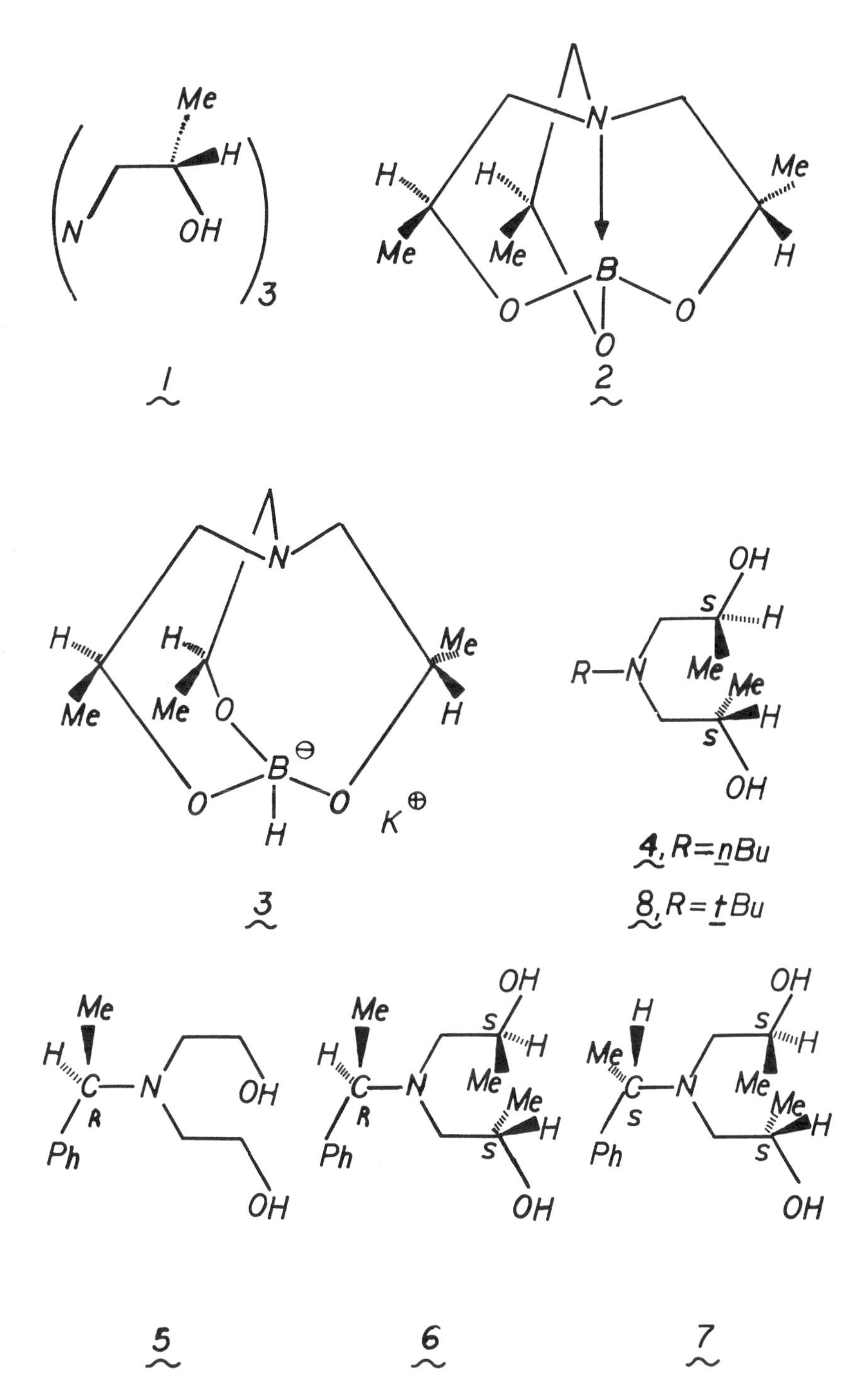

RECEIVED December 14, 1981.

Resolution by Optically Active Poly(triphenylmethyl Methacrylate)

YOSHIO OKAMOTO and HEIMEI YUKI

Osaka University, Department of Chemistry, Faculty of Engineering Science,
Toyonaka, Osaka 560, Japan

Triphenylmethyl methacrylate (TrMA) forms a highly isotactic, optically active polymer [(+)-PTrMA] when polymerized with a chiral anionic catalyst, (-)-sparteine-BuLi complex.[1] This polymer is the first optically active vinyl polymer whose chirality is due to only helicity. The high-molecular-weight (+)-PTrMA was highly crystal-

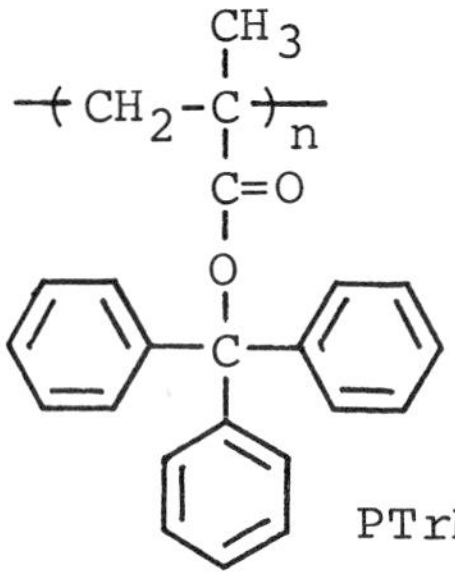

PTrMA

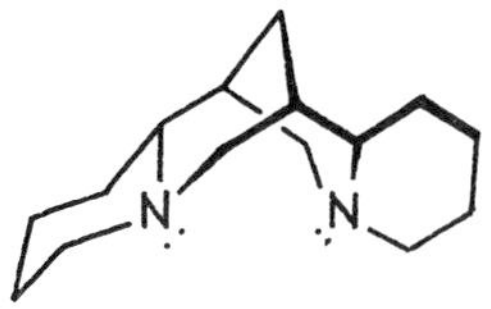

(-)-sparteine

line and insoluble in the usual organic solvents. It was ground to particles of 20-44 μm and used as a chiral support for high-performance liquid chromatography (HPLC).[2] Macroporous silica gel particles, which had been treated with dichlorodiphenylsilane, were coated with 22% (by weight) low-molecular-weight (+)-PTrMA which was soluble in THF but insoluble in methanol; this was also used as a packing material.[3] Each material was packed in stainless-steel tubing (25 × 0.46 cm) for HPLC. Various racemic compounds, particularly those having aromatic groups, were completely resolved on the columns by using methanol as eluant. However, the two columns showed different chiral recognition and some racemic compounds were much better separated by one of them. The retention volumes

0097-6156/82/0185-0280$05.00/0

of racemic compounds on the PTrMA-coated silica gel column were smaller than those on the ground PTrMA column and the theoretical plate number of the former was higher than that of the latter, indicating that more expeditious separation is possible with the former column.

The racemic compounds efficiently resolved so far are shown below.

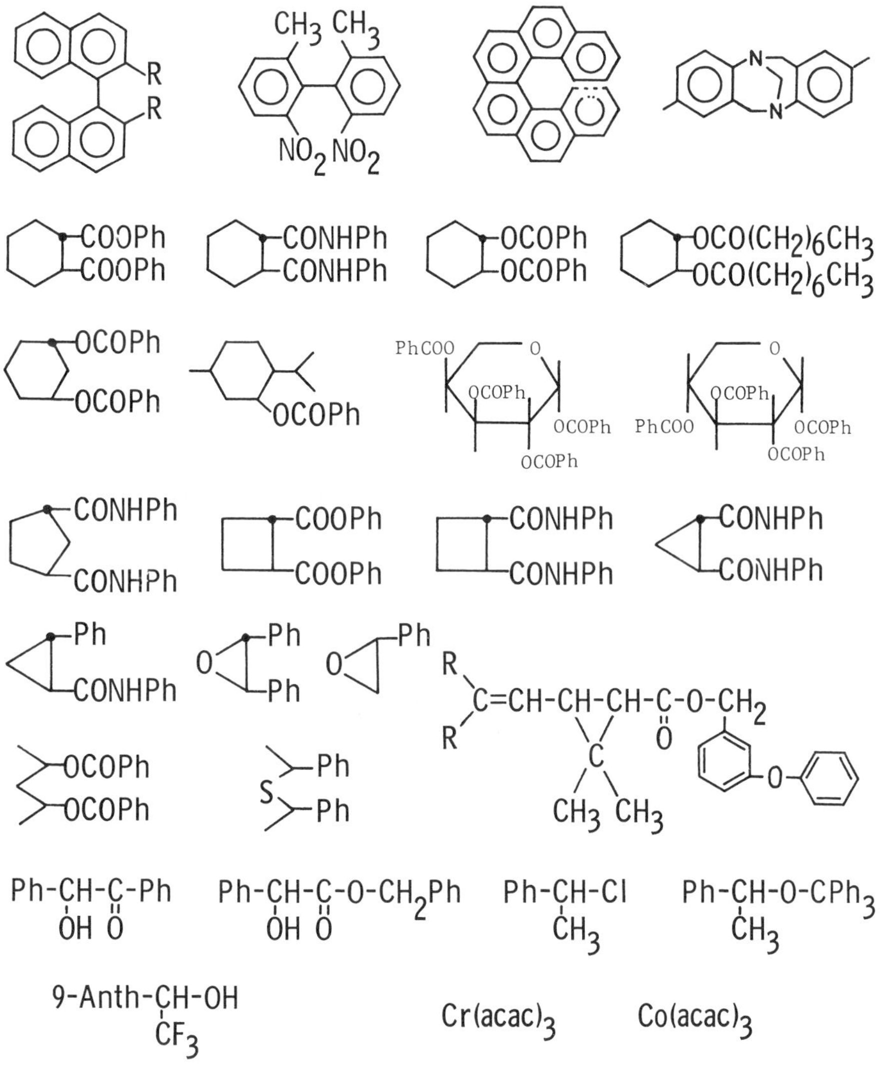

Ph-CH(OH)-C(=O)-Ph   Ph-CH(OH)-C(=O)-O-CH_2Ph   Ph-CH(CH_3)-Cl   Ph-CH(CH_3)-O-CPh_3

9-Anth-CH(CF_3)-OH   Cr(acac)$_3$ Co(acac)$_3$

Literature Cited

1. Okamoto, Y.; Suzuki, K.; Ohta, K.; Hatada, K.; Yuki, H. J. Am. Chem. Soc. 1979, 101, 4763. Okamoto, Y.; Suzuki, K.; Yuki, H. J. Polym. Sci. Polym. Chem. Ed. 1980, 18, 3043.
2. Okamoto, Y.; Okamoto, I.; Yuki, H. Chem. Lett. 1981, 835.
3. Okamoto, Y.; Honda, S.; Okamoto, I.; Yuki, H.; Murata, S.; Noyori, R.; Takaya, H. J. Am. Chem. Soc. 1981, 103, in press.

RECEIVED December 14, 1981.

Highly Active Catalysts for Enantioselective Hydrogenation of Ketones

K. TANI, K. SUWA, and S. OTSUKA

Osaka University, Department of Chemistry, Faculty of Engineering Science, Toyonaka, Osaka, Japan 560

For asymmetric hydrogenation of olefins and ketones, rhodium(I) complexes containing various optically active chelate diphosphine ligands have been reported.[1] Their chiral phosphorus ligands carry at least one aromatic substituent on the phosphorus atom. One serious drawback of these rhodium(I) complexes is their rather poor activity for hydrogenation of ketones.[2-5] The catalytic activity can be markedly improved with fully alkylated diphosphines as we report herein.

Tetraalkyl analogs of (-)-DIOP [2,3-O-isopropylidene-2,3-dihydroxy-1,4-bis(diphenylphosphino)butane][(-)-RDIOP; R=Et, i-Pr, and cyclo-C_6H_{11}] were prepared for the first time from the difluoride (1). As the catalyst precursor, either cationic complexes, [Rh(RDIOP)(diene)]ClO_4(diene=norbornadiene or 1,5-cyclooctadiene), or neutral ones formed in situ from $[Rh(C_2H_4)_2Cl]_2$ or $[Rh(C_8H_{14})_2Cl]_2$ and two equivalents of RDIOP can be used. Simple ketones, α-ketoacids, α-ketoesters, α-ketolactones, and α-ketoamides have been hydrogenated smoothly to the corresponding optically active hydroxy compounds under atmospheric pressure of hydrogen at ambient temperature with these catalysts. For example, the neutral complex prepared from (-)-CyDIOP(Cy=cyclo-C_6H_{11}) catalyzed hydrogenation of N-benzylphenylglyoxylamide to (+)-(S)-N-benzylmandelamide with an optical yield 77% and $t_{1/2}$ (25°C) of 5 min. ([Rh]=2.5 mM; [substrate]/[Rh]=200).

0097-6156/82/0185-0283$05.00/0

$$\mathbf{1} \; (\text{dioxolane with } CH_2F, CH_2F) + LiPR_2 \longrightarrow \mathbf{2} \; (\text{dioxolane with } CH_2PR_2, CH_2PR_2)$$

1

2
(-)-RDIOP

2 a : R = Et, bp 90-91°/10^{-3} Torr, $[\alpha]_D^{20}$ -25.5($CHCl_3$, c6.63)
b : R = i-Pr, bp 125-130°/10^{-3} Torr, $[\alpha]_D^{20}$ -31.0(C_6H_6, c2.97)
c : R = Cy, mp 90-92°, $[\alpha]_D^{20}$ -24.1(C_6H_6, c0.97)

Literature Cited

1. Valentine, D. S., Jr.; Scott, J. W. Synthesis 1978, 329 and references cited therein.
2. Törös, S.; Heil, B.; Kollár, L.; Markó, L. J. Organometal. Chem. 1980, 197, 85.
3. Hayashi, T.; Katsumata, A.; Konishi, M.; Kumada, M. Tetrahedron Lett. 1979, 425.
4. Achiwa, K.; Kogure, T.; Ojima, I. Chem. Lett. 1978, 297.
5. Ojima, I.; Kogure, T. J. Organometal. Chem. 1980, 195, 239.

RECEIVED December 14, 1981.

Reversibility of η^4-Cyclobutadiene Metal Formation from Complexed Alkynes

Unimolecular Isomerization of Labeled Racemic and Enantiomerically Enriched η^5-Cyclopentadienyl-η^4-cyclobutadiene Cobalt Complexes

G. VILLE, K. PETER C. VOLLHARDT, and MARK J. WINTER

University of California, Department of Chemistry, and Lawrence Berkeley Laboratory, Materials and Molecular Division, Berkeley, CA 94720

Diastereomeric 1,2-bis(trimethylsilyl)-3-alkylcyclobutadiene cyclopentadienyl cobalt complexes **1** and **2** in which the alkyl group contains a chiral center may be synthesized, separated, and equilibrated in the gas phase at 540-650° by flash pyrolysis or in solution in refluxing pristane (301°C). The process is unimolecular as shown by crossover experiments and kinetic analysis. Isomerization occurs by inversion at the four-ring, demonstrated by the pyrolysis of complexes enantiomerically enriched at the chiral carbon center and analysis of the products by optically active NMR shift reagents. No other processes but diastereoisomerization are observed in solution. In the gas phase increasing temperatures lead to increasing decomposition of starting complexes to cobalt metal and alkynes derived by retrocyclization of the cyclobutadiene ligand, in addition to positionally isomerized complexes. Such positional isomerization proceeding through B is also the mechanism of diastereoisomerization as shown by the pyrolysis of 1-triethylsilyl-2-trimethylsilyl substituted (**1d**⇄**2d**; **1e**⇄**2e**) and ^{13}C-labeled complexes. The data strongly imply that the cyclization of alkynes to cyclobutadienes in the coordination sphere of cobalt is reversible. They do not necessitate the intermediacy of a metallacycle A, which, if present, would have to be configurationally stable (i.e. the η^5-C_5H_5 ligand has to remain on one face of the other π-ligand).

0097-6156/82/0185-0285$05.00/0

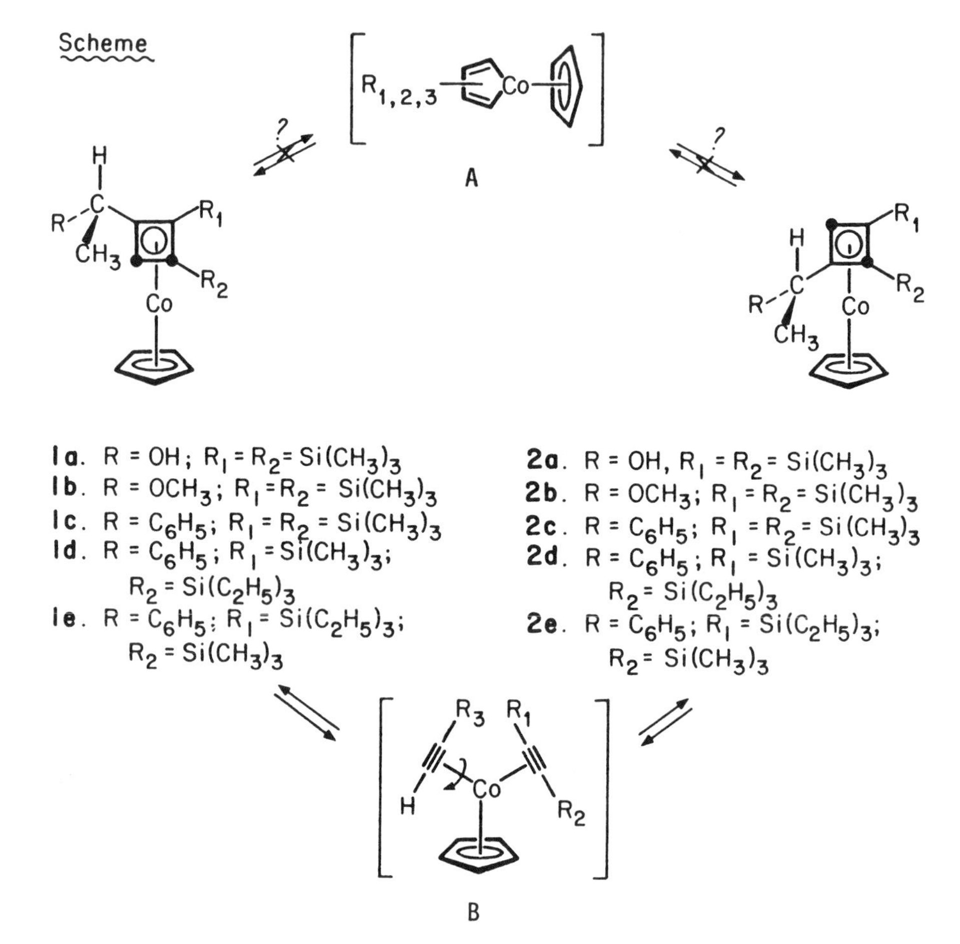

RECEIVED December 14, 1981.

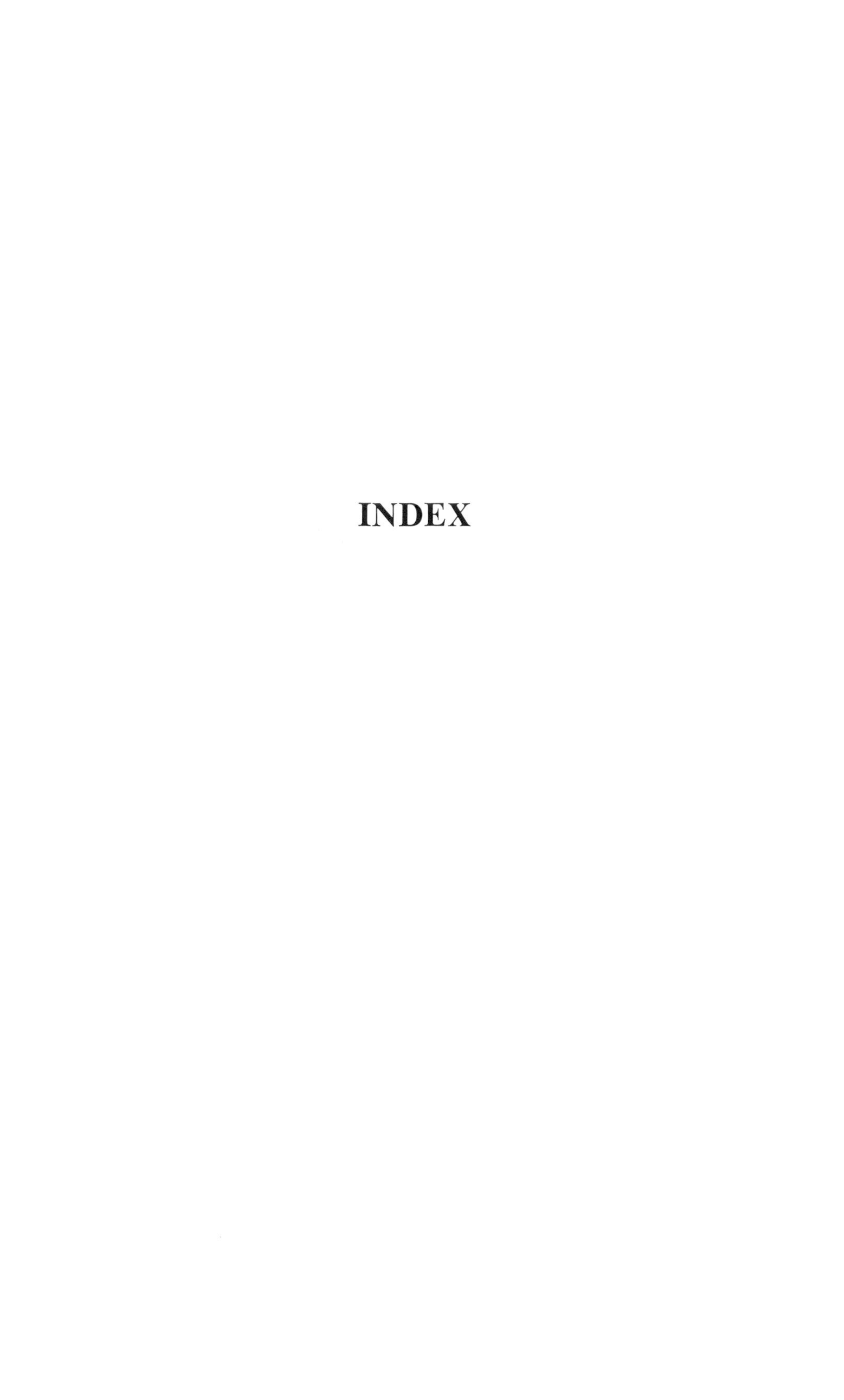

INDEX

INDEX

D

Jacket design by Kathleen Schaner
Production by Katharine Mintel and Cynthia Hale

Elements typeset by Service Composition Co., Baltimore, MD.
Printed and bound by The Maple Press Co., York, PA.